W9-BFL-050

MATH 54

An Incremental Development

MATH 54

An Incremental Development

Stephen Hake
John Saxon

SAXON PUBLISHERS, INC.

Math 54: An Incremental Development

Copyright © 1990 by Stephen Hake and John Saxon

All rights reserved.
No part of this publication may be reproduced, stored in a retrieval system, or transmitted in any form or by any means, electronic, mechanical, photocopying, recording, or otherwise, without the prior written permission of the publisher.

Printed in the United States of America

ISBN: 0-939798-21-12

Editor and Production Supervisor: Nancy Warren
Compositor: Science Topographers
Printer and Binder: R.R. Donnelley & Sons
Coordinator: Joan Coleman
Graphic Artists: Chad Threet, Scott Kirby, and David Pond

First Printing: July 1990

Saxon Publishers, Inc.
Norman, Oklahoma 73072

Contents

Preface

To the Student

We study mathematics because it is an important part of our daily lives. Our school schedule, our trip to the store, the preparation of our meals, and many of the games we play all involve mathematics. Most of the word problems you will see in this book are drawn from our daily experiences.

Mathematics is even more important in the adult world. In fact, your personal future in the adult world may depend in part upon the mathematics you have learned. This book was written with the hope that more students will learn mathematics and learn it well. For this to happen, you must use this book properly. As you work through the pages of this book, you will find similar problems presented over and over again. **Solving these problems day after day is the secret to success. Work every problem in every practice set and in every problem set. Do not skip problems. With honest effort you will experience success and true learning which will stay with you and serve you well in the future.**

Acknowledgments

We thank Shirley McQuade Davis for her ideas on teaching word problem thinking patterns.

Stephen Hake
Temple City, California

John Saxon
Norman, Oklahoma

LESSON
1

Review of Addition

Speed Test: 100 Addition Facts (Test A in Test Booklet)

When we add, we combine two groups into one group.

4 + 3 = 7 means [• •/• •] + [• •/•] = [• • • •/• • •]

The numbers that are added are called **addends**. The answer is called the **sum**. Here we show two ways to add 4 and 3

$$\begin{array}{rl} 4 & \text{addend} \\ +\ 3 & \text{addend} \\ \hline 7 & \text{sum} \end{array} \qquad \begin{array}{rl} 3 & \text{addend} \\ +\ 4 & \text{addend} \\ \hline 7 & \text{sum} \end{array}$$

Notice that if the order of the addends is changed the sum remains the same. This is true for any two numbers. **When we add two numbers, either number may be first.**

$$4 + 3 = 7 \qquad 3 + 4 = 7$$

Only two numbers can be added at one time. To add three numbers,

$$4 + 3 + 5$$

we must use two steps. We can add 4 and 3 to get 7 and then add 5. Or we can add 3 and 5 to get 8 and then add 4. Here we have used parentheses to show which two numbers we added first.

$$\begin{array}{rl} (4 + 3) + 5 & 4 + (3 + 5) \\ 7 + 5 & 4 + 8 \\ 12 & 12 \end{array}$$

From this we see that **numbers can be added in any order and the sum is the same every time.**

When we add 0 to a number, the number is not changed.

$$4 + 0 = 4 \qquad 9 + 0 = 9 \qquad 0 + 7 = 7$$

Some of the addition problems in this book will have an addend missing. We can use any letter to represent the missing addend (number). In this lesson we will use N to represent the missing addend. When an addend is missing and the sum is given, the problem is to find the missing addend.

Example (a) $\begin{array}{r} 4 \\ +\ 3 \\ \hline N \end{array}$ (b) $\begin{array}{r} 4 \\ +\ N \\ \hline 7 \end{array}$ (c) $\begin{array}{r} 4 \\ 3 \\ N \\ +\ 2 \\ \hline 14 \end{array}$ (d) $\begin{array}{r} 3 \\ N \\ +\ 5 \\ \hline 12 \end{array}$

Solution In (a) the sum of 4 and 3 is **7**. In (b) the letter N represents the missing number. This number must be **4** because 4 plus 3 equals 7. In (c) the sum of 4 + 3 + 2 is 9. The missing number N must be **5** because 9 + 5 equals 14. In (d) the missing number N must be **4** because 3 + 4 + 5 equals 12.

Practice Add or find the missing addend.

a. $4 + 3 =$ **b.** $5 + N = 13$

c. $9 + N + 7 = 16$

Problem set 1

1. $\begin{array}{r} 5 \\ +\ 6 \\ \hline \end{array}$ **2.** $\begin{array}{r} 4 \\ +\ 7 \\ \hline \end{array}$ **3.** $\begin{array}{r} 9 \\ +\ N \\ \hline 13 \end{array}$ **4.** $\begin{array}{r} 7 \\ +\ 8 \\ \hline \end{array}$

5. $\begin{array}{r} N \\ +\ 6 \\ \hline 13 \end{array}$ **6.** $\begin{array}{r} 5 \\ 2 \\ +\ N \\ \hline 12 \end{array}$ **7.** $\begin{array}{r} 4 \\ 8 \\ +\ 5 \\ \hline \end{array}$ **8.** $\begin{array}{r} 9 \\ 3 \\ +\ 7 \\ \hline \end{array}$

9. $\begin{array}{r} 8 \\ N \\ +\ 3 \\ \hline 16 \end{array}$ **10.** $\begin{array}{r} 9 \\ 7 \\ +\ 3 \\ \hline \end{array}$ **11.** $\begin{array}{r} 2 \\ 9 \\ +\ 6 \\ \hline \end{array}$ **12.** $\begin{array}{r} 3 \\ 8 \\ +\ 2 \\ \hline \end{array}$

13. $\begin{array}{r} 9 \\ 5 \\ +\ 3 \\ \hline \end{array}$ **14.** $\begin{array}{r} 2 \\ N \\ +\ 4 \\ \hline 9 \end{array}$ **15.** $\begin{array}{r} 5 \\ 3 \\ +\ N \\ \hline 9 \end{array}$ **16.** $\begin{array}{r} 2 \\ 3 \\ +\ N \\ \hline 7 \end{array}$

17. $\begin{array}{r} 5 \\ 3 \\ +\ N \\ \hline 10 \end{array}$ **18.** $\begin{array}{r} 8 \\ 4 \\ +\ 6 \\ \hline \end{array}$ **19.** $\begin{array}{r} 2 \\ N \\ +\ 7 \\ \hline 11 \end{array}$ **20.** $\begin{array}{r} 5 \\ 2 \\ +\ 6 \\ \hline \end{array}$

21. $5 + 8 + 2 + 7 + 4 + 3$

22. $9 + 3 + 11 + 2 + 1 + N + 4 = 38$

23. $\begin{array}{r} 8 \\ 7 \\ 6 \\ 5 \\ +\ 4 \\ \hline \end{array}$ **24.** $\begin{array}{r} 5 \\ 8 \\ 7 \\ 6 \\ +\ 4 \\ \hline \end{array}$ **25.** $\begin{array}{r} 2 \\ 3 \\ 5 \\ 7 \\ +\ 2 \\ \hline \end{array}$ **26.** $\begin{array}{r} 8 \\ 7 \\ 6 \\ 4 \\ +\ 5 \\ \hline \end{array}$ **27.** $\begin{array}{r} 4 \\ 7 \\ 8 \\ 4 \\ +\ 5 \\ \hline \end{array}$

LESSON 2 Counting Patterns

Speed Test: 100 Addition Facts (Test A in Test Booklet)

Counting is a math skill we learn early in life. Counting by ones, we say, "one, two, three, four, five, ..."

1, 2, 3, 4, 5, . . .

We often count by a number other than 1.

Counting by twos: 2, 4, 6, 8, 10, ...

Counting by fives: 5, 10, 15, 20, 25, ...

These are examples of counting patterns. A counting pattern is a **sequence**. The three dots mean that the sequence continues without end. A counting sequence may count up or count down. We may study a sequence to discover a rule for the sequence. Then we can find more numbers in the sequence.

Example 1 Find the next three numbers in this sequence:

10, 20, 30, 40, ___, ___, ___, ...

Solution The pattern is "counting up by tens." Counting this way, we find that the next three numbers are **50, 60, 70.**

Example 2 Find the next number in this sequence:

30, 27, 24, 21, ___

Solution The rule is "count down by threes." If we count down by threes from 21, we find that the next number in the sequence is **18.**

Practice Write the next three numbers in each sequence.

a. 10, 9, 8, 7, ___, ___, ___, ...

b. 3, 6, 9, 12, ___, ___, ___, ...

c. 80, 70, 60, 50, ___, ___, ___, ...

d. 5, 10, 15, 20, ___, ___, ___, ...

Problem set 2 Find either the sum or the missing addend.

1. $\begin{array}{r} 9 \\ 5 \\ 3 \\ +\ 9 \\ \hline \end{array}$

2. $\begin{array}{r} 4 \\ 2 \\ 4 \\ +\ 5 \\ \hline \end{array}$

3. $\begin{array}{r} 9 \\ 1 \\ 2 \\ +\ 7 \\ \hline \end{array}$

4. $\begin{array}{r} 8 \\ 3 \\ N \\ +\ 2 \\ \hline 19 \end{array}$

5. $\begin{array}{r} 4 \\ 5 \\ N \\ +\ 4 \\ \hline 22 \end{array}$

6. $\begin{array}{r} 8 \\ 2 \\ N \\ +\ 6 \\ \hline 19 \end{array}$

Write the next number in each sequence.

7. 10, 20, 30, ___, . . .

8. 22, 21, 20, ___, . . .

9. 40, 35, 30, 25, ___, . . .

10. 70, 80, 90, ___, . . .

Write the next three numbers in each sequence.

11. 6, 12, 18, ___, ___, ___, . . .

12. 3, 6, 9, ___, ___, ___, . . .

13. 4, 8, 12, ___, ___, ___, . . .

14. 45, 36, 27, ___, ___, ___, . . .

Find the missing number in each sequence.

15. 8, 12, ___, 20, . . .

16. 12, 18, ___, 30, . . .

17. 30, 25, ___, 15, . . .

18. 6, 9, ___, 15, . . .

19. How many small rectangles are shown? Count by twos.

20. How many X's are shown? Count by fours.

XX XX XX
XX XX XX

XX XX XX
XX XX XX

21. How many circles are shown? Count by fives.

22. 4 + 8 + 7 + 5 + 2 =

23. 9 + 5 + 7 + 8 + 3 =

24. 8 + 4 + 7 + 2 + 3 =

25. 2 + 9 + 7 + 5 + 4 =

26. 7 + 2 + 3 + 5 + 2 + 4 + 5 =

27. 9 + 3 + 2 + 7 + 5 + 6 + 1 =

28. 2 + 5 + 3 + 8 + 6 + 2 + 5 =

29. 3 + 5 + 8 + 2 + 2 + 3 + 5 =

LESSON 3 Digits

Speed Test: 100 Addition Facts (Test A in Test Booklet)

To write numbers we use digits. **Digits are the numerals 0, 1, 2, 3, 4, 5, 6, 7, 8, and 9**. The number 356 has three digits, and the last digit is 6. The number 67,896,094 has eight digits, and the last digit is 4.

Example 1 The number 64,000 has how many digits?

Solution The number 64,000 has **five digits**.

Example 2 What is the last digit of 2001?

Solution The last digit of 2001 is **1**.

Practice How many digits are in each number?

a. 18 **b.** 5280 **c.** 8,403,227,189

What is the last digit in each number?

d. 19 **e.** 5281 **f.** 8,403,190

Problem set 3

Write the next three numbers in each sequence.

1. 5, 10, 15, 20, ___, ___, ___, ...

2. 7, 14, 21, 28, ___, ___, ___, ...

3. 8, 16, 24, 32, ___, ___, ___, ...

4. 4, 8, 12, 16, ___, ___, ___, ...

Find the missing number in each sequence.

5. 90, ___, 70, 60, ...

6. 10, 8, ___, 4, ...

7. 6, ___, 12, 15, ...

8. 50, 45, ___, 35, ...

9. 45, 54, ___, 72, ...

10. 16, ___, 32, 40, ...

11. How many digits are in each number?
(a) 593 (b) 180 (c) 186,527,394

12. What is the last digit in each number?
(a) 3427 (b) 460 (c) 437,269

Find either the sum or the missing addend.

13. $\begin{array}{r} 4 \\ 3 \\ +\ 5 \\ \hline \end{array}$

14. $\begin{array}{r} 9 \\ 3 \\ +\ 8 \\ \hline \end{array}$

15. $\begin{array}{r} 7 \\ N \\ +\ 6 \\ \hline 22 \end{array}$

16. $\begin{array}{r} 8 \\ N \\ +\ 7 \\ \hline 19 \end{array}$

17. $\begin{array}{r} 9 \\ 8 \\ +\ N \\ \hline 19 \end{array}$

18. How many X's are shown? Count by fours.

XX XX XX XX
XX XX XX XX

XX XX XX XX
XX XX XX XX

19. How many cents are in 4 nickels? Count by fives.

5¢ 5¢

5¢ 5¢

20. How many X's are in this pattern? Count by threes.

X X X X X X X X X X
X X X X X X X X X X
X X X X X X X X X X

21.
$$\begin{array}{r} 2 \\ 7 \\ 3 \\ 5 \\ +\ 4 \\ \hline \end{array}$$

22.
$$\begin{array}{r} 5 \\ 2 \\ 3 \\ 8 \\ +\ 2 \\ \hline \end{array}$$

23.
$$\begin{array}{r} 9 \\ 8 \\ 7 \\ 4 \\ +\ 3 \\ \hline \end{array}$$

24.
$$\begin{array}{r} 8 \\ 7 \\ 2 \\ 5 \\ +\ 7 \\ \hline \end{array}$$

25.
$$\begin{array}{r} 9 \\ 3 \\ 2 \\ 5 \\ +\ 4 \\ \hline \end{array}$$

26.
$$\begin{array}{r} 8 \\ 7 \\ 6 \\ 5 \\ 4 \\ 7 \\ +\ 6 \\ \hline \end{array}$$

27.
$$\begin{array}{r} 4 \\ 3 \\ 2 \\ 6 \\ 5 \\ 4 \\ +\ 3 \\ \hline \end{array}$$

28.
$$\begin{array}{r} 9 \\ 7 \\ 2 \\ 1 \\ 4 \\ 8 \\ +\ 6 \\ \hline \end{array}$$

LESSON 4 Place Value

Speed Test: 100 Addition Facts (Test A in Test Booklet)

The greatest number we can write with just one digit is 9. To write numbers greater than 9, we use two or more digits. The number 43 has two digits. The two digits occupy **two places**. Each place has a **value**. The 4 in 43 is in the place that has a value of **ten**. The 3 in 43 is in the place that has a value of **one**. So 43 means **4 tens** and **3 ones**.

43 means

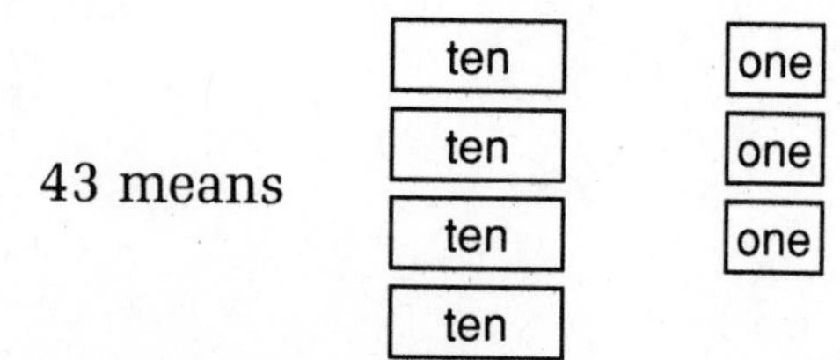

The number 243 has three digits. The 2 is in the place that has a value of **one hundred**. The 4 is in the place that has a value of **ten**. The 3 is in the place that has a value of **one**. So 243 means **2 hundreds and 4 tens and 3 ones**.

	hundred	ten	one
243 means	hundred	ten	one
		ten	one
		ten	

Example Draw a diagram to show the meaning of the number 323.

Solution The number 323 means 3 hundreds and 2 tens and 3 ones.

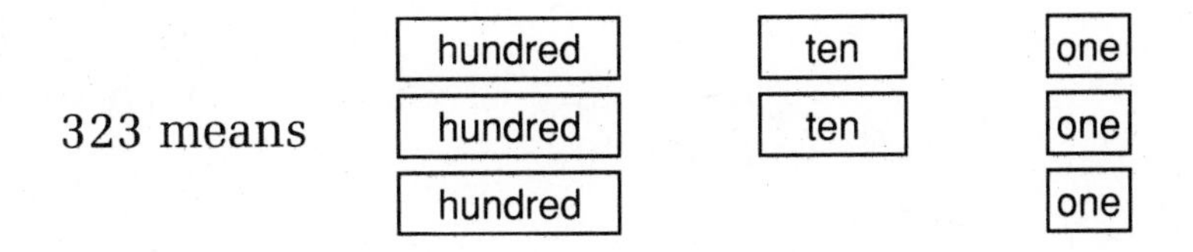

Practice Draw a diagram to show the meaning of the number 231.

Problem set 4 Find either the sum or the missing addend.

1. $\begin{array}{r} 2 \\ 4 \\ +\ N \\ \hline 12 \end{array}$ **2.** $\begin{array}{r} 4 \\ 5 \\ +\ 3 \\ \hline \end{array}$ **3.** $\begin{array}{r} 3 \\ 13 \\ +\ N \\ \hline 19 \end{array}$ **4.** $\begin{array}{r} 2 \\ 7 \\ +\ N \\ \hline 14 \end{array}$

5. $\begin{array}{r} 3 \\ N \\ +\ 2 \\ \hline 11 \end{array}$ **6.** $\begin{array}{r} 1 \\ +\ N \\ \hline 5 \end{array}$ **7.** $\begin{array}{r} N \\ +\ 2 \\ \hline 7 \end{array}$ **8.** $\begin{array}{r} 3 \\ N \\ +\ 2 \\ \hline 9 \end{array}$

9. $\begin{array}{r} 4 \\ N \\ +\ 5 \\ \hline 12 \end{array}$ **10.** $\begin{array}{r} N \\ 2 \\ +\ 3 \\ \hline 8 \end{array}$

Write the next three numbers in each sequence.

11. 30, 40, 50, ___, ___, ___, ...

12. 15, 20, 25, ___, ___, ___, ...

13. 9, 12, 15, ___, ___, ___, ...

14. 30, 24, 18, ___, ___, ___, ...

15. 12, 16, 20, ___, ___, ___, ...

16. 35, 28, 21, ___, ___, ___, ...

17. 27, 36, 45, ___, ___, ___, ...

18. How many digits are in each number?
(a) 37,432 (b) 5,934,286 (c) 453,000

19. What is the last digit in each number?
(a) 734 (b) 347 (c) 473

20. Draw a diagram for the number 322.

21. What number is indicated by this diagram?

hundred	ten	one
hundred	ten	one
hundred	ten	one
hundred		one

Find the missing number in each sequence.

22. 24, ___, 36, 42, ...

23. 36, 32, ___, 24, ...

24. How many ears are on 10 rabbits? Count by twos.

25.	26.	27.	28.	29.
2	1	9	9	4
5	8	4	3	8
3	2	2	6	3
2	3	3	5	2
3	5	8	2	5
1	7	4	4	4
+ 7	+ 2	+ 1	+ 3	+ 1

LESSON 5 Place Value through Hundreds

Speed Test: 100 Addition Facts (Test A in Test Booklet)

Places are named by the value of the place. Three-digit numbers occupy three different places.

hundreds' place	tens' place	ones' place
5	6	7

The number 567 means 5 hundreds and 6 tens and 7 ones.

Example 1 The digit 7 is in what place in 753?

Solution The 7 is in the third place from the right, which is the **hundreds' place**.

Example 2 Use three digits to write 8 hundreds plus 3 tens plus 9 ones.

Solution **839**

Practice The digit 6 is in what place in each number?

a. 16 **b.** 65 **c.** 623

Use three digits to write a number that has the given value.

d. 5 hundreds and 2 tens and 3 ones

e. 7 hundreds and 4 tens and 6 ones

Problem set 5

Find the missing addend.

1. $\begin{array}{r} 2 \\ 6 \\ +\ N \\ \hline 15 \end{array}$ **2.** $\begin{array}{r} 1 \\ N \\ +\ 7 \\ \hline 14 \end{array}$ **3.** $\begin{array}{r} 3 \\ N \\ +\ 5 \\ \hline 12 \end{array}$ **4.** $\begin{array}{r} 1 \\ N \\ +\ 6 \\ \hline 13 \end{array}$

5. $\begin{array}{r} 2 \\ 5 \\ +\ N \\ \hline 10 \end{array}$ **6.** $\begin{array}{r} 2 \\ +\ N \\ \hline 7 \end{array}$ **7.** $\begin{array}{r} N \\ +\ 5 \\ \hline 11 \end{array}$ **8.** $\begin{array}{r} 3 \\ +\ N \\ \hline 5 \end{array}$

9. $\begin{array}{r} 2 \\ N \\ +\ 3 \\ \hline 7 \end{array}$ **10.** $\begin{array}{r} 5 \\ 2 \\ +\ N \\ \hline 8 \end{array}$

Write the next three numbers in each sequence.

11. 40, 50, 60, ___, ___, ___, . . .

12. 20, 25, 30, ___, ___, ___, . . .

13. 12, 15, 18, ___, ___, ___, . . .

14. 24, 30, 36, ___, ___, ___, . . .

15. 16, 20, 24, ___, ___, ___, . . .

16. 28, 35, 42, ___, ___, ___, . . .

Find the missing number in each sequence.

17. 30, ___, 42, 48

18. 30, ___, 40, 45

19. Draw a diagram for the number 324.

20. The digit 8 is in what place in each number?
(a) 845 (b) 458 (c) 584

21. Use three digits to write the number that equals 2 hundreds plus 3 tens plus 5 ones.

22. If the pattern is continued, what would be the next number circled?

1 2 (3) 4 5 (6) 7 8 (9) 10 ...

23. Seven boys have how many elbows? Count by twos.

24.
$$\begin{array}{r} 5 \\ 8 \\ 4 \\ 7 \\ 4 \\ +\ 3 \\ \hline \end{array}$$

25.
$$\begin{array}{r} 5 \\ 7 \\ 3 \\ 8 \\ 4 \\ +\ 2 \\ \hline \end{array}$$

26.
$$\begin{array}{r} 9 \\ 7 \\ 6 \\ 5 \\ 4 \\ +\ 2 \\ \hline \end{array}$$

27.
$$\begin{array}{r} 8 \\ 7 \\ 3 \\ 5 \\ 4 \\ +\ 9 \\ \hline \end{array}$$

28.
$$\begin{array}{r} 5 \\ 4 \\ 7 \\ 3 \\ 6 \\ +\ 4 \\ \hline \end{array}$$

29.
$$\begin{array}{r} 5 \\ 2 \\ 1 \\ 7 \\ 8 \\ +\ 4 \\ \hline \end{array}$$

30.
$$\begin{array}{r} 5 \\ N \\ 2 \\ 5 \\ 7 \\ +\ 4 \\ \hline 25 \end{array}$$

31.
$$\begin{array}{r} 5 \\ 3 \\ 4 \\ 7 \\ 1 \\ +\ N \\ \hline 24 \end{array}$$

LESSON 6 Subtraction

Speed Test: 100 Addition Facts (Test A in Test Booklet)

We remember that when we add, we combine two groups into one group.

$4 + 2 = 6$ means [• •/• •] + [•/•] = [• • •/• • •]

When we subtract, we take a number away from a group. To take away 2 from 6, we subtract.

$6 - 2 = 4$ means [• • •/• • •] − [•/•] = [• •/• •]

When we subtract one number from another number, the answer is called the **difference**. If we subtract 2 from 6, the difference is 4.

$$\begin{array}{r} 6 \\ -\ 2 \\ \hline 4 \end{array} \text{ difference}$$

We can check a subtraction problem by adding the two bottom numbers. **The sum of the two bottom numbers must equal the top number.**

$$\begin{array}{r} 6 \\ -\ 2 \\ \hline 4 \end{array} \qquad 4 + 2 = 6$$

Subtraction facts should be memorized. Practice is the best way to memorize subtraction facts.

In the practice and problem sets for the rest of the book, all subtraction problems should be checked by adding.

Practice Subtract. Check by adding.

a. $\begin{array}{r} 14 \\ -\ 8 \\ \hline \end{array}$ **b.** $\begin{array}{r} 9 \\ -\ 3 \\ \hline \end{array}$ **c.** $\begin{array}{r} 15 \\ -\ 7 \\ \hline \end{array}$ **d.** $\begin{array}{r} 11 \\ -\ 4 \\ \hline \end{array}$ **e.** $\begin{array}{r} 12 \\ -\ 5 \\ \hline \end{array}$

Problem set 6

1. $\begin{array}{r} 14 \\ -\ 5 \\ \hline \end{array}$ 2. $\begin{array}{r} 15 \\ -\ 8 \\ \hline \end{array}$ 3. $\begin{array}{r} 9 \\ -\ 4 \\ \hline \end{array}$ 4. $\begin{array}{r} 11 \\ -\ 7 \\ \hline \end{array}$

5. $\begin{array}{r} 15 \\ -\ 8 \\ \hline \end{array}$ 6. $\begin{array}{r} 11 \\ -\ 6 \\ \hline \end{array}$ 7. $\begin{array}{r} 15 \\ -\ 7 \\ \hline \end{array}$ 8. $\begin{array}{r} 9 \\ -\ 6 \\ \hline \end{array}$

9. $\begin{array}{r} 13 \\ -\ 5 \\ \hline \end{array}$ 10. $\begin{array}{r} 12 \\ -\ 6 \\ \hline \end{array}$ 11. $\begin{array}{r} 8 \\ +\ N \\ \hline 17 \end{array}$ 12. $\begin{array}{r} N \\ +\ 8 \\ \hline 14 \end{array}$

13. $\begin{array}{r} 3 \\ +\ N \\ \hline 11 \end{array}$ 14. $\begin{array}{r} 1 \\ 4 \\ +\ N \\ \hline 13 \end{array}$ 15. $\begin{array}{r} 2 \\ 5 \\ +\ N \\ \hline 11 \end{array}$

Write the next three numbers in each sequence.

16. 20, 25, 30, ___, ___, ___, . . .

17. 16, 18, 20, ___, ___, ___, . . .

18. 21, 28, 35, ___, ___, ___, . . .

19. 20, 24, 28, ___, ___, ___, . . .

20. Use three digits to write the number that names 6 hundreds plus 5 tens plus 4 ones.

21. Draw a diagram for the number 326.

22. The digit 6 is in what place in each number?
(a) 625 (b) 456 (c) 263

Find the missing addend.

23. $\begin{array}{r} 2 \\ N \\ +\ 4 \\ \hline 13 \end{array}$ 24. $\begin{array}{r} N \\ 3 \\ +\ 5 \\ \hline 16 \end{array}$ 25. $\begin{array}{r} 2 \\ 3 \\ +\ N \\ \hline 9 \end{array}$ 26. $\begin{array}{r} N \\ 2 \\ +\ 5 \\ \hline 13 \end{array}$

27.	**28.**	**29.**	**30.**
4	4	9	2
7	3	8	8
8	2	2	3
6	5	5	7
4	7	4	6
3	8	6	4
+ 8	+ 4	+ 8	+ 5

LESSON 7 Writing Numbers through 999

Speed Test: 100 Subtraction Facts (Test B in Test Booklet)

To write the names of whole numbers through 999 (nine hundred ninety-nine), we need to know the following words and how to put them together.

0	zero	10	ten	20	twenty
1	one	11	eleven	30	thirty
2	two	12	twelve	40	forty
3	three	13	thirteen	50	fifty
4	four	14	fourteen	60	sixty
5	five	15	fifteen	70	seventy
6	six	16	sixteen	80	eighty
7	seven	17	seventeen	90	ninety
8	eight	18	eighteen	100	one hundred
9	nine	19	nineteen		

You may look at this chart when you are asked to write the names of numbers in the problem sets.

The names of two-digit numbers greater than 20 that do not end with zero are written with a hyphen.

Example 1 Use words to write the number 44.

Solution We use a hyphen and write **forty-four**. Notice that "forty" is spelled without a "u."

To write three-digit numbers, we first write the number of hundreds and then we write the rest of the number. **We do not use the word "and" when writing whole numbers.**

Example 2 Use words to write the number 313.

Solution First we write the number of hundreds. Then we write the rest of the number: **three hundred thirteen**.

Example 3 Use words to write the number 705.

Solution First we write the number of hundreds. Then we write the rest of the number: **seven hundred five**.

Example 4 Use digits to write the number six hundred eight.

Solution Six hundred eight means six hundreds and eight ones. There are no tens, so we write a zero in the tens' place: **608**.

Practice Use words to write each number.

a. 0 **b.** 81

c. 99 **d.** 515

e. 444 **f.** 909

Use digits to write each number.

g. Nineteen

h. Ninety-one

i. Five hundred twenty-four

j. Eight hundred sixty

Problem set 7

1. $\begin{array}{r} 2 \\ 3 \\ + N \\ \hline 13 \end{array}$ **2.** $\begin{array}{r} 3 \\ N \\ + 6 \\ \hline 14 \end{array}$ **3.** $\begin{array}{r} 5 \\ N \\ + 2 \\ \hline 11 \end{array}$ **4.** $\begin{array}{r} 2 \\ 6 \\ + N \\ \hline 15 \end{array}$

5.
$$\begin{array}{r} N \\ 4 \\ +\ 3 \\ \hline 14 \end{array}$$

6.
$$\begin{array}{r} 16 \\ -\ \ 8 \\ \hline \end{array}$$

7.
$$\begin{array}{r} 13 \\ -\ \ 7 \\ \hline \end{array}$$

8.
$$\begin{array}{r} 12 \\ -\ \ 8 \\ \hline \end{array}$$

9.
$$\begin{array}{r} 13 \\ -\ \ 5 \\ \hline \end{array}$$

10.
$$\begin{array}{r} 16 \\ -\ \ 9 \\ \hline \end{array}$$

Use digits to write each number.

11. Two hundred fourteen

12. Five hundred thirty-two

Use words to write each number.

13. 301

14. 320

15. 312

Write the next three numbers in each sequence.

16. 12, 18, 24, ___, ___, ___, ...

17. 15, 18, 21, ___, ___, ___, ...

18. 20, 24, 28, ___, ___, ___, ...

19. 30, 36, 42, ___, ___, ___, ...

20. 35, 42, 49, ___, ___, ___, ...

21. 40, 48, 56, ___, ___, ___, ...

What number is indicated by each diagram?

22. **23.**

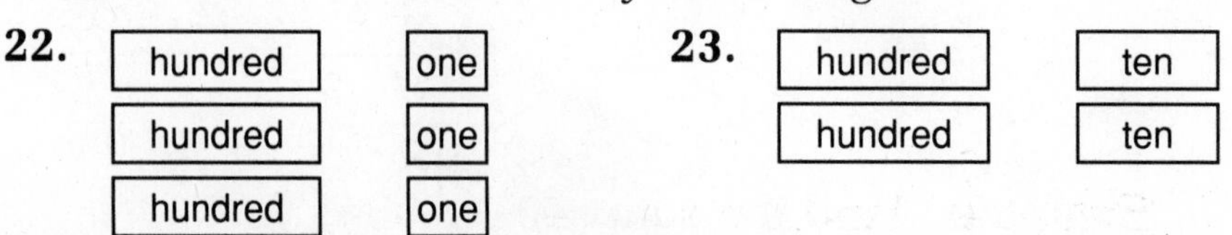

24. $$\begin{array}{r} 4 \\ 7 \\ 8 \\ 5 \\ +\ 4 \\ \hline \end{array}$$

25. $$\begin{array}{r} 2 \\ 3 \\ 5 \\ 8 \\ +\ 5 \\ \hline \end{array}$$

26. $$\begin{array}{r} 9 \\ 3 \\ 2 \\ 4 \\ +\ 8 \\ \hline \end{array}$$

27. $$\begin{array}{r} 4 \\ 5 \\ 8 \\ 7 \\ +\ 4 \\ \hline \end{array}$$

28. $$\begin{array}{r} 5 \\ 8 \\ 6 \\ 4 \\ 3 \\ 7 \\ +\ 2 \\ \hline \end{array}$$

29. $$\begin{array}{r} 9 \\ 3 \\ 5 \\ 2 \\ 7 \\ 8 \\ +\ 5 \\ \hline \end{array}$$

30. $$\begin{array}{r} 4 \\ 7 \\ 8 \\ 2 \\ 5 \\ 7 \\ +\ 4 \\ \hline \end{array}$$

LESSON 8 Adding Two-Digit Numbers, Part 1

Speed Test: 100 Addition Facts (Test A in Test Booklet)

We remember that the answer we get when we add is called the **sum**. The sum of numbers that have two digits may be found by adding one column of digits at a time. We add the digits in the right-hand column first. Then we add the digits in the left-hand column.

$$\begin{array}{r} \text{Add ones} \\ \downarrow \\ 24 \\ +\ 15 \\ \hline 9 \end{array} \qquad \begin{array}{r} \text{Add tens} \\ \downarrow\ \ \ \\ 24 \\ +\ 15 \\ \hline 39 \end{array}$$

Example Find the sum: $46 + 33$

Solution First write one number above the other number. We line up the last digit in each number vertically. Then we add. The sum of 46 and 33 is **79**.

$$\begin{array}{r} 46 \\ +\ 33 \\ \hline 79 \end{array}$$

Practice a. 53 + 26 b. 14 + 75 c. 36 + 42

d. 27 + 51 e. 15 + 21 f. 32 + 23

Problem set 8 Use digits to write each number.

1. Three hundred forty-three

2. Three hundred seven

3. Six hundred fourteen

Use words to write each number.

4. 592

5. 603

6. 713

7. $\begin{array}{r} 2 \\ 4 \\ +\ N \\ \hline 12 \end{array}$ 8. $\begin{array}{r} 1 \\ N \\ +\ 6 \\ \hline 10 \end{array}$ 9. $\begin{array}{r} 1 \\ N \\ +\ 7 \\ \hline 14 \end{array}$ 10. $\begin{array}{r} 2 \\ 6 \\ +\ N \\ \hline 13 \end{array}$

11. $\begin{array}{r} N \\ 5 \\ +\ 4 \\ \hline 15 \end{array}$ 12. $\begin{array}{r} 85 \\ +\ 14 \\ \hline \end{array}$ 13. $\begin{array}{r} 22 \\ +\ 16 \\ \hline \end{array}$ 14. $\begin{array}{r} 40 \\ +\ 38 \\ \hline \end{array}$

15. $\begin{array}{r} 25 \\ +\ 14 \\ \hline \end{array}$ 16. $\begin{array}{r} 42 \\ +\ 26 \\ \hline \end{array}$ 17. $\begin{array}{r} 17 \\ -\ 8 \\ \hline \end{array}$ 18. $\begin{array}{r} 14 \\ -\ 6 \\ \hline \end{array}$

19. $\begin{array}{r} 13 \\ -\ 9 \\ \hline \end{array}$ 20. $\begin{array}{r} 17 \\ -\ 5 \\ \hline \end{array}$ 21. $\begin{array}{r} 14 \\ -\ 8 \\ \hline \end{array}$

Write the next three numbers in each sequence.

22. 12, 15, 18, ___, ___, ___, ...

23. 28, 35, 42, ___, ___, ___, ...

24. 21, 28, 35, ___, ___, ___, ...

25. 16, 24, 32, ___, ___, ___, ...

26. $\begin{array}{r} 5 \\ 8 \\ 7 \\ 6 \\ 4 \\ +\ 3 \\ \hline \end{array}$ **27.** $\begin{array}{r} 9 \\ 7 \\ 6 \\ 4 \\ 8 \\ +\ 7 \\ \hline \end{array}$ **28.** $\begin{array}{r} 2 \\ 5 \\ 7 \\ 3 \\ 5 \\ +\ 4 \\ \hline \end{array}$ **29.** $\begin{array}{r} 9 \\ 3 \\ 8 \\ 4 \\ 7 \\ +\ 6 \\ \hline \end{array}$ **30.** $\begin{array}{r} 4 \\ 8 \\ 7 \\ 4 \\ 5 \\ +\ 6 \\ \hline \end{array}$

LESSON 9 Adding Two-Digit Numbers, Part 2

Speed Test: 100 Addition Facts (Test A in Test Booklet)

When we add numbers, we can write only one digit in the ones' place. We find that it is often necessary to **regroup** to write the answer.

Example 1 Add: 28 + 46

Solution We write one number above the other number. **We line up the last digits.** We add the ones' digits:

$$\begin{array}{r} 28 \\ +\ 46 \\ \hline \textcircled{14} \end{array}$$

The number 14 is 1 ten plus 4 ones. We write the digit 4 in the ones' column and **carry** the 1 to the tens' column.

$$\begin{array}{r} {}^{1} \\ 28 \\ +\ 46 \\ \hline 4 \end{array}$$

Then we add 1 + 2 + 4 and get 7, which we write in the tens' column. The sum is **74**.

$$\begin{array}{r} {}^{1} \\ 28 \\ +\ 46 \\ \hline \mathbf{74} \end{array}$$

Example 2 Add: 24 + 9

Solution We write one number above the other number. **We line up the last digits.** Then we add 4 and 9 to get 13. We write the 3 below the line and carry the 1 to the top of the tens' column. Then we add 1 and 2 and write 3 below the line.

$$\begin{array}{r} 2\ 4 \\ +\quad 9 \\ \hline \textcircled{13} \end{array} \qquad \begin{array}{r} {}^{1} \\ 24 \\ +\ \ 9 \\ \hline \mathbf{33} \end{array}$$

The sum is **33**.

Practice

a. $\begin{array}{r} 36 \\ +\ 29 \\ \hline \end{array}$ **b.** $\begin{array}{r} 47 \\ +\ \ 8 \\ \hline \end{array}$ **c.** $\begin{array}{r} 57 \\ +\ 13 \\ \hline \end{array}$

d. 68 + 24 **e.** 59 + 8 **f.** 46 + 25

Problem set 9

Use digits to write each number.

1. Six hundred thirteen

2. Five hundred forty-two

3. Nine hundred one

Use words to write each number.

4. 502

5. 720

6. 941

7. $\begin{array}{r} 2 \\ 4 \\ + N \\ \hline 11 \end{array}$

8. $\begin{array}{r} 5 \\ N \\ + 2 \\ \hline 13 \end{array}$

9. $\begin{array}{r} N \\ 4 \\ + 7 \\ \hline 15 \end{array}$

10. $\begin{array}{r} 2 \\ 7 \\ + N \\ \hline 16 \end{array}$

11. $\begin{array}{r} N \\ 5 \\ + 3 \\ \hline 15 \end{array}$

12. $\begin{array}{r} 47 \\ + 18 \\ \hline \end{array}$

13. $\begin{array}{r} 27 \\ + 69 \\ \hline \end{array}$

14. $\begin{array}{r} 49 \\ + 25 \\ \hline \end{array}$

15. $\begin{array}{r} 33 \\ + 8 \\ \hline \end{array}$

16. $\begin{array}{r} 42 \\ + 38 \\ \hline \end{array}$

17. $\begin{array}{r} 9 \\ - 7 \\ \hline \end{array}$

18. $\begin{array}{r} 13 \\ - 6 \\ \hline \end{array}$

19. $\begin{array}{r} 17 \\ - 8 \\ \hline \end{array}$

20. $\begin{array}{r} 12 \\ - 6 \\ \hline \end{array}$

21. $\begin{array}{r} 18 \\ - 9 \\ \hline \end{array}$

Write the next three numbers in each sequence.

22. 30, 35, 40, ___, ___, ___, . . .

23. 30, 36, 42, ___, ___, ___, . . .

24. 28, 35, 42, ___, ___, ___, . . .

25. What digit is in the hundreds' place in 843?

26. 28 + 6

27. 47 + 28

28. 35 + 27

29. 4 + 5 + 7 + 8 + 4 + 6 + 3

30. 8 + 3 + 7 + 4 + 2 + 3 + 5

LESSON 10 Even and Odd Numbers

Speed Test: 100 Addition Facts (Test A in Test Booklet)

Even numbers The numbers we say when we count by twos are **even numbers.** Notice that every even number ends with either 2, 4, 6, 8, or 0.

2, 4, 6, 8, 10, 12, 14, 16, 18, 20, 22, 24, 26, . . .

The list of even numbers goes on and on. We do not begin with 0 when we count by twos. However, the number 0 is an even number.

Example 1 Which one of these numbers is an even number?

463 285 456

Solution **A number is an even number if the last digit is even.** The last digits of these numbers are 3, 5, and 6. Of these, the only even digit is 6, so the even number is **456**.

Odd numbers If a whole number is not an even number, then it is an **odd number.** We can make a list of odd numbers by beginning with the number 1. Then we add 2 to get the next odd number, add 2 more to get the next odd number, and so on. The sequence of odd numbers is

1, 3, 5, 7, 9, 11, . . .

An even number of objects can be separated into two groups and it "comes out even." Six is an even number. Here we show six dots separated into two equal groups.

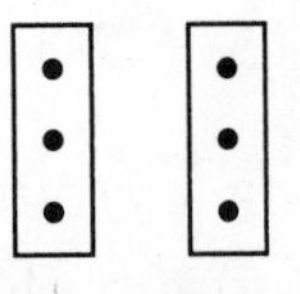

If we try to separate an odd number of objects into two groups, it will not "come out even." There will always be one extra object. Five is an odd number. Five will not separate into two equal groups because one is left over.

[: :] [: :] • ← one dot left over

Example 2 Write an odd three-digit number that has a 7 in the hundreds' place and a 5 in the tens' place.

Solution The first two digits are 7 and 5.

7 5 _

The last digit can be either 1, 3, 5, 7, or 9. So any of these numbers will do.

751, 753, 755, 757, or 759

Practice Write "even" or "odd" for each number.

a. 563

b. 328

c. 99

d. 0

e. Write an even three-digit number that has a 6 in the hundreds' place and a 3 in tens' place.

Problem set 10

1. Use digits to write five hundred forty-two.

2. Use digits to write six hundred nineteen.

3. Use digits to write three hundred two.

Use words to write each number.

4. 903

5. 746

6. 813

7. Write an odd three-digit number that has a 5 in the hundreds' place and a 0 in the tens' place.

8. Write an even three-digit number that has a 9 in the hundreds' place and a 2 in the tens' place.

9. $\begin{array}{r} 4 \\ N \\ +\ 3 \\ \hline 14 \end{array}$ **10.** $\begin{array}{r} N \\ 4 \\ +\ 2 \\ \hline 13 \end{array}$ **11.** $\begin{array}{r} 5 \\ N \\ +\ 7 \\ \hline 14 \end{array}$ **12.** $\begin{array}{r} N \\ 3 \\ +\ 2 \\ \hline 11 \end{array}$

13. $\begin{array}{r} 2 \\ 7 \\ +\ N \\ \hline 16 \end{array}$ **14.** $\begin{array}{r} 14 \\ -\ 7 \\ \hline \end{array}$ **15.** $\begin{array}{r} 17 \\ -\ 8 \\ \hline \end{array}$ **16.** $\begin{array}{r} 11 \\ -\ 6 \\ \hline \end{array}$

17. $\begin{array}{r} 15 \\ -\ 7 \\ \hline \end{array}$ **18.** $\begin{array}{r} 16 \\ -\ 9 \\ \hline \end{array}$ **19.** $\begin{array}{r} 42 \\ +\ 8 \\ \hline \end{array}$ **20.** $\begin{array}{r} 17 \\ +49 \\ \hline \end{array}$

21. $\begin{array}{r} 25 \\ +\ 38 \\ \hline \end{array}$ **22.** $\begin{array}{r} 19 \\ +\ 34 \\ \hline \end{array}$ **23.** $\begin{array}{r} 15 \\ +\ 68 \\ \hline \end{array}$

Write the next three numbers in each sequence.

24. 18, 21, 24, ___, ___, ___, . . .

25. 18, 24, 30, ___, ___, ___, . . .

26. 18, 27, 36, ___, ___, ___, . . .

27. $\begin{array}{r} 2 \\ 3 \\ 5 \\ 7 \\ 8 \\ 4 \\ +\ 5 \\ \hline \end{array}$ **28.** $\begin{array}{r} 9 \\ 4 \\ 7 \\ 8 \\ 6 \\ 5 \\ +\ 3 \\ \hline \end{array}$ **29.** $\begin{array}{r} 9 \\ 2 \\ 4 \\ 8 \\ 7 \\ 6 \\ +\ 2 \\ \hline \end{array}$ **30.** $\begin{array}{r} 4 \\ 7 \\ 4 \\ 6 \\ 5 \\ 2 \\ +\ 3 \\ \hline \end{array}$

LESSON 11 "Some and Some More" Problems, Part 1

Speed Test: 100 Addition Facts (Test A in Test Booklet)

To solve word problems, we use ideas. Many ideas form a **thinking pattern.** Some problems use the idea of **putting things together.** Look at this story.

John had 5 marbles. He got 7 more marbles.
Now John has 12 marbles.

There is a pattern to this story. John had **some** marbles. Then he got **some more** marbles. When he put the marbles together, he found out the **total** number of marbles. We will call these problems "**some and some more**" problems.

Pattern	Problem	
Some	5 marbles	
+ Some more	+ 7 marbles	
Total	12 marbles	← largest number

The some and some more pattern is an addition pattern. **We remember that the bottom number in an addition pattern is the largest number.** A some and some more pattern is shown in the box below. There are six lines and a plus sign. Three of the lines are for numbers. Three of the lines are for words. All the words are the same.

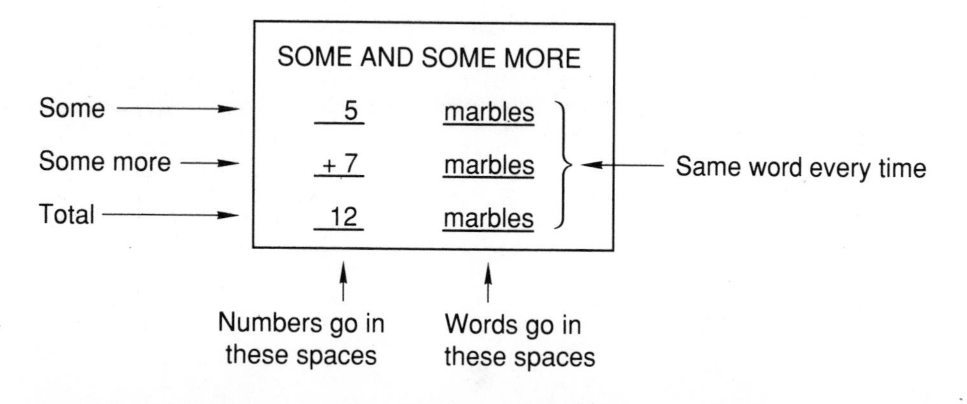

Example 1 Mickey saw 15 rabbits. Then he saw 7 more rabbits. How many rabbits did he see in all?

Solution This problem has the idea of some and some more. To practice the pattern, we draw six lines and a plus sign.

Some	___	______
Some more	+ ___	______
Total	___	______

Next we write rabbits in the three spaces on the right.

Some	___	rabbits
Some more	+ ___	rabbits
Total	___	rabbits

Now we put in the numbers. In the beginning there were 15. Then he saw 7 more.

Some	15	rabbits
Some more	+ 7	rabbits
Total	___	rabbits

We add the 15 and 7 to find the total number.

Some	15	rabbits
Some more	+ 7	rabbits
Total	22	rabbits

He saw **22 rabbits** in all.

Example 2 Joan found 5 beans on the floor, 3 beans behind the stove, and 6 beans in the cupboard. How many beans did she find in all?

Solution This problem puts beans together. There are 5 beans, 3 beans, and 6 beans to put together. This is a some, and some more, and some more problem.

Some	5	beans
Some more	3	beans
Some more	+ 6	beans
Total	___	beans

If we add 5 beans and 3 beans and 6 beans, we find the total is 14 beans. She found **14 beans** in all.

Practice Draw the lines and the plus sign to make the some and some more pattern for each problem. Then work the problem.

a. Diane ran 5 laps in the morning. She ran 8 laps in the afternoon. How many laps did she run in all?

b. Nathan found 3 pencils on the table, 5 pencils behind his desk, and 11 pencils under his bed. Altogether, how many pencils did Nathan find?

Problem set 11

1. James had 17 balls. Then his brother gave him 8 more balls. How many balls does James have now?

2. Samantha kept rabbits in cages. She had 5 rabbits in one cage. She had 3 rabbits in another cage. She had 4 rabbits in another cage. How many rabbits did Samantha have in all?

3. Use digits to write seven hundred fourteen.

4. Use digits to write three hundred forty-seven.

Use words to write each number.

5. 706

6. 532

7. Write an even three-digit number that has 5 in the tens' place and 7 in the hundreds' place.

8. Write an odd three-digit number that has 6 in the tens' place and 2 in the hundreds' place.

9. $\begin{array}{r} 4 \\ N \\ +\ 6 \\ \hline 15 \end{array}$

10. $\begin{array}{r} N \\ 7 \\ +\ 4 \\ \hline 16 \end{array}$

11. $\begin{array}{r} 9 \\ N \\ +\ 3 \\ \hline 12 \end{array}$

12. $\begin{array}{r} 5 \\ 2 \\ +\ N \\ \hline 13 \end{array}$

13. $\begin{array}{r} N \\ 4 \\ +\ 3 \\ \hline 14 \end{array}$

14. $\begin{array}{r} 18 \\ -\ 9 \\ \hline \end{array}$

15. $\begin{array}{r} 17 \\ -\ 5 \\ \hline \end{array}$

16. $\begin{array}{r} 12 \\ -\ 6 \\ \hline \end{array}$

17. $\begin{array}{r} 14 \\ -\ 7 \\ \hline \end{array}$

18. $\begin{array}{r} 13 \\ -\ 4 \\ \hline \end{array}$

19. $\begin{array}{r} 28 \\ +\ 9 \\ \hline \end{array}$

20. $\begin{array}{r} 15 \\ +\ 8 \\ \hline \end{array}$

21. $\begin{array}{r} 27 \\ +\ 16 \\ \hline \end{array}$

22. $\begin{array}{r} 35 \\ +\ 38 \\ \hline \end{array}$

23. $\begin{array}{r} 47 \\ +\ 49 \\ \hline \end{array}$

Write the next three numbers in each sequence.

24. 9, 12, 15, ___, ___, ___, . . .

25. 14, 21, 28, ___, ___, ___, . . .

26. 24, 32, 40, ___, ___, ___, . . .

27. $\begin{array}{r} 4 \\ 3 \\ 7 \\ 8 \\ 5 \\ 6 \\ +\ 4 \\ \hline \end{array}$

28. $\begin{array}{r} 2 \\ 5 \\ 8 \\ 7 \\ 4 \\ 8 \\ +\ 2 \\ \hline \end{array}$

29. $\begin{array}{r} 9 \\ 3 \\ 2 \\ 4 \\ 6 \\ 5 \\ +\ 7 \\ \hline \end{array}$

30. $\begin{array}{r} 4 \\ 8 \\ 2 \\ 5 \\ 7 \\ 3 \\ +\ 6 \\ \hline \end{array}$

LESSON 12 Number Lines

Speed Test: 100 Addition Facts (Test A in Test Booklet)

To make a number line, We use a ruler and draw a straight line on the paper. The line we draw is called a **line**

segment. Then we put marks on the segment. The distances from one mark to the next mark are all the same.

Then we put 0 next the leftmost mark. Then we count by ones, two, tens, or some other number to name the other marks. If we count by ones, we get

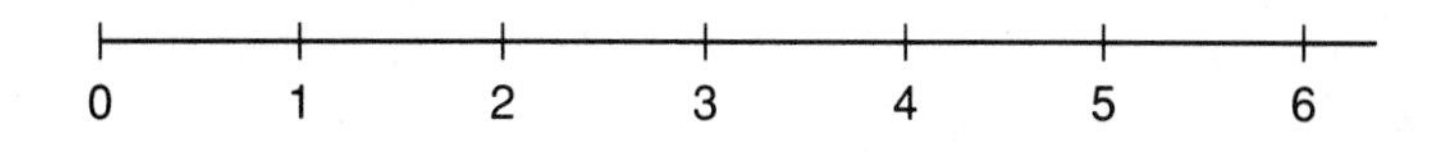

The distance between two marks on this line is 1 unit.

Number lines are very useful. Number lines may be used to help us measure things like distance, temperature, and time. We will practice reading different types of number lines.

Example 1 The arrow is pointing to what number on this number line?

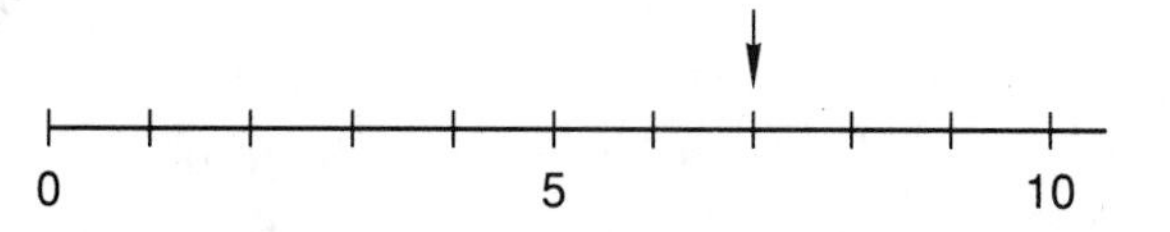

Solution First we must find the distance between two marks. There are 5 spaces between 0 and 5. So each space equals 1 unit. The arrow is 2 spaces to the right of 5. The arrow is pointing to the number **7**.

Example 2 The arrow is pointing to what number on this number line?

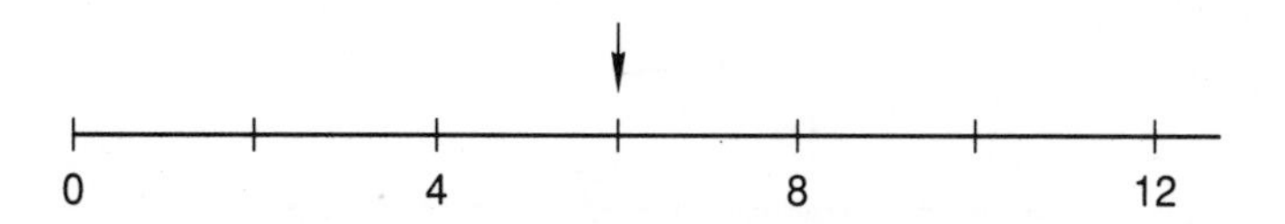

Solution First we must find the distance between two marks. There are 2 spaces between 0 and 4 and 2 spaces between 4 and 8. So the marks on this number line are 2 units apart. We can count by twos to name the marks on this line. The arrow points to the number **6**.

Practice The arrow is pointing to what number on each number line below?

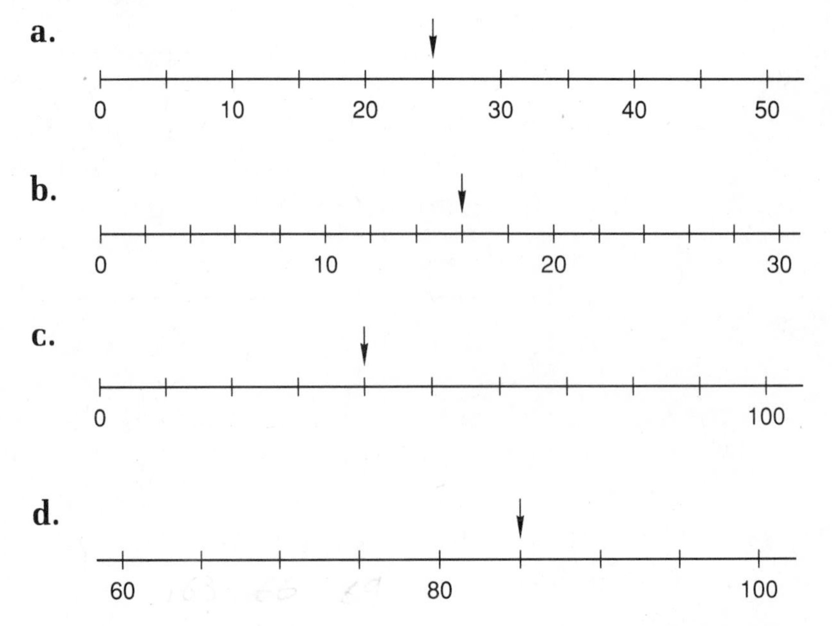

Problem set 12

1. Sandra saw 4 horses at the fair. Then she saw 13 horses on a farm. How many horses did Sandra see in all?

2. Mickey saw 14 gnomes in the forest and 23 gnomes in the valley. How many gnomes did Mickey see in all?

3. Use digits to write six hundred forty-two.

4. Use digits to write three hundred twelve.

Use words to write each number.

5. 713

6. 941

7. Write an odd three-digit number that has 5 in the hundreds' place and 7 in the tens' place.

8. To which number is the arrow pointing?

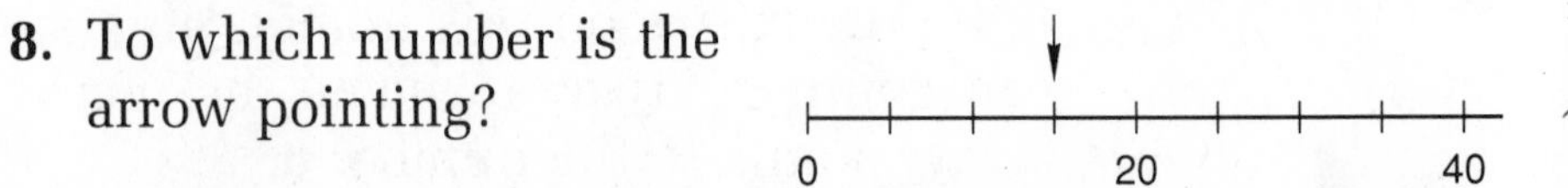

We have been using N to represent missing addends. We do not have to use N. We can use any letter we wish to use. In problems 9–13, we use A, B, N, M, and K to represent the missing addends. Find each missing addend.

9. $\begin{array}{r} 5 \\ B \\ +\ 7 \\ \hline 18 \end{array}$ **10.** $\begin{array}{r} N \\ 5 \\ +\ 3 \\ \hline 15 \end{array}$ **11.** $\begin{array}{r} 7 \\ A \\ +\ 4 \\ \hline 12 \end{array}$ **12.** $\begin{array}{r} M \\ 2 \\ +\ 8 \\ \hline 14 \end{array}$

13. $\begin{array}{r} 6 \\ K \\ +\ 3 \\ \hline 17 \end{array}$ **14.** $\begin{array}{r} 14 \\ -\ 7 \\ \hline \end{array}$ **15.** $\begin{array}{r} 12 \\ -\ 8 \\ \hline \end{array}$ **16.** $\begin{array}{r} 13 \\ -\ 6 \\ \hline \end{array}$

17. $\begin{array}{r} 12 \\ -\ 3 \\ \hline \end{array}$ **18.** $\begin{array}{r} 8 \\ -\ 4 \\ \hline \end{array}$ **19.** $\begin{array}{r} 28 \\ +\ 47 \\ \hline \end{array}$ **20.** $\begin{array}{r} 39 \\ +\ 53 \\ \hline \end{array}$

21. $\begin{array}{r} 74 \\ +\ 18 \\ \hline \end{array}$ **22.** $\begin{array}{r} 93 \\ +\ 39 \\ \hline \end{array}$ **23.** $\begin{array}{r} 28 \\ +\ 45 \\ \hline \end{array}$

Write the next three numbers in each sequence.

24. 12, 15, 18, ___, ___, ___, . . .

25. 30, 36, 42, ___, ___, ___, . . .

26. 16, 24, 32, ___, ___, ___, . . .

27. $\begin{array}{r} 4 \\ 3 \\ 5 \\ 8 \\ 7 \\ 6 \\ +\ 2 \\ \hline \end{array}$ **28.** $\begin{array}{r} 4 \\ 3 \\ 6 \\ 7 \\ 8 \\ 4 \\ +\ 3 \\ \hline \end{array}$ **29.** $\begin{array}{r} 5 \\ 7 \\ 6 \\ 4 \\ 8 \\ 3 \\ +\ 2 \\ \hline \end{array}$ **30.** $\begin{array}{r} 9 \\ 5 \\ 3 \\ 8 \\ 7 \\ 4 \\ +\ 6 \\ \hline \end{array}$

LESSON 13 More Missing Digits

Speed Test: 100 Addition Facts (Test A in Test Booklet)

Thinking patterns are important in mathematics. We remember that addition problems have a pattern. **The bottom number in an addition pattern is the largest number** (unless the middle number is zero). The bottom number in an addition pattern is always the sum of the two top numbers.

$$\begin{array}{rrl} & 4 & \text{addend} \\ & +\ 3 & \text{addend} \\ \hline \text{largest number} \rightarrow & 7 & \text{sum} \end{array} \qquad 7 = 4 + 3$$

Subtraction problems also have a pattern. **The top number in a subtraction pattern is the largest number** (unless the middle number is zero). The top number in a subtraction pattern is always the sum of the two bottom numbers.

$$\begin{array}{rrl} \text{largest number} \rightarrow & 7 & \\ & -\ 3 & \\ \hline & 4 & \text{difference} \end{array} \qquad 7 = 4 + 3$$

We remember that the bottom number in a subtraction pattern is called the **difference**.

Sometimes we know the bottom number in a subtraction pattern and one of the top numbers is missing. We can find the missing number in a subtraction pattern by asking a question about addition.

$$\begin{array}{r} 14 \\ -\ N \\ \hline 6 \end{array}$$

We ask, "What number added to 6 equals 14?" If we add 8 to 6, we get 14. So N equals **8**.

If digits are missing in an addition problem or a subtraction problem, we can use the pattern to help us make a guess. Then we check our guess.

Example 1 Find the missing number.

$$\begin{array}{r} 17 \\ -\ N \\ \hline 8 \end{array}$$

Solution The largest number in a subtraction problem is the top number. So 8 plus N must equal 17. This means N equals **9**.

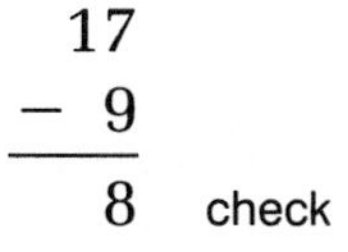

Example 2 Find the missing number.

$$\begin{array}{r} 19 \\ -\ N \\ \hline 14 \end{array}$$

Solution To find N, we ask a question. What number added to 14 equals 19? The answer is **5**. Now we check.

$$\begin{array}{r} 19 \\ -\ 5 \\ \hline 14 \end{array} \quad \text{check}$$

Practice Find each missing digit. Check your answer.

a. $\begin{array}{r} 14 \\ -\ N \\ \hline 6 \end{array}$ **b.** $\begin{array}{r} N \\ -\ 5 \\ \hline 2 \end{array}$ **c.** $\begin{array}{r} 19 \\ -\ N \\ \hline 12 \end{array}$ **d.** $\begin{array}{r} 12 \\ -\ N \\ \hline 5 \end{array}$

Problem set 13

1. Jimmy found forty-two acorns in the forest. Then he found thirty-seven acorns in his back yard. How many acorns did he find in all?

2. At first thirty-five fairies were flying about. Later twenty-seven more fairies began to fly about. In all, how many fairies were flying about?

3. Use digits to write the number seven hundred fifteen.

Use words to write each number.

4. 603

5. 222

6. Write the largest three-digit number that has 6 in the ones' place and 4 in the tens' place.

7. To which number is the arrow pointing?

40 60 80

8. $\begin{array}{r} 5 \\ N \\ +\ 6 \\ \hline 15 \end{array}$ **9.** $\begin{array}{r} A \\ 2 \\ +\ 5 \\ \hline 15 \end{array}$ **10.** $\begin{array}{r} 7 \\ 2 \\ +\ N \\ \hline 15 \end{array}$ **11.** $\begin{array}{r} 4 \\ A \\ +\ 2 \\ \hline 15 \end{array}$

12. $\begin{array}{r} B \\ 5 \\ +\ 6 \\ \hline 15 \end{array}$ **13.** $\begin{array}{r} 16 \\ -\ 8 \\ \hline \end{array}$ **14.** $\begin{array}{r} 14 \\ -\ 7 \\ \hline \end{array}$ **15.** $\begin{array}{r} 12 \\ -\ A \\ \hline 7 \end{array}$

16. $\begin{array}{r} B \\ -\ 6 \\ \hline 6 \end{array}$ **17.** $\begin{array}{r} 13 \\ -\ C \\ \hline 8 \end{array}$ **18.** $\begin{array}{r} 48 \\ +\ 16 \\ \hline \end{array}$ **19.** $\begin{array}{r} 37 \\ +\ 14 \\ \hline \end{array}$

20. $\begin{array}{r} 26 \\ +\ 35 \\ \hline \end{array}$ **21.** $\begin{array}{r} 8 \\ +\ N \\ \hline 12 \end{array}$ **22.** $\begin{array}{r} 6 \\ -\ N \\ \hline 2 \end{array}$

Write the next three numbers in each sequence.

23. 28, 35, 42, ___, ___, ___, . . .

24. 18, 21, 24, ___, ___, ___, . . .

25. 15, 20, 25, ___, ___, ___, . . .

26.	**27.**	**28.**	**29.**	**30.**
2	2	7	9	5
3	5	3	8	2
5	7	8	5	4
7	3	5	7	8
4	5	4	6	7
8	8	3	4	6
+ 2	+ 4	+ 2	+ 6	+ 2

LESSON 14 Comparing Numbers

Facts Practice: 100 Addition Facts (Test A in Test Booklet)

When we compare two numbers, we decide whether one number is greater than, equal to, or less than another number.

> Three **is equal to** three.
> Three **is less than** four.
> Four **is greater than** three.

The three sentences above compare two numbers. We may use digits and symbols to write the same sentences. If two numbers are equal, we use the equals sign.

$3 = 3$ is read "Three is equal to three."

If numbers are not equal, we use the greater than/less than symbol.

$$> \qquad <$$

The pointed end points to the smaller number. We read from left to right. If the pointed end comes first, we say "less than."

$3 < 4$ is read "Three is less than four."

If the open end comes first, we say "greater than."

$4 > 3$ is read "Four is greater than three."

Example 1 Use digits and symbols to write "five is greater than four."

Solution The pointed end must point to the smaller number.

$$\mathbf{5 > 4}$$

Example 2 Compare the following numbers by replacing the circle with the proper comparison symbol.

$$12 \bigcirc 21$$

Solution We use the greater than/less than symbol. Since 12 is less than 21, we let the pointed end point to 12.

$$12 < 21$$

We read this by saying "12 is less than 21."

Example 3 Which of these numbers is the greatest?

123 231 213

Solution All of these numbers have the same digits, but their values are different because the place values are different. The greatest (largest) of these three numbers is **231**.

Example 4 Arrange these numbers in order from the least to the greatest.

640 406 460

Solution The least (smallest) is 406. The greatest is 640. Arranging the numbers in order, we write:

406 460 640

Practice Write each statement using symbols instead of words.

a. Thirteen is less than thirty.

b. Forty is greater than fourteen.

Replace each circle with the proper symbol.

c. 432 ◯ 324 d. 212 ◯ 221 e. 316 ◯ 316

f. Arrange the numbers 132, 123, and 213 in order from the least to the greatest.

Problem set 14

1. Forty-two students had red hair. Thirty-seven students did not have red hair. How many students were there in all?

2. Seventeen brave students rode the loop-o-plane at the fair. Forty-seven others just watched. How many students were there altogether?

3. Use digits to write the number six hundred forty-two.

4. Use words to write the number 502.

5. Use symbols to write "fourteen is greater than seven."

Compare:

6. 16 ◯ 12 7. 141 ◯ 204

8. To which number is the arrow pointing?

9. $\begin{array}{r} 4 \\ N \\ +\ 5 \\ \hline 13 \end{array}$

10. $\begin{array}{r} N \\ 6 \\ +\ 2 \\ \hline 9 \end{array}$

11. $\begin{array}{r} 5 \\ 2 \\ +\ P \\ \hline 14 \end{array}$

12. $\begin{array}{r} 3 \\ 5 \\ +\ K \\ \hline 16 \end{array}$

13. $\begin{array}{r} A \\ 4 \\ +\ 3 \\ \hline 12 \end{array}$

14. $\begin{array}{r} 17 \\ -\ 9 \\ \hline \end{array}$

15. $\begin{array}{r} 12 \\ -\ 6 \\ \hline \end{array}$

16. $\begin{array}{r} 9 \\ -\ A \\ \hline 4 \end{array}$

17. $\begin{array}{r} 7 \\ -\ B \\ \hline 3 \end{array}$ **18.** $\begin{array}{r} C \\ -\ 5 \\ \hline 7 \end{array}$ **19.** $\begin{array}{r} 68 \\ +\ 27 \\ \hline \end{array}$ **20.** $\begin{array}{r} 49 \\ +\ 24 \\ \hline \end{array}$

21. $\begin{array}{r} 37 \\ +\ 28 \\ \hline \end{array}$ **22.** $\begin{array}{r} 5 \\ +\ N \\ \hline 12 \end{array}$ **23.** $\begin{array}{r} 12 \\ -\ N \\ \hline 4 \end{array}$

Write the next three numbers in each sequence.

24. 18, 27, 36, ___, ___, ___, ...

25. 32, 40, 48, ___, ___, ___, ...

26. 20, 24, 28, ___, ___, ___, ...

27. $\begin{array}{r} 2 \\ 5 \\ 7 \\ 6 \\ 5 \\ +\ 4 \\ \hline \end{array}$ **28.** $\begin{array}{r} 2 \\ 5 \\ 7 \\ 3 \\ 4 \\ +\ 2 \\ \hline \end{array}$ **29.** $\begin{array}{r} 8 \\ 4 \\ 2 \\ 7 \\ 5 \\ +\ 4 \\ \hline \end{array}$ **30.** $\begin{array}{r} 5 \\ 2 \\ 7 \\ 3 \\ 7 \\ +\ 2 \\ \hline \end{array}$

LESSON 15 Adding Three-Digit Numbers

Speed Test: 100 Addition Facts (Test A in Test Booklet)

When we add three-digit numbers, we add the digits in the ones' column first. Next we add the digits in the tens' column. Then we add the digits in the hundreds' column.

Add ones	Add tens	Add hundreds
↓	↓	↓
$\begin{array}{r} 124 \\ +\ 315 \\ \hline 9 \end{array}$	$\begin{array}{r} 124 \\ +\ 315 \\ \hline 39 \end{array}$	$\begin{array}{r} 124 \\ +\ 315 \\ \hline 439 \end{array}$

As we add, we remember that we can write only one digit in each column. If we add a column of numbers and get a two-digit answer, we write the last digit of the sum in that column. We carry the first digit of the sum to the next column.

Example 1 Add: $\begin{array}{r} 456 \\ +\ 374 \\ \hline \end{array}$

Solution We begin by adding the digits in the ones' column. Then we move one column at a time to the left, carrying the first digit of two-digit answers. The sum is **830**.

$$\begin{array}{r} {}^{1}{}^{1} \\ 456 \\ +\ 374 \\ \hline 830 \end{array}$$

Example 2 Add: 579 + 24

Solution We write the numbers in columns so that the last digits are in line, one above the other. We treat the empty place below the 5 as though it were a zero. The sum is **603**.

$$\begin{array}{r} {}^{1}{}^{1} \\ 579 \\ +\ \ 24 \\ \hline 603 \end{array}$$

Practice

a. $\begin{array}{r} 579 \\ +\ 186 \\ \hline \end{array}$ **b.** $\begin{array}{r} 408 \\ +\ 243 \\ \hline \end{array}$ **c.** $\begin{array}{r} 498 \\ +\ \ 89 \\ \hline \end{array}$

d. 458 + 336 **e.** 56 + 569

Problem set 15

1. Seventy-seven students ran in circles and waved their arms. Nineteen students watched in amazement. How many students were there in all?

2. The king counted seventy-five on the way to the fair. The queen counted thirty-eight on the way home. How many did the king and queen count in all?

3. Use words to write the number 913.

4. Use digits to write the number seven hundred forty-three.

5. Use symbols and digits to write "seventy-five is less than ninety-eight."

Compare:

6. 413 ◯ 314

7. 256 ◯ 652

8. To which number is the arrow pointing?

9. $475 + 332$

10. $714 + 226$

11. $943 + 187$

12. $576 + 228$

13. $996 + 48$

14. $4 + N + 6 = 15$

15. $9 + A + 6 = 17$

16. $N + 3 + 7 = 16$

17. $8 + 5 + K = 17$

18. $17 - 8$

19. $13 - 7$

20. $N - 8 = 7$

21. $14 - N = 6$

22. $16 - A = 9$

23. $N - 9 = 7$

24. $49 + 76$

25. $94 + 58$

26. $29 + 98$

27. $14 + N = 16$

28. $8 - N = 2$

Write the next three numbers in each sequence.

29. 25, 30, 35, ___, ___, ___, ...

30. 28, 35, 42, ___, ___, ___, ...

LESSON 16 Some and Some More Problems, Part 2

Speed Test: 100 Addition Facts (Test A in Test Booklet)

We have been working problems that have a some and some more thinking pattern. In the problems we have worked thus far both the "some" number and the "some more" number were given. We add these numbers to find the total. If we have 5 marbles and get 7 more marbles, we have 12 marbles in all.

	SOME AND SOME MORE	
Some →	5	marbles
Some more →	+ 7	marbles
Total →	12	marbles

To review, we look at this thinking pattern carefully. The largest number is the total number. The largest number is on the bottom of the pattern.

Sometimes a some and some more problem gives us the total number and one other number. Then we must find the missing number. These problems are just like addition problems that have a missing addend.

Example 1 Walter had 8 marbles. Then Sam gave him some more marbles. He has 17 marbles now. How many marbles did Sam give him?

Solution **If we can recognize the pattern, we can solve the problem.** Walter had some marbles. Then he got some more marbles. This problem has a some and some more pattern. We know the "some" number. We know the total number. We put these numbers in the pattern. In a some and some more pattern, the bottom number is the largest number. **The sum of the two top numbers must equal the bottom number.**

Some	8	marbles
Some more	$+ N$	marbles
Total	17	marbles

We see that this time one of the top numbers is missing. One way to find the missing number is to ask an addition question.

8 plus what number equals 17?

$$\begin{array}{r} 8 \\ + N \\ \hline 17 \end{array}$$

Another way is to ask a subtraction question.

17 minus 8 equals what number?

$$\begin{array}{r} 17 \\ - \ 8 \\ \hline N \end{array}$$

Both answers are 9.

The key to the problem is recognizing the pattern. Then finding the answer is easy!

Example 2 Jamie had some pies. Then Frank gave her 5 more pies. Now she has 12 pies. How many pies did Jamie have at first?

Solution First we must find the pattern. Does this problem have a some and some more pattern? Let's see. **We remember that in a some and some more pattern the largest number is on the bottom.**

Some	N	pies
Some more	$+ \ 5$	pies
Total	12	pies

Yes. This problem is a some and some more problem. Finding the answer is easy now. We can find the missing number by asking an addition question or by asking a subtraction question.

5 plus what number equals 12? answer 7
12 minus what number equals 5?answer 7

At first Jamie had **7 pies**.

Practice Draw the lines to make a some and some more pattern for each problem. Then work the problem.

a. Lucille had 4 marigolds. Jim gave her some more marigolds. Now she has 17 marigolds. How many marigolds did Jim give Lucille?

b. Sid had some agates. Then he found 8 more agates. Now he has 15 agates. How many agates did he have at first?

Problem set 16

1. There were 11 towheads in the hall. There were 32 towheads in the room. How many towheads were there in all?

2. Christy saw 8 birds in the back yard. Then she saw some more birds in the front yard. She saw 17 birds in all. How many birds did she see in the front yard?

3. Think of any even number and think of any odd number. Then add the numbers. Is the sum even or odd? Try another even number and another odd number. Is the sum even or odd?

4. Use words to write the number 904.

5. Use digits to write five hundred eleven.

Compare:

6. 314 ◯ 143

7. 906 ◯ 609

8. To which number is the arrow pointing?

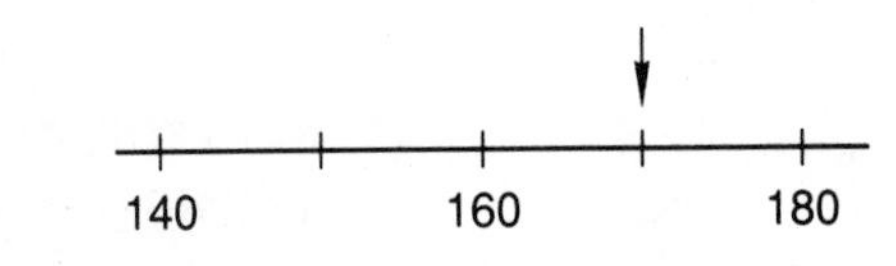

9. $\begin{array}{r} 693 \\ +\ 148 \\ \hline \end{array}$ **10.** $\begin{array}{r} 108 \\ +\ 699 \\ \hline \end{array}$ **11.** $\begin{array}{r} 524 \\ +\ 219 \\ \hline \end{array}$ **12.** $\begin{array}{r} 416 \\ +\ 328 \\ \hline \end{array}$

13. $\begin{array}{r} 532 \\ +\ 198 \\ \hline \end{array}$ **14.** $\begin{array}{r} 4 \\ +\ A \\ \hline 8 \end{array}$ **15.** $\begin{array}{r} 7 \\ +\ P \\ \hline 13 \end{array}$ **16.** $\begin{array}{r} 5 \\ -\ K \\ \hline 1 \end{array}$

17. $\begin{array}{r} 7 \\ +\ P \\ \hline 12 \end{array}$ **18.** $\begin{array}{r} 15 \\ -\ X \\ \hline 9 \end{array}$ **19.** $\begin{array}{r} 4 \\ N \\ +\ 6 \\ \hline 15 \end{array}$ **20.** $\begin{array}{r} C \\ 5 \\ +\ 7 \\ \hline 15 \end{array}$

21. $\begin{array}{r} N \\ 3 \\ +\ 6 \\ \hline 21 \end{array}$ **22.** $\begin{array}{r} 4 \\ +\ N \\ \hline 12 \end{array}$ **23.** $\begin{array}{r} 8 \\ -\ N \\ \hline 2 \end{array}$ **24.** $\begin{array}{r} 7 \\ -\ X \\ \hline 3 \end{array}$

25. $\begin{array}{r} N \\ -\ 2 \\ \hline 5 \end{array}$ **26.** $\begin{array}{r} 7 \\ +\ O \\ \hline 13 \end{array}$ **27.** $\begin{array}{r} N \\ +\ 4 \\ \hline 11 \end{array}$ **28.** $\begin{array}{r} 13 \\ -\ N \\ \hline 9 \end{array}$

Write the next three numbers in each sequence.

29. 35, 30, 25, ___, ___, ___, ...

30. 42, 35, 28, ___, ___, ___, ...

LESSON 17 Subtracting Two-Digit Numbers

> **Speed Test:** 100 Subtraction Facts (Test B in Test Booklet)

We remember that the answer to a subtraction problem is called the difference. When we subtract two-digit numbers, we subtract in the ones' column first. Then we subtract in the tens' column.

Subtract ones	Subtract tens
$\begin{array}{r} 65 \\ -\ 23 \\ \hline 2 \end{array}$	$\begin{array}{r} 65 \\ -23 \\ \hline 42 \end{array}$

Example 1 Find the difference: 68 − 43

Solution We write the first number above the second number. Next we subtract the bottom digit from the top digit in the ones' column. Then we subtract the bottom digit from the top digit in the tens' column. When we subtract 43 from 68, the difference is **25**.

$$\begin{array}{r} 68 \\ -\ 43 \\ \hline 25 \end{array}$$

Example 2 Find XY:

$$\begin{array}{r} 47 \\ -\ XY \\ \hline 25 \end{array}$$

Solution To find XY, we subtract 25 from 47 and get **22**. To check our answer, we replace XY with 22 and check the answer.

$$\begin{array}{r} 47 \\ -\ 25 \\ \hline 22 \end{array} \qquad \begin{array}{r} 47 \\ -\ 22 \\ \hline 25 \end{array}$$

Practice

a. $\begin{array}{r} 48 \\ -\ 24 \\ \hline \end{array}$

b. $\begin{array}{r} 56 \\ -\ 33 \\ \hline \end{array}$

c. $\begin{array}{r} 97 \\ -\ XY \\ \hline 53 \end{array}$

d. $\begin{array}{r} MC \\ -\ 23 \\ \hline 54 \end{array}$

Problem set 17

1. Forty-two red ones were on the first wave. Seventeen red ones were on the second wave. How many red ones were on the first two waves?

2. Mindy saw four green ones in the first hour. In the second hour she saw some more green ones. She saw eleven green ones in all. How many green ones did she see in the second hour?

3. Use the digits 1, 2, and 3 once each to name an even number less than 200.

4. Use words to write the number 915.

5. Use digits to write seven hundred thirteen.

Compare:

6. 218 ◯ 821

7. 704 ◯ 407

8. To which number is the arrow pointing?

30 50

9. $\begin{array}{r} 346 \\ +\ 298 \\ \hline \end{array}$

10. $\begin{array}{r} 499 \\ +\ 275 \\ \hline \end{array}$

11. $\begin{array}{r} 421 \\ +\ 389 \\ \hline \end{array}$

12. $\begin{array}{r} 506 \\ +\ 210 \\ \hline \end{array}$

13. $\begin{array}{r} 438 \\ +\ 296 \\ \hline \end{array}$

14. $\begin{array}{r} 17 \\ -\ A \\ \hline 9 \end{array}$

15. $\begin{array}{r} 7 \\ +\ B \\ \hline 14 \end{array}$

16. $\begin{array}{r} 5 \\ -\ C \\ \hline 2 \end{array}$

17. $\begin{array}{r} 8 \\ +\ D \\ \hline 15 \end{array}$

18. $\begin{array}{r} 15 \\ -\ K \\ \hline 9 \end{array}$

19. $\begin{array}{r} 3 \\ N \\ +\ 2 \\ \hline 13 \end{array}$

20. $\begin{array}{r} 47 \\ -\ 25 \\ \hline \end{array}$

21. $\begin{array}{r} 47 \\ -\ VW \\ \hline 16 \end{array}$

22. $\begin{array}{r} 28 \\ -\ XY \\ \hline 13 \end{array}$

23. $\begin{array}{r} 75 \\ +\ ST \\ \hline 87 \end{array}$

24. $\begin{array}{r} 24 \\ +\ EE \\ \hline 67 \end{array}$

25. $\begin{array}{r} GH \\ +\ 72 \\ \hline 86 \end{array}$

26. $\begin{array}{r} 37 \\ +\ HK \\ \hline 49 \end{array}$

27. $\begin{array}{r} LM \\ +\ 33 \\ \hline 56 \end{array}$

28. $\begin{array}{r} 27 \\ +\ NP \\ \hline 48 \end{array}$

Write the next three numbers in each sequence.

29. 81, 72, 63, ___, ___, ___, . . .

30. 35, 28, 21, ___, ___, ___, . . .

LESSON 18 Subtracting Two-Digit Numbers • Regrouping

Speed Test: 100 Subtraction Facts (Test B in Test Booklet)

Sometimes silly counting is helpful. If we count up by ones from 27, we can say 31 a new way.

twenty-eight
twenty-nine
twenty-ten
twenty-eleven

This shows us that twenty-eleven is the same as 31. The number 31 means 3 tens plus 1 one.

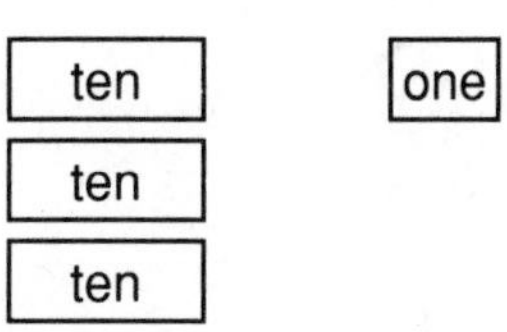

The number 31 also means 2 tens plus 11 ones.

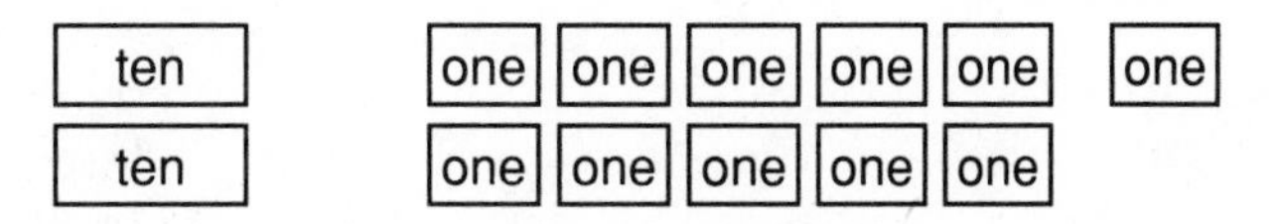

We use this idea to help us subtract. We say that we are **regrouping** or **borrowing** or **exchanging.** We cannot subtract 8 from 1 in this subtraction problem.

$$\begin{array}{r} 31 \\ -\ 18 \\ \hline ? \end{array}$$

But if we trade 1 ten for 10 ones, we can write 31 as twenty-eleven. Then we can subtract.

$$\begin{array}{r} \overset{2}{\cancel{3}}\,{}^{1}1 \\ -\ 1\ 8 \\ \hline 1\ 3 \end{array}$$

Example 1 Find the difference: 56 − 29

Solution We write the first number on top.

$$\begin{array}{r} 56 \\ -\ 29 \\ \hline ? \end{array}$$

Now to subtract we borrow 1 ten from the tens' column. This lets us write 56 as 40 and 16. Then we subtract.

$$\begin{array}{r} \overset{4}{\cancel{5}}\,{}^{1}6 \\ -\ 2\ 9 \\ \hline \mathbf{2\ 7} \end{array}$$

Example 2 Find XY:

$$\begin{array}{r} XY \\ +\ 28 \\ \hline 63 \end{array}$$

Solution To find the missing number in this addition pattern, we can subtract.

$$\begin{array}{r} \overset{5}{\cancel{6}}\,{}^{1}3 \\ -\ 2\ 8 \\ \hline \mathbf{3\ 5} \end{array}$$

Now we check to see if the sum of 35 and 28 is 63.

$$\begin{array}{r} 35 \\ +\ 28 \\ \hline 63 \end{array} \quad \text{check}$$

Example 3 Find AB:

$$\begin{array}{r} 67 \\ -\ AB \\ \hline 48 \end{array}$$

Solution To find the answer, we can subtract 48 from 67.

$$\begin{array}{r} \overset{5}{\not{6}}\,{}^{1}7 \\ -\ 4\ 8 \\ \hline \mathbf{1\ 9} \end{array}$$

To check we remember that the sum of the two bottom numbers in a subtraction pattern must equal the top number.

$$\begin{array}{r} 48 \\ +\ 19 \\ \hline 67 \end{array} \quad \text{check}$$

Practice **a.** 63 − 36 **b.** 40 − 13 **c.** 34 − 27

d. $\begin{array}{r} 70 \\ -\ 42 \\ \hline \end{array}$ **e.** $\begin{array}{r} XY \\ -\ 19 \\ \hline 27 \end{array}$ **f.** $\begin{array}{r} 43 \\ -\ XY \\ \hline 17 \end{array}$

Problem set 18

1. Jimmy found six hundred eighteen acorns under one tree. He found one hundred seventeen acorns under another tree. How many acorns did Jimmy find in all?

2. One day Richard the Lion-Hearted had sixteen knights. On the second day some more knights came. Then he had seventy-two knights. How many knights came on the second day?

3. Use the digits 3, 6, and 7 once each to write an even number less than 400.

4. Use words to write the number 605.

5. The smallest odd two-digit whole number is 11. What is the smallest even two-digit whole number?

Compare:

6. 75 ◯ 57 **7.** 318 ◯ 813

8. To which number is the arrow pointing?

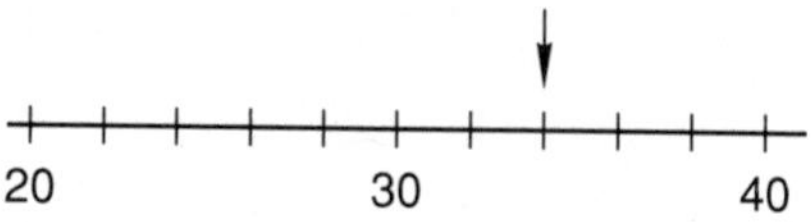

9. $\begin{array}{r} 426 \\ +\ 298 \\ \hline \end{array}$ **10.** $\begin{array}{r} 278 \\ +\ 456 \\ \hline \end{array}$ **11.** $\begin{array}{r} 721 \\ +\ 189 \\ \hline \end{array}$ **12.** $\begin{array}{r} 409 \\ +\ 198 \\ \hline \end{array}$

13. $\begin{array}{r} 421 \\ +\ 149 \\ \hline \end{array}$ **14.** $\begin{array}{r} 18 \\ -\ A \\ \hline 9 \end{array}$ **15.** $\begin{array}{r} 8 \\ +\ B \\ \hline 12 \end{array}$ **16.** $\begin{array}{r} C \\ -\ 4 \\ \hline 1 \end{array}$

17. $\begin{array}{r} D \\ +\ 7 \\ \hline 12 \end{array}$ **18.** $\begin{array}{r} 14 \\ -\ E \\ \hline 8 \end{array}$ **19.** $\begin{array}{r} 46 \\ -\ 28 \\ \hline \end{array}$ **20.** $\begin{array}{r} 35 \\ -\ 16 \\ \hline \end{array}$

21. $\begin{array}{r} 22 \\ -\ 18 \\ \hline \end{array}$ **22.** $\begin{array}{r} 44 \\ -\ 28 \\ \hline \end{array}$ **23.** $\begin{array}{r} 64 \\ -\ NP \\ \hline 36 \end{array}$ **24.** $\begin{array}{r} 14 \\ -\ N \\ \hline 6 \end{array}$

25. $\begin{array}{r} N \\ -\ 5 \\ \hline 8 \end{array}$ **26.** $\begin{array}{r} 9 \\ -\ N \\ \hline 2 \end{array}$ **27.** $\begin{array}{r} N \\ -\ 5 \\ \hline 6 \end{array}$ **28.** $\begin{array}{r} 7 \\ -\ N \\ \hline 3 \end{array}$

Write the next three numbers in each sequence.

29. 80, 75, 70, ___, ___, ___, . . .

30. 14, 21, 28, ___, ___, ___, . . .

LESSON **19**

Expanded Form of Three-Digit Numbers

> **Speed Test:** 100 Subtraction Facts (Test B in Test Booklet)

We can practice the idea of place value by writing numbers in **expanded form.** We remember that the value of a digit in a number depends on the location of the digit.

The value of the 5 in 356 is 5 tens, which is 50.
The value of the 6 in 625 is 6 hundreds, which is 600.
The value of the 8 in 368 is 8 ones, which is 8.

To write a number in expanded form, we write the value of the digits. We separate these values with a plus sign. The value of 365 is 3 hundreds plus 6 tens plus 5 ones. So, in expanded form, 365 is

$$300 + 60 + 5$$

Example 1 Write 275 in expanded form.

Solution We add the values of the digits and get

$$\mathbf{200 + 70 + 5}$$

Example 2 Write 407 in expanded form.

Solution There are no tens, so we write

$$\mathbf{400 + 7}$$

Practice Write these numbers in expanded form.

a. 86 **b.** 325

c. 507 **d.** 225

Problem set 19

1. Twenty-three horses grazed in the pasture close to the barn. Eighty-nine horses were grazing in all the pastures. How many horses were grazing in pastures not close to the barn?

2. Three hundred seventy-five students stood silently in the hall. The other one hundred seven students in the hall were shouting and jumping up and down. Altogether, how many students were in the hall?

3. Use the digits 1, 2, and 8 once each to write an odd number less than 300.

4. Write 782 in expanded form.

5. The largest odd three-digit whole number is 999. What is the smallest three-digit even number?

Compare:

6. 918 ◯ 819

7. 047 ◯ 704

8. To which number is the arrow pointing?

300 400 500

9. $\begin{array}{r} 576 \\ +\ 128 \\ \hline \end{array}$

10. $\begin{array}{r} 243 \\ +\ 578 \\ \hline \end{array}$

11. $\begin{array}{r} 186 \\ +\ 285 \\ \hline \end{array}$

12. $\begin{array}{r} 329 \\ +\ 186 \\ \hline \end{array}$

13. $\begin{array}{r} 406 \\ +\ 318 \\ \hline \end{array}$

14. $\begin{array}{r} 17 \\ -\ A \\ \hline 9 \end{array}$

15. $\begin{array}{r} 8 \\ +\ B \\ \hline 14 \end{array}$

16. $\begin{array}{r} C \\ -\ 7 \\ \hline 2 \end{array}$

17. $\begin{array}{r} D \\ +\ 12 \\ \hline 17 \end{array}$

18. $\begin{array}{r} 15 \\ -\ E \\ \hline 7 \end{array}$

19. $\begin{array}{r} 46 \\ -\ 18 \\ \hline \end{array}$

20. $\begin{array}{r} 42 \\ -\ 16 \\ \hline \end{array}$

21. $\begin{array}{r} 25 \\ -\ 19 \\ \hline \end{array}$

22. $\begin{array}{r} 42 \\ -\ 28 \\ \hline \end{array}$

24. $\begin{array}{r} 64 \\ -\ NP \\ \hline 27 \end{array}$

24. $\begin{array}{r} AB \\ -\ 34 \\ \hline 15 \end{array}$

25. $\begin{array}{r} 68 \\ -\ CD \\ \hline 34 \end{array}$

26. $\begin{array}{r} EF \\ +\ 48 \\ \hline 64 \end{array}$

27. $\begin{array}{r} 62 \\ -\ GH \\ \hline 28 \end{array}$

28. $\begin{array}{r} KL \\ -\ 46 \\ \hline 36 \end{array}$

Write the next three numbers in each sequence.

29. 80, 65, 50, ___, ___, ___, . . .

30. 16, 20, 24, ___, ___, ___, . . .

LESSON 20 Adding Columns of Numbers

Speed Test: 100 Addition Facts (Test A in Test Booklet)

Sometimes the sum of the digits in a column is a number greater than 19. When this happens, we must carry a number that is greater than 1. If the sum of the digits is 32, we carry the 3 to the next column.

Example 1 Add 28, 16, 39, and 29.

Solution We write the numbers one above the other. We add the digits in the ones' column. The sum is 32. We carry the 3 to the tens' column and add. The answer is **112**.

$$\begin{array}{r} 2\ 8 \\ 1\ 6 \\ 3\ 9 \\ +\ 2\ 9 \\ \hline (32) \end{array} \qquad \begin{array}{r} {}^{3} \\ 28 \\ 16 \\ 39 \\ +\ 29 \\ \hline \mathbf{112} \end{array}$$

Example 2 Add: 227 + 88 + 6

Solution We line up the last digits of the numbers. Then we add the digits in the ones' column and get 21.

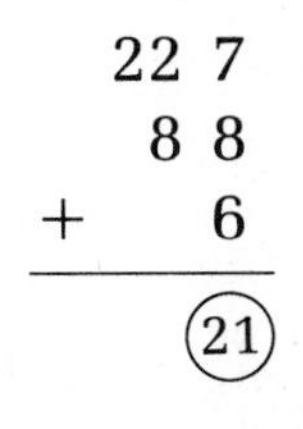

The number 21 is 2 tens plus 1 one. We record the 1 and carry the 2 to the tens' column and add. The sum is 12 tens.

$$\begin{array}{r} {}^{2} \\ 2\ 2\ 7 \\ 8\ 8 \\ 6 \\ \hline (12)\,1 \end{array}$$

We record a 2 in the tens' place and carry the 1 to the next column. Then we add in the hundreds' column.

$$\begin{array}{r} {}^{1}{}^{2} \\ 227 \\ 88 \\ 6 \\ \hline \mathbf{321} \end{array}$$

Practice

a. $\begin{array}{r} 47 \\ 29 \\ 46 \\ +\ 95 \\ \hline \end{array}$

b. $\begin{array}{r} 28 \\ 47 \\ +\ 65 \\ \hline \end{array}$

c. $\begin{array}{r} 38 \\ 22 \\ 31 \\ +\ 46 \\ \hline \end{array}$

d. $\begin{array}{r} 438 \\ 76 \\ +\quad 5 \\ \hline \end{array}$

e. 15 + 24 + 11 + 25 + 36

Problem set 20

1. One doctor put in twenty-four stitches. A second doctor put in some more stitches. There were seventy-five stitches in all. How many stitches did the second doctor put in?

2. Four hundred seven roses were in front. Three hundred sixty-two roses were behind. How many roses were there in all?

3. Use the digits 9, 2, and 8 once each to write an even number less than 300.

4. Use words to write the number 813.

5. The largest two-digit even number is 98. What is the smallest two-digit odd number?

6. To which number is the arrow pointing?

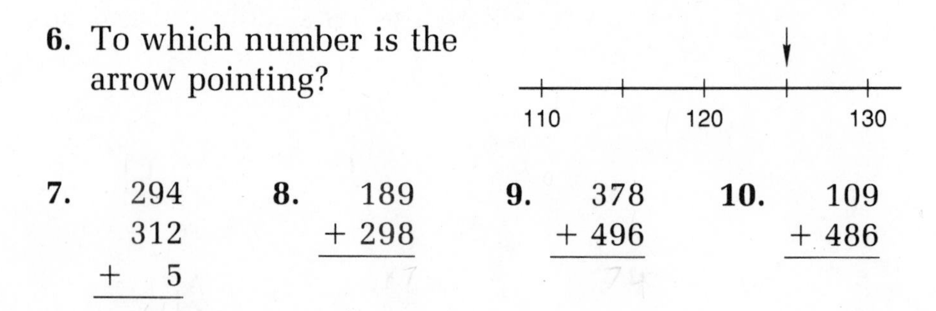

7. $\begin{array}{r} 294 \\ 312 \\ +\quad 5 \\ \hline \end{array}$

8. $\begin{array}{r} 189 \\ +\ 298 \\ \hline \end{array}$

9. $\begin{array}{r} 378 \\ +\ 496 \\ \hline \end{array}$

10. $\begin{array}{r} 109 \\ +\ 486 \\ \hline \end{array}$

11. $$\begin{array}{r} 14 \\ 28 \\ 35 \\ 16 \\ +\ 227 \\ \hline \end{array}$$

12. $$\begin{array}{r} 14 \\ -\ A \\ \hline 7 \end{array}$$

13. $$\begin{array}{r} 8 \\ +\ B \\ \hline 14 \end{array}$$

14. $$\begin{array}{r} C \\ -\ 13 \\ \hline 5 \end{array}$$

15. $$\begin{array}{r} 11 \\ -\ D \\ \hline 9 \end{array}$$

16. $$\begin{array}{r} E \\ -\ 5 \\ \hline 8 \end{array}$$

17. $$\begin{array}{r} 38 \\ -\ 29 \\ \hline \end{array}$$

18. $$\begin{array}{r} 57 \\ -\ 38 \\ \hline \end{array}$$

19. $$\begin{array}{r} 62 \\ -\ 28 \\ \hline \end{array}$$

20. $$\begin{array}{r} 93 \\ -\ 47 \\ \hline \end{array}$$

21. $$\begin{array}{r} 28 \\ 35 \\ 46 \\ 91 \\ +\ 26 \\ \hline \end{array}$$

22. $$\begin{array}{r} 34 \\ +\ AB \\ \hline 81 \end{array}$$

23. $$\begin{array}{r} MN \\ +\ 22 \\ \hline 47 \end{array}$$

24. $$\begin{array}{r} 48 \\ -\ PO \\ \hline 25 \end{array}$$

25. $$\begin{array}{r} CD \\ -\ 46 \\ \hline 15 \end{array}$$

26. $$\begin{array}{r} XY \\ -\ 15 \\ \hline 17 \end{array}$$

Write the next three numbers in each sequence.

27. 105, 90, 75, ___, ___, ___, . . .

28. 48, 44, 40, ___, ___, ___, . . .

29. 12, 15, 18, ___, ___, ___, . . .

30. 20, 25, 30, ___, ___, ___, . . .

LESSON 21 Reading Scales

Speed Test: 100 Subtraction Facts (Test B in Test Booklet)

A **scale** is a type of number line often used for measuring. Scales are found on rulers, gauges, thermometers, speedometers, and many other instruments. The trick to reading a scale is to discover the distance between the marks on the scale. Then we can find the values of all the marks on the scale.

Example 1 What temperature is shown on the thermometer?

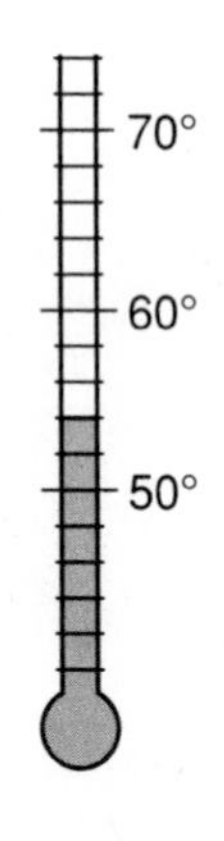

Solution There are five spaces between 50° and 60° on this scale. So each space equals 2°. Thus we must count by twos. If we count up by twos from 50°, we see that the temperature is **54°**.

Example 2 This speedometer shows speed in miles per hours (mph). The speedometer shows that the car is going how fast?

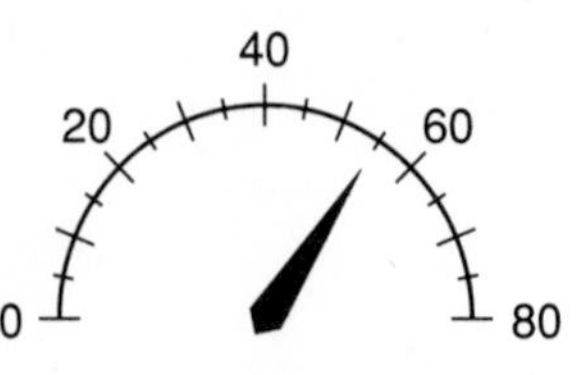

Solution There are 4 spaces between 40 and 60. Each space must equal 5. If we count up by fives from 40, we see that the pointer points to 55. The car is going **55 mph**.

Practice The arrow marks what number on each of these scales?

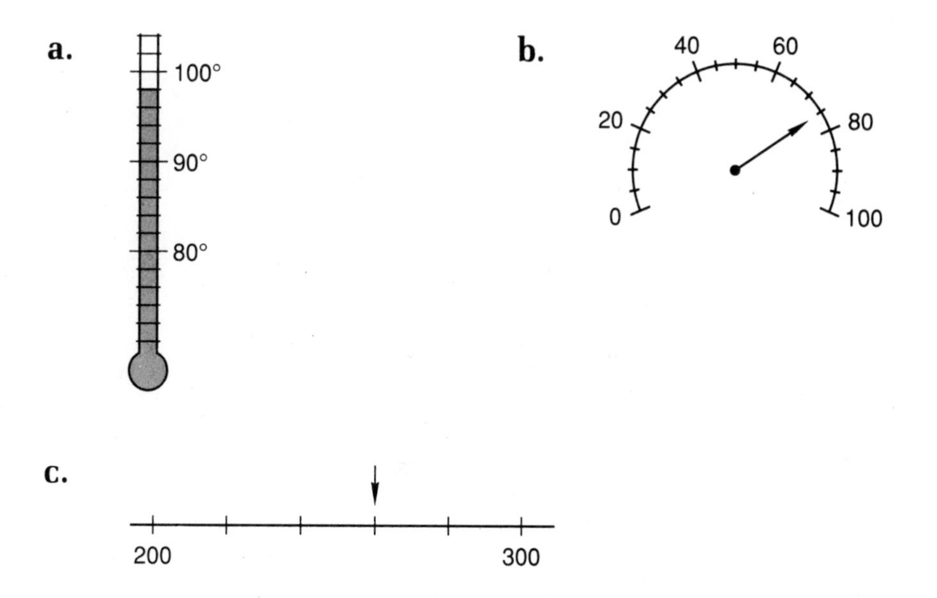

Problem set 21

1. The first flower had twenty-four petals. The second flower had more petals. There were fifty-six petals in all. How many petals did the second flower have?

2. Four hundred seventy-two strutted proudly in front. Two hundred seven more walked sadly behind them. How many were there in all?

3. Use the digits 2, 5, and 7 once each to write an even number that is greater than 700.

4. The tens' digit was 4. The ones' digit was 9. The number was between 200 and 300. What was the number?

5. Which of these numbers is an odd number that is greater than 750?
   (a) 846 (b) 864 (c) 903 (d) 309

6. To which number is the arrow pointing?

   400 500

7. $$\begin{array}{r} 392 \\ +\ 278 \\ \hline \end{array}$$

8. $$\begin{array}{r} 439 \\ +\ 339 \\ \hline \end{array}$$

9. $$\begin{array}{r} 774 \\ +\ 174 \\ \hline \end{array}$$

10. $$\begin{array}{r} 389 \\ +\ 398 \\ \hline \end{array}$$

11. $$\begin{array}{r} 13 \\ 25 \\ 46 \\ 25 \\ +\ 29 \\ \hline \end{array}$$

12. $$\begin{array}{r} 18 \\ -\ A \\ \hline 12 \end{array}$$

13. $$\begin{array}{r} 8 \\ +\ B \\ \hline 16 \end{array}$$

14. $$\begin{array}{r} C \\ -\ 5 \\ \hline 3 \end{array}$$

15. $$\begin{array}{r} 13 \\ -\ D \\ \hline 9 \end{array}$$

16. $$\begin{array}{r} 17 \\ -\ E \\ \hline 12 \end{array}$$

17. $$\begin{array}{r} 42 \\ -\ 29 \\ \hline \end{array}$$

18. $$\begin{array}{r} 22 \\ -\ 18 \\ \hline \end{array}$$

19. $$\begin{array}{r} 62 \\ -\ 48 \\ \hline \end{array}$$

20. $$\begin{array}{r} 82 \\ -\ 58 \\ \hline \end{array}$$

21. $$\begin{array}{r} 28 \\ 36 \\ 57 \\ +\ 47 \\ \hline \end{array}$$

22. $$\begin{array}{r} 35 \\ -XY \\ \hline 14 \end{array}$$

23. $$\begin{array}{r} 45 \\ +\ MP \\ \hline 54 \end{array}$$

24. $$\begin{array}{r} 75 \\ -\ KL \\ \hline 47 \end{array}$$

25. $$\begin{array}{r} BC \\ -\ 47 \\ \hline 35 \end{array}$$

26. $$\begin{array}{r} DE \\ +\ 15 \\ \hline 37 \end{array}$$

Write the next three numbers in this sequence:

27. 210, 190, 170, ___, ___, ___, . . .

28. Write 498 in expanded form.

29. Compare: 423 ◯ 432

30. $4 + 7 + 3 + 5 + 2 + 3 + 2 + 7 + 8$

LESSON 22 Reading Time from a Clock

Speed Test: 100 Subtraction Facts (Test B in Test Booklet)

The scale on a clock is actually two number lines in one. One number line marks hours and is usually numbered. The other number line marks minutes and is not numbered.

To tell time, we read the location of the short hand on the hour number line. We also read the location of the long hand on the minute number line.

To write the time of day, we write the hour followed by a colon. Then we write two digits to show the number of minutes after the hour. We use the abbreviation a.m. for the 12 hours before noon and p.m. for the 12 hours after noon.

Example If it is evening, what time is shown by the clock?

Solution It is after 9 p.m. It is 20 minutes after 9:00 p.m. It is **9:20 p.m.**

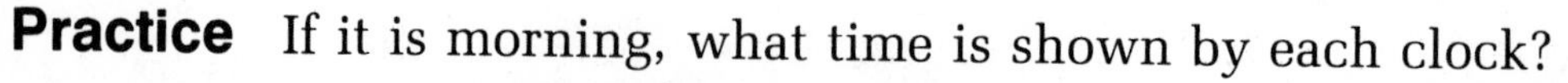

Practice If it is morning, what time is shown by each clock?

Problem set 22

1. On the first day Sarah sharpened fifty-one pencils. Then she sharpened some more pencils on the second day. She sharpened seventy-six pencils in all. How many pencils did she sharpen on the second day?

2. Two hundred seventy came on Monday. Three hundred two came on Tuesday. One hundred eleven came on Wednesday. How many came in all?

3. The hundreds' digit was 5. The ones' digit was 7. The number was greater than 540 and less than 550. What was the number?

4. Write 905 in expanded form.

5. Use digits and a comparison symbol to write "One hundred twenty is greater than one hundred twelve."

6. If it is morning, what time is shown on this clock?

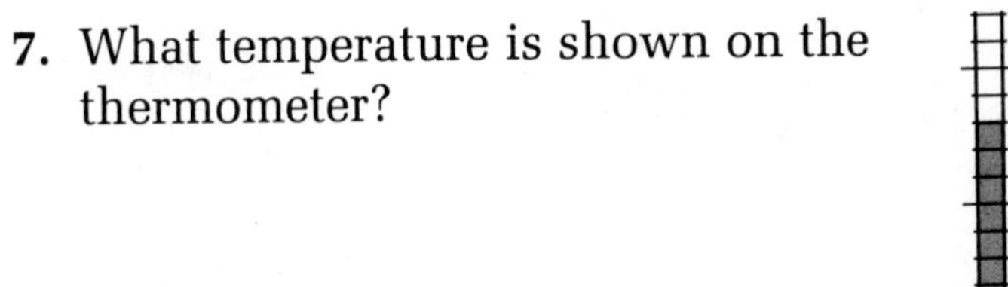

7. What temperature is shown on the thermometer?

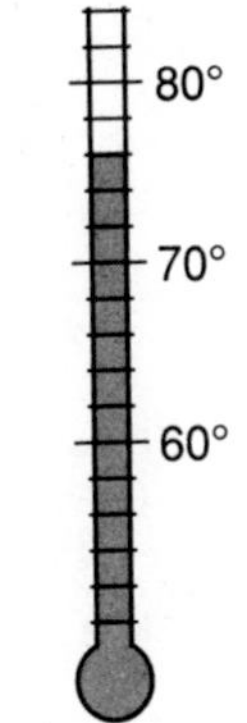

8. $468 + 293$

9. $468 + 185$

10. $187 + 698$

11. $14 - A = 7$

12. $8 + B = 16$

13. $C - 8 = 7$

14. $14 - D = 9$

15. $E - 5 = 4$

16. $44 - 28$

17. $23 - 18$

18. $62 - 43$

19. $74 - 58$

20. $45 - CP = 21$

21. $13 + AB = 37$

22. $EF - 45 = 32$

23. $75 - GH = 54$

24. $KP - 47 = 35$

25. $NQ - 15 = 37$

26. $14 + 28 + 36 + 84$

27. $25 + 28 + 46 + 88$

28. $93 + 44 + 26 + 54$

29. $25 + 37 + 93 + 24$

30. Write the next three numbers in this sequence:

60, 75, 90, ___, ___, ___, . . .

LESSON 23 Reading a Centimeter Scale

Speed Test: 100 Subtraction Facts (Test B in Test Booklet)

A **centimeter** is a unit used for measuring length. A centimeter is this long:

———

To measure lengths in centimeters, we use a centimeter scale. We match the length of the object we are measuring with the marks on the scale.

Example 1 How long is this nail?

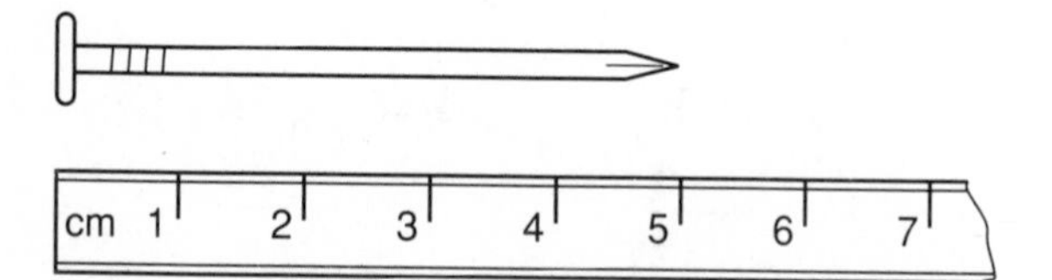

Solution The head of the nail is lined up on the "zero" of the centimeter scale. The point of the nail is closest to the 5-centimeter mark of the scale, so the nail is 5 centimeters long. We abbreviate this by writing **5 cm**.

Practice Find the length of each object to the nearest centimeter.

a.

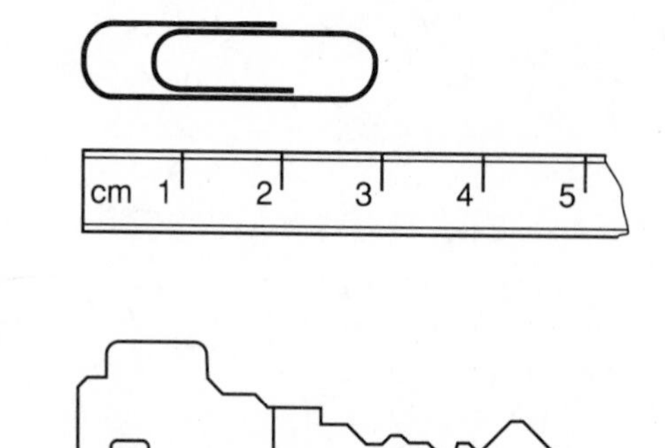

b.

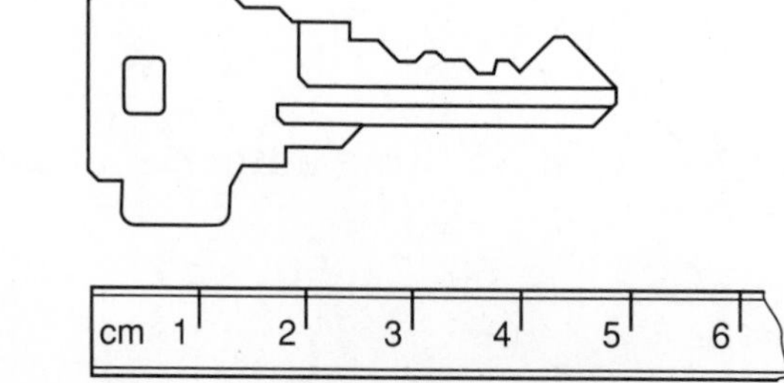

Problem set 23

1. Martine saw a "whole bunch" of them one day. She saw forty-eight of them on the second day. If she saw seventy-two of them in all, how many were in the whole bunch?

2. Four hundred seventy-six stood quietly in one line. Three hundred ninety-seven stood quietly in another line. Altogether, how many stood quietly in line?

3. The ones' digit was 5. The hundreds' digit was 6. The number was between 640 and 650. What was the number?

4. Write 509 in expanded form.

5. Use digits and a comparison symbol to compare 493 and 439.

6. What temperature is shown on this thermometer?

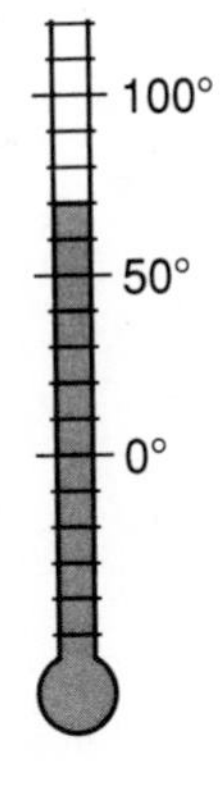

7. It is afternoon. What time is it?

8. How long is the pin?

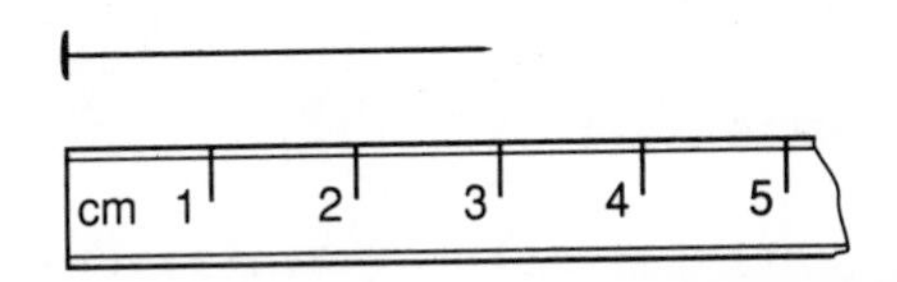

9.	10.	11.	12.
476 + 285	185 + 499	568 + 397	478 + 196

13. $\begin{array}{r} 17 \\ -\ A \\ \hline 9 \end{array}$ **14.** $\begin{array}{r} 14 \\ -\ B \\ \hline 14 \end{array}$ **15.** $\begin{array}{r} 13 \\ -\ C \\ \hline 6 \end{array}$ **16.** $\begin{array}{r} 35 \\ -\ 28 \\ \hline \end{array}$

17. $\begin{array}{r} 23 \\ -\ 15 \\ \hline \end{array}$ **18.** $\begin{array}{r} 63 \\ -\ 36 \\ \hline \end{array}$ **19.** $\begin{array}{r} 74 \\ -\ 59 \\ \hline \end{array}$ **20.** $\begin{array}{r} KM \\ +\ 22 \\ \hline 45 \end{array}$

21. $\begin{array}{r} 47 \\ -\ FK \\ \hline 34 \end{array}$ **22.** $\begin{array}{r} 41 \\ -\ NP \\ \hline 24 \end{array}$ **23.** $\begin{array}{r} 75 \\ -\ AC \\ \hline 43 \end{array}$ **24.** $\begin{array}{r} BF \\ -\ 42 \\ \hline 44 \end{array}$

25. $\begin{array}{r} GK \\ -\ 15 \\ \hline 32 \end{array}$ **26.** $\begin{array}{r} 28 \\ 36 \\ 44 \\ +\ 58 \\ \hline \end{array}$ **27.** $\begin{array}{r} 49 \\ 28 \\ 32 \\ +\ 55 \\ \hline \end{array}$

28. $\begin{array}{r} 93 \\ 86 \\ 75 \\ +\ 98 \\ \hline \end{array}$ **29.** $\begin{array}{r} 44 \\ 28 \\ 37 \\ +\ 25 \\ \hline \end{array}$

30. Write the next three numbers in this sequence:

48, 40, 32, ___, ___, ___, ...

LESSON 24 Triangles, Rectangles, Squares, and Circles

Speed Test: 100 Addition Facts (Test A in Test Booklet)

In this lesson we will discuss drawing triangles, rectangles, squares, and circles.

Example 1 Draw a triangle that has all sides the same length.

Solution You may need to practice on scratch paper until you find out how to draw this triangle. A triangle has three sides. If you start out like this,

the third side will be too long. A triangle that has all sides the same length looks like this:

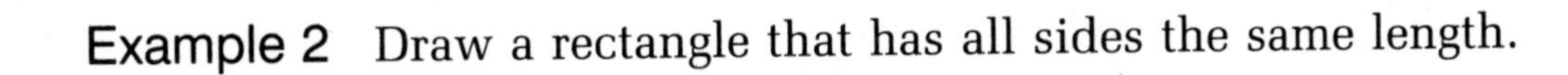

Example 2 Draw a rectangle that has all sides the same length.

Solution A rectangle does not have to be longer than it is wide. A rectangle does have "square corners" and four sides. A rectangle with all sides the same length looks like this:

This may look like a square to you. It **is** a square. It is also a rectangle. **A square is a special kind of rectangle.**

Example 3 Draw a circle that is about 1 centimeter across the center.

Solution First we draw a line segment one centimeter long. Then draw a circle that goes around the segment.

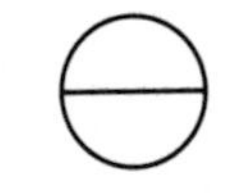

Circle drawing improves with practice.

Practice a. Draw a triangle that has two equal sides.

b. Draw a rectangle that is about twice as long as it is wide.

c. Draw a square. Inside the square draw a circle that just touches all four sides of the square.

Problem set 24

1. Roger had four hundred seventeen marbles. Harry had two hundred twenty-two marbles. How many marbles did Roger and Harry have in all?

2. Susie put forty-seven jacks into the pile. Jane put all of her jacks in the pile. Then there were seventy-two jacks in the pile. How many jacks did Jane put in?

3. The hundreds' digit was 6. The ones' digit was 5. The number was greater than 640 and less than 650. What was the number?

4. Write seven hundred fifty-three in expanded form.

5. Use digits and a comparison symbol to compare four hundred seventy-seven and five hundred forty-two.

6. The arrow is pointing to what number on this scale?

0 200 400 600

7. How long is the pencil?

cm 1 2 3 4 5 6 7 8 9 10 11 12 13

8. $\begin{array}{r} 493 \\ +\ 278 \\ \hline \end{array}$

9. $\begin{array}{r} 486 \\ +\ 378 \\ \hline \end{array}$

10. $\begin{array}{r} 524 \\ +\ 109 \\ \hline \end{array}$

11. $\begin{array}{r} 327 \\ +\ 486 \\ \hline \end{array}$

12. $\begin{array}{r} 128 \\ +\ 549 \\ \hline \end{array}$

13. Draw a triangle that has all sides the same length.

14. Draw a square. Then draw a circle around the square so that each "corner" of the square just touches the circle.

15. $\begin{array}{r} 17 \\ -\ A \\ \hline 9 \end{array}$

16. $\begin{array}{r} 45 \\ -\ 29 \\ \hline \end{array}$

17. $\begin{array}{r} 15 \\ -\ B \\ \hline 6 \end{array}$

18. $\begin{array}{r} 62 \\ -\ 45 \\ \hline \end{array}$

19. $\begin{array}{r} 24 \\ +\ CD \\ \hline 45 \end{array}$

20. $\begin{array}{r} 14 \\ -\ AB \\ \hline 2 \end{array}$

21. $\begin{array}{r} XY \\ -\ 36 \\ \hline 56 \end{array}$

22. $\begin{array}{r} 75 \\ -\ MP \\ \hline 48 \end{array}$

23. $\begin{array}{r} 46 \\ 35 \\ 27 \\ +\ 39 \\ \hline \end{array}$

24. $\begin{array}{r} 14 \\ 28 \\ 77 \\ +\ 23 \\ \hline \end{array}$

25. $\begin{array}{r} 14 \\ 23 \\ 38 \\ +\ 64 \\ \hline \end{array}$

26. $\begin{array}{r} 15 \\ 24 \\ 36 \\ +\ 99 \\ \hline \end{array}$

27. $\begin{array}{r} 22 \\ 33 \\ 44 \\ +\ 55 \\ \hline \end{array}$

28. $\begin{array}{r} 30 \\ 49 \\ 78 \\ +\ 46 \\ \hline \end{array}$

29. $\begin{array}{r} 32 \\ 91 \\ 43 \\ +\ 75 \\ \hline \end{array}$

30. Write the next three numbers in this sequence:

28, 35, 42, ___, ___, ___, . . .

LESSON 25 Rounding Whole Numbers to the Nearest Ten

Speed Test: 100 Subtraction Facts (Test B in Test Booklet)

One of the sentences below uses an **exact number**. The other sentence uses a **rounded number**. Can you tell which sentence uses the rounded number?

- The radio cost about $70.
- The radio cost $68.47.

The first sentence uses a rounded number. Rounded numbers are often used in place of exact numbers because they are easy to understand and easy to work with.

To round an exact number to the nearest ten, we choose the closest number that ends in zero. The number line can help us understand rounding.

60 61 62 63 64 65 66 67 68 69 70

We see that 67 is between 60 and 70. Since 67 is closer to 70 than it is to 60, we say that 67 is "about 70." When we say this, we have rounded 67 to the nearest ten.

Example Round 82 to the nearest ten.

Solution Rounding to the nearest ten means we round to a number we would say when counting by tens (10, 20, 30, 40, and so on). We will use a number line marked off in tens to picture this problem.

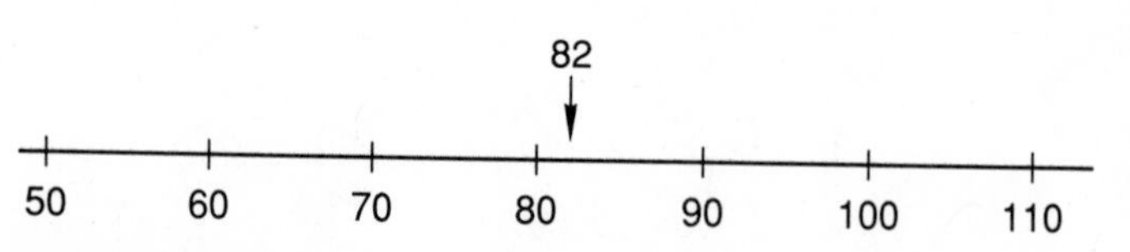

We see that 82 is between 80 and 90. Since 82 is closer to 80 than it is to 90, we round 82 to **80**.

Practice Round each number to the nearest ten. For each problem draw a number line to show your work.

a. 78 **b.** 43 **c.** 61 **d.** 49

Problem set 25

1. Forty-four were in front. Seventy-six were in the middle. Twenty-three brought up the rear. How many were there in all?

2. The smallest three-digit whole number was subtracted from two hundred twenty-four. What was the answer?

3. Amelda had twenty-six books. Then Grant gave her some more books. Now she has ninety-four books. How many books did Grant give to Amelda?

4. The hundreds' digit was 7. The ones' digit was 7. The number was between 740 and 750. What was the number?

5. Write four hundred eleven in expanded form.

6. Use digits and a comparison symbol to write five hundred eighty is greater than five hundred eighteen.

7. Round 71 to the nearest ten.

8. How long is the AA battery?

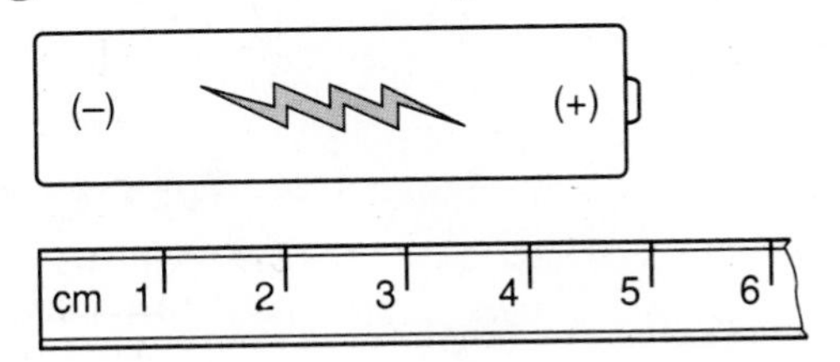

9. Draw a triangle that has all sides the same length. Then draw a circle around the triangle that just touches the "corners" of the triangle.

10. $\begin{array}{r} 387 \\ +\ 196 \\ \hline \end{array}$ **11.** $\begin{array}{r} 496 \\ +\ 508 \\ \hline \end{array}$ **12.** $\begin{array}{r} 762 \\ +\ 109 \\ \hline \end{array}$ **13.** $\begin{array}{r} 458 \\ +\ 385 \\ \hline \end{array}$

14. $\begin{array}{r} 129 \\ +\ 548 \\ \hline \end{array}$ **15.** $\begin{array}{r} 43 \\ -\ 27 \\ \hline \end{array}$ **16.** $\begin{array}{r} 42 \\ -\ 15 \\ \hline \end{array}$ **17.** $\begin{array}{r} 52 \\ -\ XY \\ \hline 27 \end{array}$

18. $\begin{array}{r} BX \\ -\ 29 \\ \hline 18 \end{array}$ **19.** $\begin{array}{r} 73 \\ -\ AC \\ \hline 46 \end{array}$ **20.** $\begin{array}{r} 14 \\ 28 \\ 47 \\ +\ 33 \\ \hline \end{array}$ **21.** $\begin{array}{r} 21 \\ 38 \\ 46 \\ +\ 93 \\ \hline \end{array}$

22. $\begin{array}{r} 37 \\ 45 \\ 36 \\ +\ 24 \\ \hline \end{array}$ **23.** $\begin{array}{r} MK \\ +\ 54 \\ \hline 76 \end{array}$ **24.** $\begin{array}{r} 15 \\ -\ N \\ \hline 6 \end{array}$

Write the next three numbers in each sequence.

25. 14, 21, 28, ___, ___, ___, . . .

26. 45, 50, 55, ___, ___, ___, . . .

27. 28, 35, 42, ___, ___, ___, . . .

28. $\begin{array}{r} 4 \\ 12 \\ 15 \\ +\ 24 \\ \hline \end{array}$ **29.** $\begin{array}{r} 15 \\ 27 \\ 46 \\ +\ 23 \\ \hline \end{array}$ **30.** $\begin{array}{r} 93 \\ 45 \\ 26 \\ +\ 43 \\ \hline \end{array}$

LESSON 26 Metric Units of Length

Speed Test: 100 Subtraction Facts (Test B in Test Booklet)

The centimeter is a unit we use to measure short distances. To measure longer distances, we use the **meter** or the **kilometer**.

A meter is exactly 100 centimeters. Perhaps you have a meterstick in your classroom. A meterstick is a ruler that is 1 meter long. Notice it is also 100 centimeters long. If you took a **big step**, your step would be about **1 meter**.

From the word "meter" comes the word "metric." The **metric system** of measurement uses meters, fractions of meters, and multiples of meters to measure distances.

A much larger unit of length in the metric system is the kilometer. **Kilo-** means 1000. **A kilometer is 1000 meters.** If you took **1000 big steps,** you would walk about **1 kilometer**.

We often use the abbreviations cm, m, and km to stand for centimeters, meters, and kilometers.

Example A door is about how many meters tall?

Solution To estimate, you may think of a meter as one *big* step and imagine how many steps it would take to "walk up" a door. If you have a meterstick, you can measure. Most doors are about **2 meters tall**.

Practice

a. If Simon said, "Take 5 giants steps," then about how many meters would you walk?

b. One meter is how many centimeters?

c. One kilometer is how many meters?

Problem set 26

1. When the bell rang, three hundred forty-two came. When the bell rang again, two hundred fourteen came. How many came in all?

2. The small horn sounded 46 times. The big horn sounded many times. If the horns sounded a total of 93 times, how many times did the big horn sound?

3. The ones' digit was 5. The hundreds' digit was 3. The number was between 370 and 380. What was the number?

4. Write seven hundred forty-two in expanded form.

5. Round sixty-eight to the nearest ten.

6. A door is about how many meters wide?

7. What temperature is shown on the thermometer?

90°

80°

8. Draw a rectangle that is 2 centimeters long and 1 centimeter wide.

9. Use digits and a comparison symbol to write three hundred twenty-seven is less than three hundred thirty.

10. $\begin{array}{r} 43 \\ -\ 27 \\ \hline \end{array}$

11. $\begin{array}{r} 278 \\ +\ 569 \\ \hline \end{array}$

12. $\begin{array}{r} 61 \\ -\ 37 \\ \hline \end{array}$

13. $\begin{array}{r} 72 \\ -\ 29 \\ \hline \end{array}$

14. $\begin{array}{r} 389 \\ +\ 256 \\ \hline \end{array}$

15. $\begin{array}{r} 47 \\ -\ PQ \\ \hline 14 \end{array}$

16. $\begin{array}{r} 46 \\ +\ RS \\ \hline 74 \end{array}$

17. 486 + 276

18. 91 − 76

19. 42 + 23 + 86 + 98

20. 95 + 87 + 24 + 38

21. $\begin{array}{r} 18 \\ -\ AC \\ \hline 6 \end{array}$

22. $\begin{array}{r} BD \\ -\ 36 \\ \hline 36 \end{array}$

23. $\begin{array}{r} 55 \\ -\ MK \\ \hline 31 \end{array}$

24. $\begin{array}{r} 75 \\ -\ 49 \\ \hline \end{array}$

25. $\begin{array}{r} 83 \\ -\ 27 \\ \hline \end{array}$

Write the next three numbers in each sequence.

26. 24, 30, 36, ___, ___, ___, . . .

27. 28, 32, 36, ___, ___, ___, . . .

28. 80, 65, 50, ___, ___, ___, . . .

29. 24 + 26 + 35 + 27 + 15

30. 36 + 15 + 43 + 55 + 23

LESSON 27 Naming Fractions

Speed Test: 100 Subtraction Facts (Test B in Test Booklet)

Part of a whole can be named with a fraction. A fraction is written with two numbers. The bottom number of a fraction is called the **denominator**. The denominator tells how many equal parts are in the whole. The top number of a fraction is called the **numerator**. The numerator tells how many of the parts are being counted. Some fractions and their names are given here.

$\frac{1}{2}$	one half	$\frac{3}{5}$	three fifths
$\frac{1}{3}$	one third	$\frac{5}{6}$	five sixths
$\frac{2}{3}$	two thirds	$\frac{7}{8}$	seven eighths
$\frac{1}{4}$	one fourth	$\frac{1}{10}$	one tenth

Example What fraction of the circle is shaded?

Solution There are 4 equal parts, and 3 are shaded. Therefore, the fraction of the circle that is shaded is three fourths, which we write

$$\mathbf{\frac{3}{4}}$$

Practice What fraction of each shape is shaded?

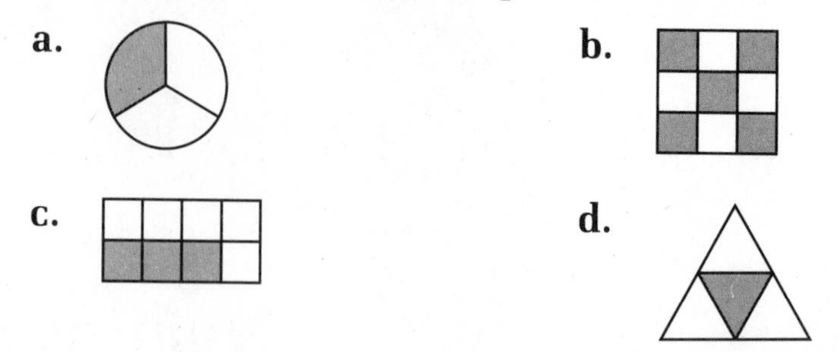

Problem set 27

1. The first four odd whole numbers are 1, 3, 5, and 7. What is their sum?

2. James was 49 inches tall at the beginning of summer. He grew 2 inches over the summer. How tall was James at the end of summer?

3. Use the digits 1, 2, and 3 once each to make an odd number less than 200.

Write the next three numbers in each sequence.

4. 80, 72, 64, ___, ___, ___, ...

5. 60, 54, 48, ___, ___, ___, ...

Find the missing number in each sequence.

6. 18, ___, 36, 45, ...

7. 18, ___, 22, 24, ...

8. What is the place value of the 9 in 891?

9. Write 106 in expanded form.

Use words to write each of these numbers.

10. 106

11. 160

12. Use digits and a comparison symbol to write that eighteen minus nine equals five plus four.

13. Round 28 to the nearest ten.

14. A bicycle is about how many meters long?

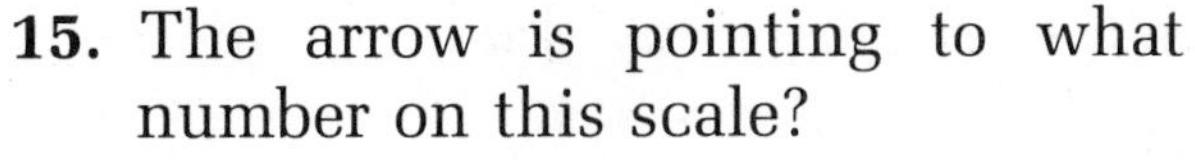

15. The arrow is pointing to what number on this scale?

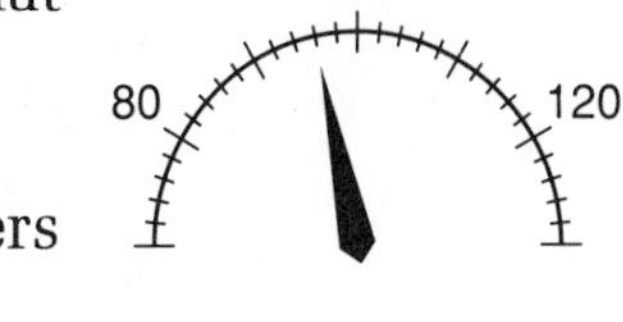

16. Draw a circle that is 2 centimeters across the center.

17. What fraction of the rectangle is shaded?

18. 51 − 43
19. 70 − 44
20. 37 − 9

21. 26 − 17
22. 61 − 35
23. 41 − 27

24. 879 + 64
25. 76 + 34 + 51 + 25

26. $\begin{array}{r} N \\ +\ 13 \\ \hline 17 \end{array}$
27. $\begin{array}{r} NX \\ -\ 42 \\ \hline 27 \end{array}$
28. $\begin{array}{r} 31 \\ -\ PO \\ \hline 14 \end{array}$
29. $\begin{array}{r} XY \\ -\ 24 \\ \hline 41 \end{array}$

30. 24 + 16 + 32 + 47 + 28 + 37

LESSON **28**

Parallel and Perpendicular Lines

Speed Test: 100 Subtraction Facts (Test B in Test Booklet)

Lines that go in the same direction and stay the same distance apart are **parallel lines**.

Pairs of parallel lines

When lines cross, we say they **intersect**.

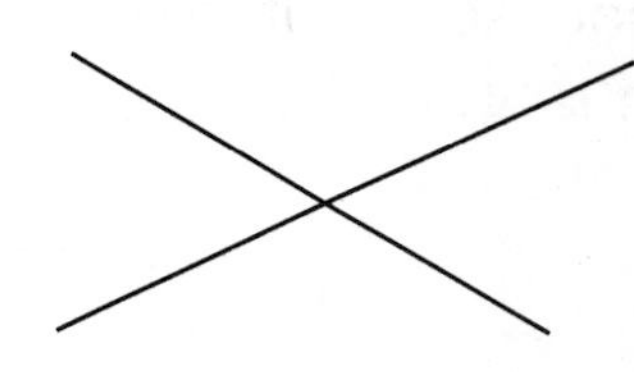

Pairs of intersecting lines

Intersecting lines that form "square corners" are **perpendicular.**

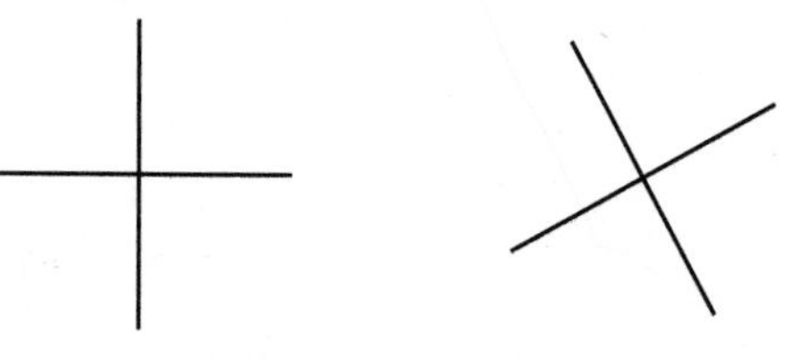

Pairs of perpendicular lines

Example (a) Draw a pair of parallel lines.
(b) Draw a pair of perpendicular lines.

Solution (a) A pair of lines that are always the same distance apart are parallel.

(b) A pair of lines that intersect and form square corners are perpendicular lines.

Practice **a.** Draw two lines that intersect but are not perpendicular.

b. Draw two lines that are perpendicular.

c. Are train tracks parallel or perpendicular?

Problem set 28

1. Twenty-eight were in the front of the line. Forty-two were next in line. Then there were eighty-eight more. Sixty-five stood sadly at the back of the line. How many were in line?

2. The reds entered first. Then came 37 blues. If there were 76 in all, how many were red?

3. Use the digits 1, 2, and 3 once each to make an odd number greater than 300.

Write the next three numbers in each sequence.

4. 40, 36, 32, ___, ___, ___, . . .

5. 30, 27, 24, ___, ___, ___, . . .

6. Use digits and a comparison symbol to write that six hundred thirty-eight is less than six hundred eighty-three.

7. Round 92 to the nearest ten.

8. A nickel is 2 centimeters across the center. If 10 nickels are laid in a line, how long would the line be?

9. How long is this rectangle?

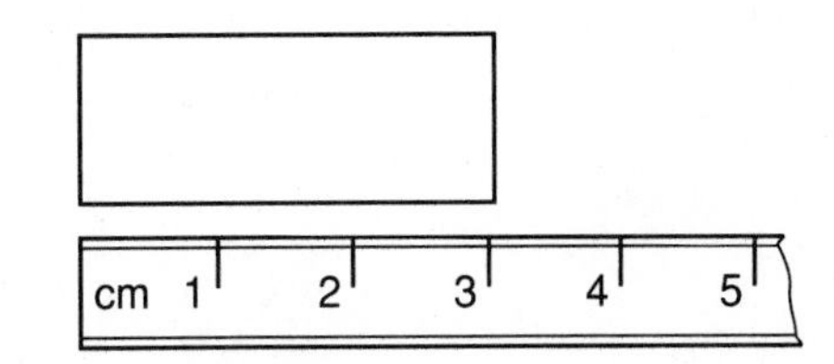

10. Draw a pair of perpendicular lines.

11. What fraction of this triangle is shaded?

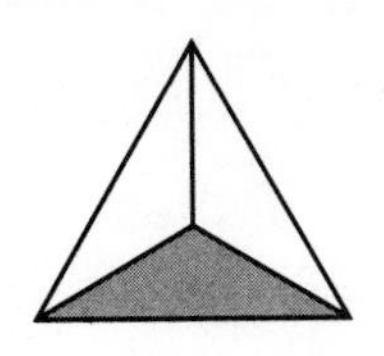

12. If it is afternoon, what time is it?

13. $\begin{array}{r} 83 \\ -\ 27 \\ \hline \end{array}$

14. $\begin{array}{r} 42 \\ -\ 27 \\ \hline \end{array}$

15. $\begin{array}{r} 72 \\ -\ 36 \\ \hline \end{array}$

16. $\begin{array}{r} 428 \\ +\ 196 \\ \hline \end{array}$

17. $\begin{array}{r} 436 \\ +\ 295 \\ \hline \end{array}$

18. $\begin{array}{r} 58 \\ +\ AK \\ \hline 87 \end{array}$

19. $\begin{array}{r} 67 \\ -\ XB \\ \hline 48 \end{array}$

20. $\begin{array}{r} PK \\ -\ 22 \\ \hline 22 \end{array}$

21. $\begin{array}{r} XY \\ -\ 18 \\ \hline 49 \end{array}$

22. $\begin{array}{r} 63 \\ -\ 48 \\ \hline \end{array}$

23. 24 + 48 + 68 + 72

24. 95 + 21 + 14 + 26

25. 42 − 27

26. 55 − 48

27. 31 − 14

28. 4 + 3 + 2 + 5 + 7 + 6 + 3 + 8 + 5 + 4

29. 21 + 25 + 43 + 64

30. 5 + 22 + 43 + 25

LESSON 29 Right Angles

Speed Test: 100 Subtraction Facts (Test B in Test Booklet)

Two pairs of intersecting lines are shown below. The intersecting lines on the left are not perpendicular. The intersecting lines on the right are perpendicular.

Lines make corners when they intersect. The corners in the left-hand figure on the last page are not square corners. The corners in the right-hand figure are square corners. **Perpendicular lines make square corners.** The corners are called **angles**. Square corners are called **right angles**.

Example 1 Which of these angles is a right angle?

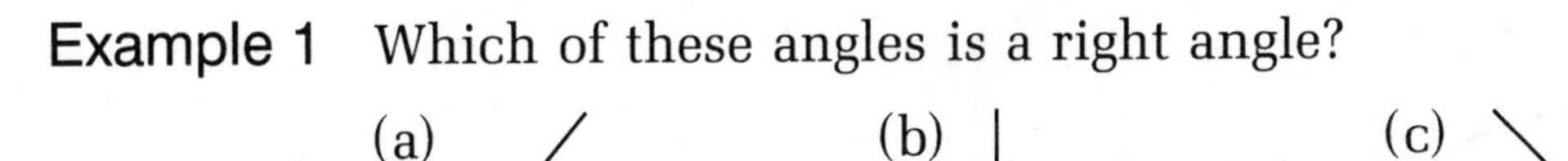

Solution Angle (b) makes a square corner, so **angle (b) is a right angle**.

Example 2 Which shape below has four right angles?

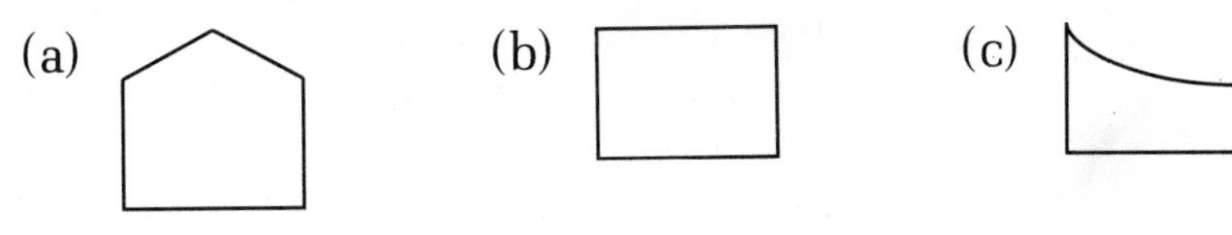

Solution The shape labeled (b) has four square corners. Thus **shape (b) has four right angles**.

Example 3 Draw a triangle that has one right angle.

Solution We begin by drawing two lines to make a right angle. Then we draw the third side.

Practice **a.** A triangle has how many angles?

b. Which of these angles is not a right angle?

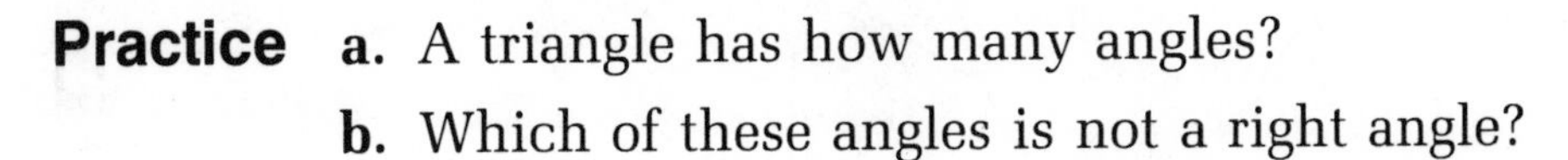

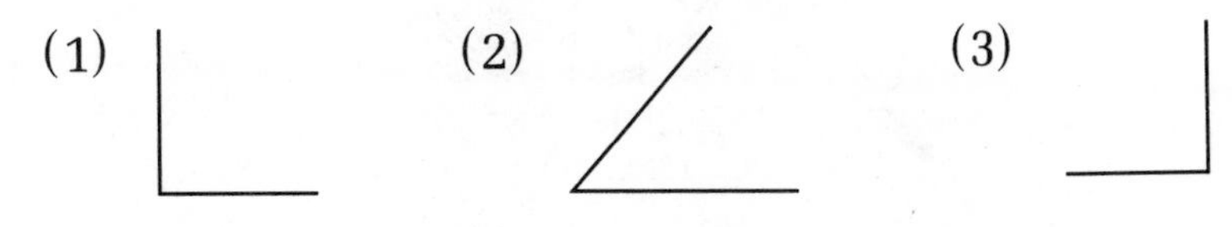

c. If two lines intersect and form right angles, then are the lines parallel or perpendicular?

Problem set 29

1. Two hundred seventy-five sat in the first row. Three hundred sixty-two sat in the second row. How many sat in the first two rows?

2. There were 47 apples in the big tree. There was a total of 82 apples in the big tree and in the little tree. How many apples were in the little tree?

3. All the students lined up in two equal rows. Which could not be the total number of students?

   36 45 60

Find the missing numbers in each sequence.

4. 9, 18, ___, ___, 45, ___, ...

5. 7, 14, ___, ___, 35, ___, ...

6. Compare: 15 − 9 ◯ 13 − 8

7. Round 77 to the nearest ten.

8. A professional basketball player is about how many meters tall?

9. If it is morning, what time is shown on this clock?

10. Which street is parallel to Elm?

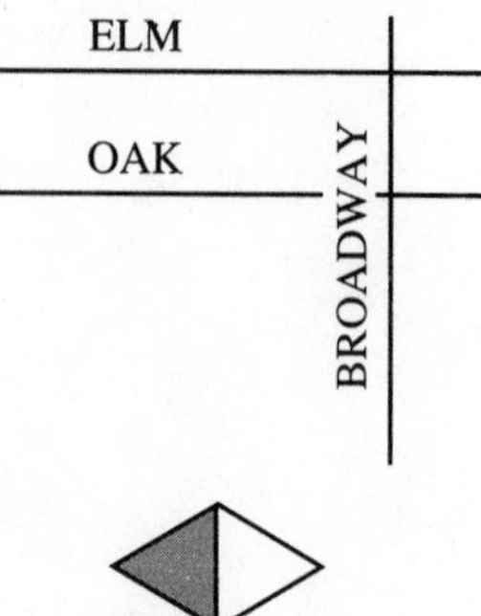

11. What fraction of this shape is shaded?

12. Draw a rectangle that is 3 centimeters long and 2 centimeters wide.

13. Which of these angles is a right angle?

(a) (b) (c)

14. $31 - 14$

15. $468 + 247$

16. $52 - 37$

17. $497 + 258$

18. $61 - 27$

19. $AB + 65 = 87$

20. $87 - CD = 59$

21. $MN - 32 = 12$

22. $36 - AC = 19$

23. $46 - E = 37$

24. 16 + 43 + 59 + 61

25. 448 + 57 + 65 + 77

26. 46 − 28

27. 41 − 32

28. 76 − 58

29. 243 + 25 + 37 + 88

30. 416 + 35 + 27 + 43 + 5

LESSON 30 "Some Went Away" Word Problems

Speed Test: 100 Addition Facts (Test A in Test Booklet)

The thinking pattern for some word problems is a subtraction pattern. One kind of problem that has a subtraction pattern has the thought that **some went away**. Read this story.

> John had 7 marbles. Then he lost 3 marbles. He had 4 marbles left.

This story is about subtraction. John had **some** marbles. Then **some marbles went away**. How many marbles are left?

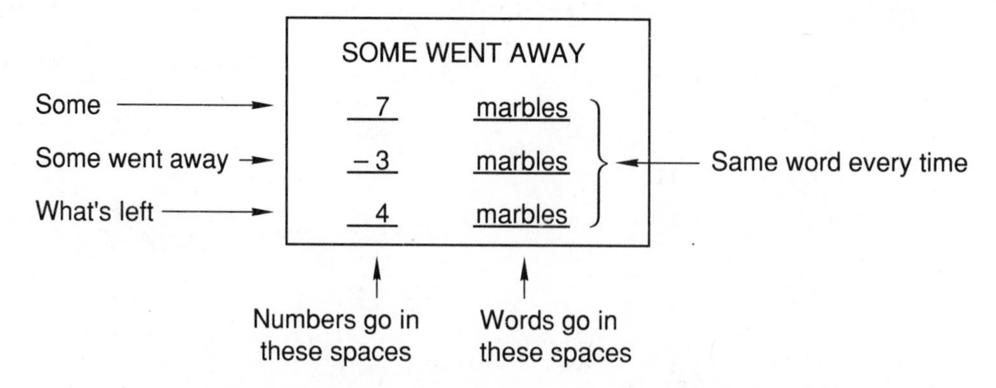

In a subtraction pattern the top number is the largest number. The top number is the sum of the other two numbers. If we know two numbers in the pattern, we can find the missing number.

Example 1 Jimmy had some marbles. Then he lost 15 marbles. He has 22 marbles left. How many marbles did he have in the beginning?

Solution He had some marbles. Then **some marbles went away**. The pattern for a "some went away" problem is a subtraction pattern. To make a subtraction pattern, we use six lines and a minus sign.

$$\begin{array}{rll} & \underline{\quad\;} & \underline{\qquad\qquad} \\ - & \underline{\quad\;} & \underline{\qquad\qquad} \\ & \underline{\quad\;} & \underline{\qquad\qquad} \end{array}$$

The same word goes in the three spaces on the right. This problem is about marbles, so the word is "marbles."

$$\begin{array}{rll} & \underline{\quad\;} & \underline{\text{marbles}} \\ - & \underline{\quad\;} & \underline{\text{marbles}} \\ & \underline{\quad\;} & \underline{\text{marbles}} \end{array}$$

He began with some marbles. Then 15 marbles went away. **In a some went away problem the number that went**

away has the minus sign. Then he had 22 marbles left.

Some	N	marbles
Some went away	$-$ 15	marbles
What's left	22	marbles

We can find the missing number in this subtraction pattern by asking an **addition question**.

22 marbles + 15 marbles = how many marbles?

The answer is **37 marbles**.

Now we check the pattern.

$$\begin{array}{r} 37 \text{ marbles} \\ -\ 15 \text{ marbles} \\ \hline 22 \text{ marbles} \end{array} \quad \text{check}$$

Example 2 Celia had 42 marbles. She lost some marbles. She has 29 marbles left. How many marbles did she lose?

Solution **Some marbles went away**. Some went away problems have a subtraction pattern. Since some marbles went away, the minus sign goes with some marbles.

Some	42	marbles
Some went away	$-$some	marbles
What's left	29	marbles

Now we must find the missing number in the pattern. To find the middle number, we subtract. Let's subtract 29 from 42.

$$\begin{array}{r} 42 \\ -\ 29 \\ \hline \mathbf{13} \end{array}$$

Now let's see if 13 makes the pattern correct.

$$\begin{array}{r} 42 \text{ marbles} \\ -\ \mathbf{13\ marbles} \\ \hline 29 \text{ marbles} \end{array}$$

Now we check.

$$42 - 13 = 29 \quad \text{check}$$

$$29 + 13 = 42 \quad \text{check}$$

Example 3 Rebecca had 65 marbles. Then she lost 13 marbles. How many marbles does she have left?

Solution This problem has the thought that **some went away**. This problem has a subtraction pattern. The number that went away has the minus sign.

Some	65	marbles
Some went away	− 13	marbles
What's left	how many	marbles

Now we have the pattern. To find the missing number, we subtract 13 from 65 and get 52.

$$\begin{array}{r} 65 \text{ marbles} \\ -\ 13 \text{ marbles} \\ \hline \mathbf{52\ marbles} \end{array}$$

To check, we note that 13 plus 52 equals 65. This means that our pattern is correct.

Practice

a. Marko had 42 marbles. Then he lost some marbles. Now he has 26 marbles. How many marbles did he lose?

b. Karen lost 42 marbles. Now she has 26 marbles. How many marbles did she have in the beginning?

c. Barbara had 75 cents. Then she spent 27 cents. Now how many cents does Barbara have?

Problem set 30

1. Micky had 75 rocks. Then she lost some rocks. Now she has 27 rocks. How many rocks did she lose?

2. Sixty-three birds sat in the tree. Then fourteen birds flew away. How many birds remained in the tree?

3. There were many cats in the alley at noon. Seventy-five cats ran away. Forty-seven cats remained. How many cats were in the alley at noon?

4. The ones' digit was 7. The tens' digit was 6. The number was between 400 and 500. What was the number?

5. Find the missing numbers in this sequence:

5, 10, ___, ___, 25, ___, ...

6. Use digits and a comparison symbol to write that seven hundred sixty-two is less than eight hundred twenty-six.

7. Round 78 to the nearest 10.

8. Use a centimeter scale to measure the length of the cover of this book.

9. If it is afternoon, what time is it?

10. Which street is perpendicular to Elm?

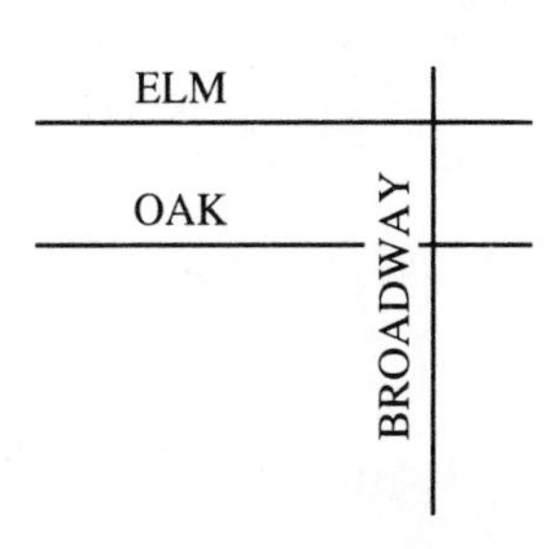

11. What fraction of this shape is shaded?

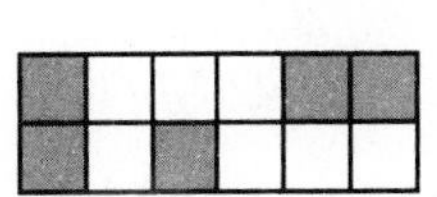

12. Draw a square whose sides are 2 cm long.

13. To what number is the arrow pointing?

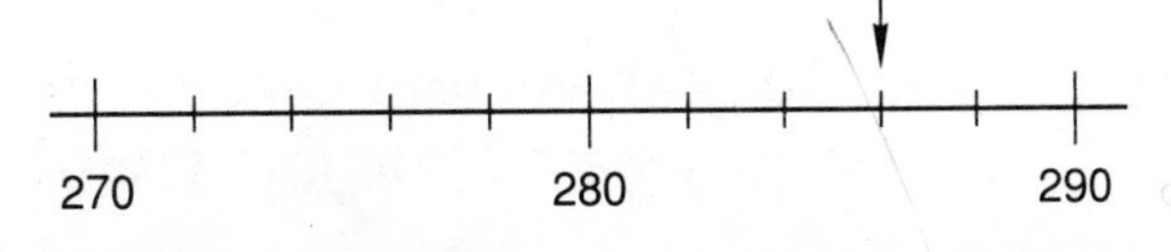

14. $\begin{array}{r} 52 \\ -\ 14 \\ \hline \end{array}$ **15.** $\begin{array}{r} 476 \\ +\ 177 \\ \hline \end{array}$ **16.** $\begin{array}{r} 62 \\ -\ 38 \\ \hline \end{array}$ **17.** $\begin{array}{r} 497 \\ +\ 258 \\ \hline \end{array}$

18. $\begin{array}{r} 61 \\ -\ 32 \\ \hline \end{array}$ **19.** $\begin{array}{r} 55 \\ +\ AB \\ \hline 87 \end{array}$ **20.** $\begin{array}{r} CD \\ -\ 23 \\ \hline 58 \end{array}$ **21.** $\begin{array}{r} XY \\ -\ 14 \\ \hline 32 \end{array}$

22. $\begin{array}{r} 36 \\ -\ FG \\ \hline 18 \end{array}$ **23.** $\begin{array}{r} 46 \\ -\ PK \\ \hline 28 \end{array}$

24. 44 + 61 + 32 + 265 **25.** 42 + 25 + 37 + 49

26. 42 − 37 **27.** 52 − 24 **28.** 73 − 59

29. 48 + 63 + 55 + 222 **30.** 463 + 55 + 84 + 29

LESSON 31 Drawing Pictures of Fractions

Speed Test: 100 Subtraction Facts (Test B in Test Booklet)

We can understand fractions better if we learn to draw pictures that represent fractions.

Example 1 Draw a rectangle and shade two thirds of it.

Solution On the left we draw a rectangle. In the center we divide the rectangle into **3 equal parts**.

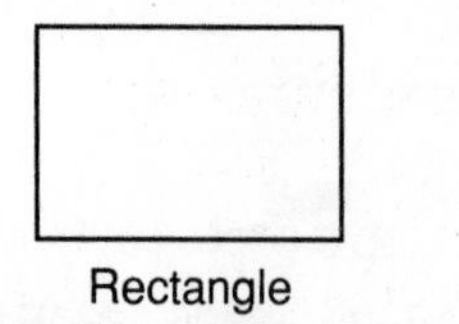

Rectangle

3 equal parts

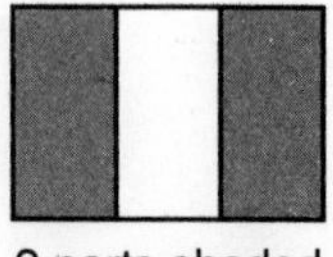

2 parts shaded

On the right we shade any two of the equal parts. We have shaded $\frac{2}{3}$ of the rectangle.

There are many ways to divide the rectangle into 3 parts. Here is another way.

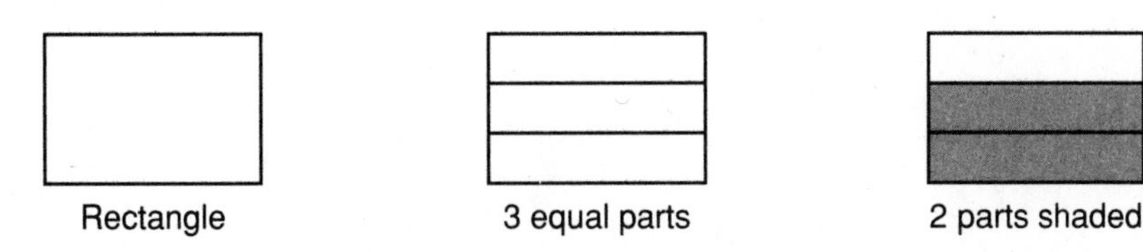

Again we have shaded $\frac{2}{3}$ of the rectangle.

Example 2 Draw a circle and shade one fourth of it.

Solution First we draw a circle. Then we divide the circle into 4 equal parts. Then we shade any 1 of the parts. We have shaded $\frac{1}{4}$ of the circle.

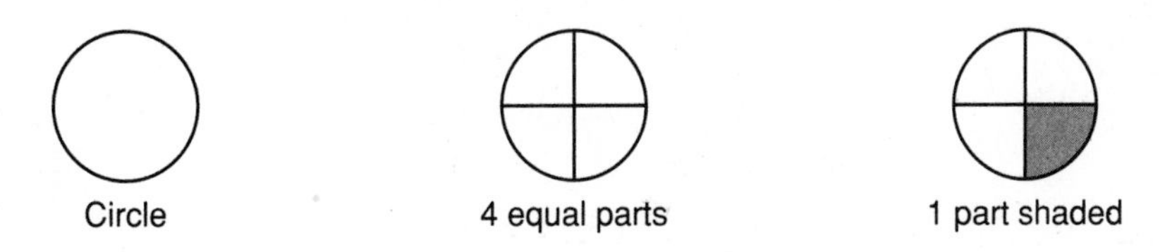

Practice

a. Draw a square and shade one half of it.

b. Draw a rectangle and shade one third of it.

c. Draw a circle and shade three fourths of it.

d. Draw a circle and shade two thirds of it.

Problem set 31

1. Mary had 42 pebbles. She threw some in the lake. Then she had 27 pebbles left. How many pebbles did she throw in the lake?

2. Demosthenes had a bag of pebbles when the sun came up. He put 17 pebbles in his mouth. Then there were 46 pebbles left in the bag. How many pebbles were in the bag when the sun came up?

3. Franklin saw one hundred forty-two the first time. Eleanor looked the other way and saw some more. If they saw three hundred seventeen in all, how many did Eleanor see?

4. Use the digits 4, 5, and 6 once each to write an even number less than 500.

Find the missing numbers in each sequence.

5. 14, 21, ___, 35, ___, ___, . . .

6. 16, 24, ___, 40, ___, ___, . . .

7. Use digits and a comparison symbol to show that twenty-seven plus eleven equals eleven plus twenty-seven.

8. Round 19 to the nearest ten.

9. One meter equals how many centimeters?

10. If it is before noon, what time is shown by this clock?

11. Which street makes a right angle with Oak?

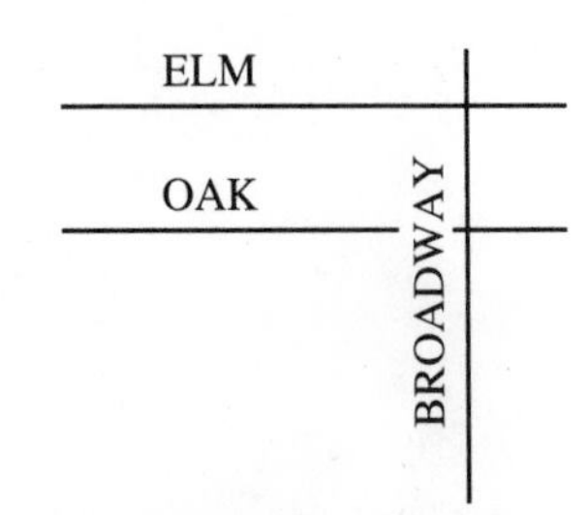

12. What fraction of this figure is shaded?

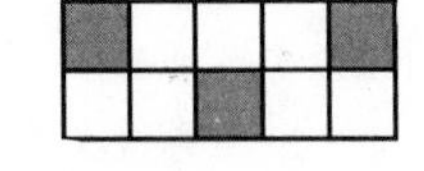

13. This scale shows weight in pounds. How many pounds does the arrow show?

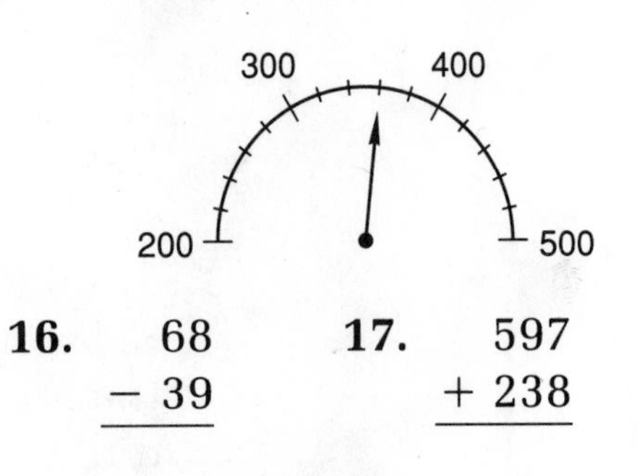

14.
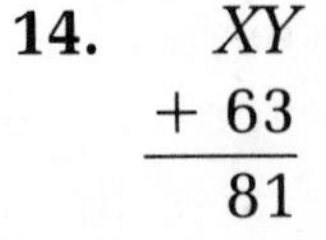

15. $\begin{array}{r} 486 \\ +\ 277 \\ \hline \end{array}$

16. $\begin{array}{r} 68 \\ -\ 39 \\ \hline \end{array}$

17. $\begin{array}{r} 597 \\ +\ 238 \\ \hline \end{array}$

18. $\begin{array}{r} 71 \\ -\ 42 \\ \hline \end{array}$ **19.** $\begin{array}{r} 87 \\ -\ MN \\ \hline 65 \end{array}$ **20.** $\begin{array}{r} 87 \\ -\ BC \\ \hline 48 \end{array}$ **21.** $\begin{array}{r} DE \\ -\ 14 \\ \hline 25 \end{array}$

22. $\begin{array}{r} 36 \\ -\ FG \\ \hline 17 \end{array}$ **23.** $\begin{array}{r} HK \\ -\ 28 \\ \hline 17 \end{array}$

24. 467 + 289 + 5 **25.** 49 + 23 + 65 + 76

26. 42 − 29 **27.** 72 − 37 **28.** 41 − 19

29. 426 + 85 + 24 + 35

30. 4 + 7 + 15 + 21 + 5 + 4 + 3

LESSON 32 Multiplication as Repeated Addition · More on Time

Speed Test: 100 Subtraction Facts (Test B in Test Booklet)

Multiplication as addition Suppose we want to find the total number of dots shown in these four groups.

There are several ways we can find the total number of dots. One way is to count the number of dots one by one. Another way is to recognize that there are 5 dots in each group and that there are four groups. We can find the answer by adding four 5s.

$$5 + 5 + 5 + 5 = 20$$

We can also use multiplication to show that we want to add 5 four times.

$$4 \times 5 = 20 \quad \text{or} \quad \begin{array}{r} 5 \\ \times\ 4 \\ \hline 20 \end{array}$$

If we find the answer this way, we are using multiplication. We call the × a **multiplication sign.** We read 4 × 5 as "4 times 5."

Example 1 Change this addition problem to a multiplication problem.

$$6 + 6 + 6 + 6 + 6$$

Solution We see five 6s. We can change this addition problem to a multiplication problem by writing either

$$5 \times 6 \quad \text{or} \quad \begin{array}{r} 6 \\ \times\ 5 \\ \hline \end{array}$$

More on time Elapsed time is time that has gone by. To find elapsed time, we must find the difference between two times.

Example 2 If it is afternoon, what time will it be in 3 hours and 20 minutes?

Solution First we count forward on the clock face 20 minutes. From that point we count forward 3 hours.

Step 1. From 1:45 p.m. we count forward 20 minutes. That makes it 2:05 p.m.
Step 2. From 2:05 p.m. we count forward 3 hours. That makes it **5:05 p.m.**

Example 3 If it is afternoon, what time was it 4 hours and 25 minutes ago?

Solution First count back the number of minutes. Then count back hours.

Step 1. We count back 25 minutes from 1:15 p.m. That makes it 12:50 p.m.

Step 2. From 12:50 p.m. we count back 4 hours. That makes it **8:50 a.m.**

Practice Change each addition problem to a multiplication problem.

a. 3 + 3 + 3 + 3

b. 9 + 9 + 9

c. 7 + 7 + 7 + 7 + 7 + 7

d. 5 + 5 + 5 + 5 + 5 + 5 + 5 + 5

Use this clock to answer questions (e) and (f).

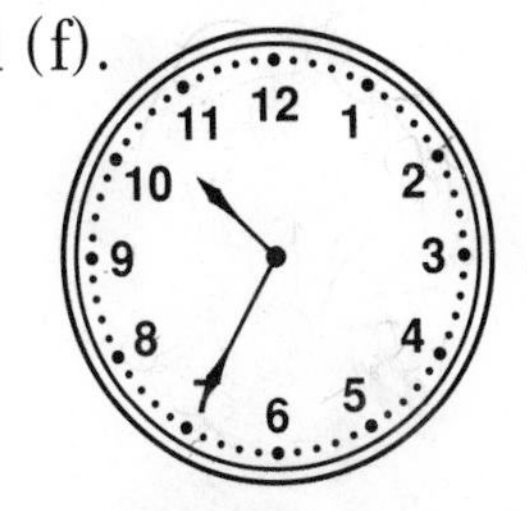

e. If it is morning, what time will it be in 2 hours and 35 minutes?

f. If it is morning, what time was it 6 hours and 40 minutes ago?

Problem set 32

1. Just before high noon, Nancy saw seventy-eight elves playing in the valley. At high noon, there were only forty-two elves playing in the valley. How many elves had left the valley by high noon?

2. According to the ancient Greeks, dryads were tree fairies. Penelope went one way and saw forty-six dryads. Perseus went the other way and saw some more dryads. Altogether they saw seventy-three dryads. How many dryads did Perseus see?

3. List the even numbers between 31 and 39.

Find the missing numbers in each sequence.

4. 8, ___, ___, 20, 24, ___, . . .

5. 24, ___, ___, 15, 12, ___, . . .

6. Write 265 in expanded form.

7. Use words to write 613.

8. Round 63 to the nearest 10.

9. Compare: 392 ◯ 329

10. The arrow is pointing to what number?

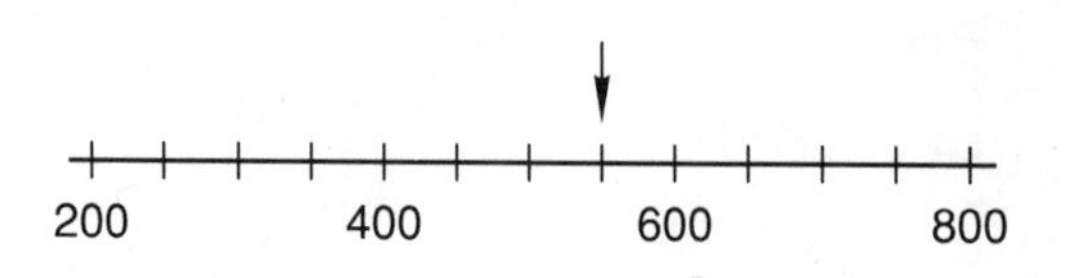

11. Draw a square and shade one fourth of it.

12. What fraction of this figure is shaded?

13. It is afternoon. What will be the time 3 hours and 10 minutes from now?

14. $\begin{array}{r} 67 \\ -\ 29 \\ \hline \end{array}$

15. $\begin{array}{r} 483 \\ +\ 378 \\ \hline \end{array}$

16. $\begin{array}{r} 71 \\ -\ 39 \\ \hline \end{array}$

17. $\begin{array}{r} 588 \\ +\ 239 \\ \hline \end{array}$

18. $\begin{array}{r} 71 \\ -\ 43 \\ \hline \end{array}$

19. $\begin{array}{r} 66 \\ +\ EF \\ \hline 87 \end{array}$

20. $\begin{array}{r} 87 \\ -\ OR \\ \hline 67 \end{array}$

21. $\begin{array}{r} AB \\ -\ 14 \\ \hline 27 \end{array}$

22. $\begin{array}{r} CD \\ -\ 19 \\ \hline 36 \end{array}$

23. $\begin{array}{r} 71 \\ -\ KD \\ \hline 39 \end{array}$

24. 467 + 288

25. 83 + 67 + 98 + 47 + 28

26. 43 − 28 **27.** 63 − 36 **28.** 51 − 19

29. Change this addition problem to a multiplication problem:

$$9 + 9 + 9 + 9$$

30. 5 + 7 + 3 + 6 + 4 + 8 + 7 + 2 + 5 + 6 + 4 + 8

LESSON 33 The Multiplication Table

Speed Test: 100 Subtraction Facts (Test B in Test Booklet)

If we count by zeros, we get a string of zeros, as we show in the first row below. If we count by ones, twos, and threes, we get the sequences shown in the next three rows.

zeros	0	0	0	0	0	0	0	0	0	0	0	0
ones	1	2	3	4	5	6	7	8	9	10	11	12*
twos	2	4	6	8	10	12	14	16	18	20	22	24
threes	3	6	9	12	15	18	21	24	27	30	33	36

If we count by fours, fives, sixes, sevens, eights, nines,

*Multiplication by tens, elevens, and twelves is considered in Appendix A.

tens, elevens, and twelves and put all these sequences in a box, we have a **multiplication table.**

Multiplication Table

	0	1	2	3	4	5	6	7	8	9	10	11	12
0	0	0	0	0	0	0	0	0	0	0	0	0	0
1	0	1	2	3	4	5	6	7	8	9	10	11	12
2	0	2	4	6	8	10	12	14	16	18	20	22	24
3	0	3	6	9	12	15	18	21	24	27	30	33	36
4	0	4	8	12	16	20	24	28	32	36	40	44	48
5	0	5	10	15	20	25	30	35	40	45	50	55	60
6	0	6	12	18	24	30	36	42	48	54	60	66	72
7	0	7	14	21	28	35	42	49	56	63	70	77	84
8	0	8	16	24	32	40	48	56	64	72	80	88	96
9	0	9	18	27	36	45	54	63	72	81	90	99	108
10	0	10	20	30	40	50	60	70	80	90	100	110	120
11	0	11	22	33	44	55	66	77	88	99	110	121	132
12	0	12	24	36	48	60	72	84	96	108	120	132	144

From a multiplication table we can find the answer to multiplications such as 3×4. We find a 3 from the first row and a 4 from the first column and look for the number where that row and column meet.

	0	1	2	3	4 ↓	5	6	7	8	9	10	11	12
0	0	0	0	0	0	0	0	0	0	0	0	0	0
1	0	1	2	3	4	5	6	7	8	9	10	11	12
2	0	2	4	6	8	10	12	14	16	18	20	22	24
→ 3	0	3	6	9	(12)	15	18	21	24	27	30	33	36
4	0	4	8	12	16	20	24	28	32	36	40	44	48
5	0	5	10	15	20	25	30	35	40	45	50	55	60
6	0	6	12	18	24	30	36	42	48	54	60	66	72
7	0	7	14	21	28	35	42	49	56	63	70	77	84
8	0	8	16	24	32	40	48	56	64	72	80	88	96
9	0	9	18	27	36	45	54	63	72	81	90	99	108
10	0	10	20	30	40	50	60	70	80	90	100	110	120
11	0	11	22	33	44	55	66	77	88	99	110	121	132
12	0	12	24	36	48	60	72	84	96	108	120	132	144

Each of the two numbers multiplied is called a **factor**. The answer to a multiplication problem is called a **product**. In this problem, 3 and 4 are factors and 12 is the product.

Practice Use the multiplication table to find the following products.

a. $\begin{array}{r} 9 \\ \times\ 3 \\ \hline \end{array}$ **b.** $\begin{array}{r} 3 \\ \times\ 9 \\ \hline \end{array}$ **c.** $\begin{array}{r} 6 \\ \times\ 4 \\ \hline \end{array}$ **d.** $\begin{array}{r} 4 \\ \times\ 6 \\ \hline \end{array}$

e. $\begin{array}{r} 7 \\ \times\ 8 \\ \hline \end{array}$ **f.** $\begin{array}{r} 8 \\ \times\ 7 \\ \hline \end{array}$ **g.** $\begin{array}{r} 5 \\ \times\ 8 \\ \hline \end{array}$ **h.** $\begin{array}{r} 8 \\ \times\ 5 \\ \hline \end{array}$

Problem set 33

1. Hansel ate seventy-two pieces of gingerbread. Then Gretel ate three hundred forty-two pieces of gingerbread. How many pieces of gingerbread did they eat in all?

2. Sherry needs $35 to buy a baseball glove. She has saved $18. How much more money does she need?

3. All of them walked to the meeting. Then forty-two got mad and went home. Only twenty-three were left at the meeting. How many walked to the meeting?

Find the missing numbers in each sequence.

4. 12, ___, ___, 30, 36, ___, . . .

5. 36, ___, ___, 24, 20, ___, . . .

6. Use words to write 265.

7. Round 28 to the nearest ten.

8. A right triangle has one right angle. Draw a right triangle. Let one side be 2 cm long and let another side be 4 cm long.

9. It is morning. What time will it be 6 hours and 10 minutes from now?

10. What fraction of the group is shaded?

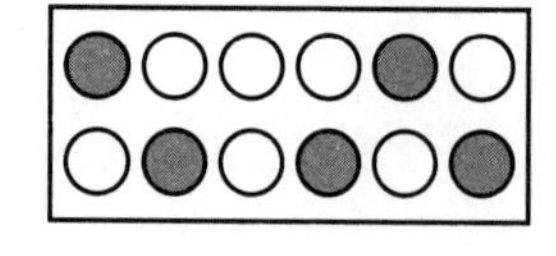

11. Write 417 in expanded form.

12. How long is this arrow?

13. $\begin{array}{r} 76 \\ -\ 29 \\ \hline \end{array}$

14. $\begin{array}{r} 286 \\ +\ 388 \\ \hline \end{array}$

15. $\begin{array}{r} 73 \\ -\ 39 \\ \hline \end{array}$

16. $\begin{array}{r} 587 \\ +\ 243 \\ \hline \end{array}$

17. $\begin{array}{r} 61 \\ -\ 45 \\ \hline \end{array}$

18. $\begin{array}{r} AN \\ +\ 48 \\ \hline 87 \end{array}$

19. $\begin{array}{r} 87 \\ -\ XY \\ \hline 57 \end{array}$

20. $\begin{array}{r} CD \\ -\ 14 \\ \hline 37 \end{array}$

21. $\begin{array}{r} 46 \\ -\ AC \\ \hline 19 \end{array}$

22. $\begin{array}{r} 41 \\ -\ N \\ \hline 35 \end{array}$

23. 489 + 299

24. 87 + 26 + 35 + 48 + 72

25. 78 − 43

26. 77 − 29

27. 53 − 19

28. 35 − 20

29. Change this addition problem to a multiplication problem:

$$6 + 6 + 6 + 6 + 6 + 6 + 6$$

30. $4 + 3 + 6 + 5 + 2 + 3 + 7 + 4 + 5 + 2 + 8 + 9$

LESSON **34**

Multiplication Facts (0, 1, 2, 5)

Speed Test: Multiplication Facts 0, 1, 2, 5 (Test C in Test Booklet)

We must memorize all the multiplication facts. Sixty-four of the 100 multiplication facts in the multiplication table shown in this lesson have 0, 1, 2, or 5 as one of the factors. These facts are the easiest to learn.

Zero times any number equals zero.

$0 \times 5 = 0$ $5 \times 0 = 0$ $7 \times 0 = 0$ $0 \times 7 = 0$

One times any number equals the number.

$1 \times 5 = 5$ $5 \times 1 = 5$ $7 \times 1 = 7$ $1 \times 7 = 7$

Two times any number is an even number.

$2 \times 5 = 10$ $2 \times 7 = 14$ $2 \times 6 = 12$ $2 \times 8 = 16$

Until we have memorized the facts, we can find multiples of 2 by counting by twos. So 2×6 is the sixth number we say when counting by twos: 2, 4, 6, 8, 10, 12. **Counting by twos is not a substitute for memorizing the facts.**

Five times any number equals a number that ends in 0 or in 5.

$5 \times 1 = 5$ $5 \times 3 = 15$ $5 \times 8 = 40$ $5 \times 7 = 35$

Until we have memorized the facts, we can find multiples of 5 by counting by fives. The sixth number we say when counting by fives is 30, so $5 \times 6 = 30$.

Practice Do the Facts Practice Multiplication Facts 0, 1, 2, 5 (Test C in Test Booklet)

Problem set 34

1. Two hundred sixty-two were in the vanguard. Three hundred seventy-five were in the main body. One hundred seven were in the rear guard. How many soldiers were there in all?

2. Ninety-two blackbirds squawked noisily in the tree. Then some flew away. Four and twenty blackbirds remained. That was enough for a pie, so the cook was happy. How many blackbirds flew away?

3. Robill threw 42 rocks. Then Buray threw some rocks. They threw 83 rocks in all. How many rocks did Buray throw?

Find the missing numbers in each sequence.

4. 8, ___, ___, 32, 40, ___, ...

5. 14, ___, ___, 35, 42, ...

6. Use the digits 4, 5, and 6 once each to write a three-digit odd number that is less than 640.

7. Use digits and a comparison symbol to write that two hundred nine is greater than one hundred ninety.

8. It is afternoon. What time will it be in 6 hours and 15 minutes?

9. Draw a rectangle and shade two thirds of it.

10. 2×8 11. 5×7 12. 2×7 13. 5×8

14. $\begin{array}{r} 83 \\ -\ 19 \\ \hline \end{array}$ 15. $\begin{array}{r} 286 \\ +\ 387 \\ \hline \end{array}$ 16. $\begin{array}{r} 72 \\ -\ 38 \\ \hline \end{array}$ 17. $\begin{array}{r} 587 \\ +\ 279 \\ \hline \end{array}$

18. $\begin{array}{r} 51 \\ -\ 23 \\ \hline \end{array}$ **19.** $\begin{array}{r} 88 \\ -\ AN \\ \hline 37 \end{array}$ **20.** $\begin{array}{r} 86 \\ -\ NM \\ \hline 47 \end{array}$ **21.** $\begin{array}{r} FG \\ -\ 14 \\ \hline 47 \end{array}$

22. $\begin{array}{r} 19 \\ +\ NQ \\ \hline 46 \end{array}$ **23.** $\begin{array}{r} 72 \\ -\ N \\ \hline 66 \end{array}$

24. 489 + 278 + 7

25. 44 + 55 + 66 + 77 + 88

26. 87 − 48 **27.** 67 − 19 **28.** 44 − 19

29. Change this addition problem to a multiplication problem:

$$8 + 8 + 8$$

30. 2 + 5 + 3 + 7 + 8 + 5 + 6 + 4 + 3 + 2 + 1 + 5 + 4

LESSON
35

Multiplication Patterns

Speed Test: Multiplication Facts 0, 1, 2, 5 (Test C in Test Booklet)

We can draw patterns that help us understand multiplication. The patterns have the shapes of rectangles. The patterns have rows and columns. One of the numbers in the multiplication pattern is the number of rows. The other number is the number of columns.

To make our first pattern we will use X's to show the multiplication of 3 times 4. We will use 3 rows and 4 columns. The total number of X's is the answer.

←4→

3 ↕

X X X X
X X X X
X X X X

$3 \times 4 = 12$

We do not have to use X's. Another way is to draw a rectangle with small squares. We will use 3 rows of small squares. There are 4 columns of small squares.

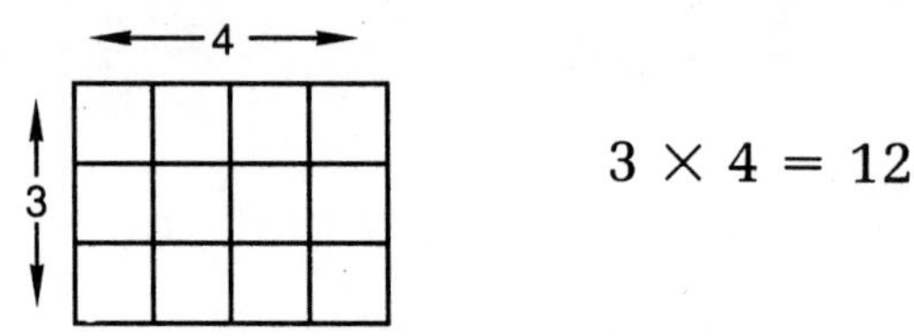

$3 \times 4 = 12$

Example 1 Draw a pattern of X's to show the multiplication of 4×5.

Solution One of the numbers is the number of rows, and the other number is the number of columns. It does not matter which is which. We show both patterns.

←4→

5 ↕

X X X X
X X X X
X X X X
X X X X
X X X X

$4 \times 5 = 20$

←5→

4 ↕

X X X X X
X X X X X
X X X X X
X X X X X

$4 \times 5 = 20$

Example 2 Draw a rectangle to show the multiplication of 2×4.

Solution We will draw a rectangle that is 2 small squares wide and 4 small squares long.

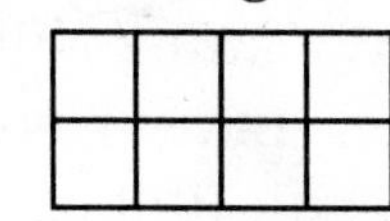

$2 \times 4 = 8$

Practice a. Draw a pattern of X's to show the multiplication of 6×3.

b. Draw a rectangle to show the multiplication of 2×5.

Write a multiplication problem for each pattern below.

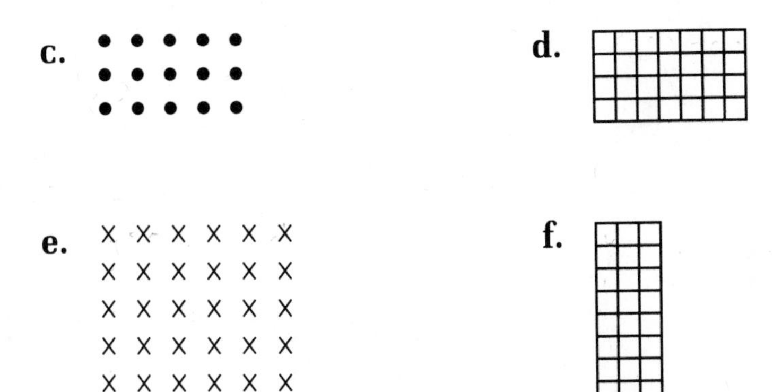

Problem set 35

1. Twenty-eight boys sat in the first row. Thirty-five boys sat in the second row. Fifty boys sat in the third row. How many boys sat in the first three rows?

2. Forty crayons were in the box. Fran took some crayons out of the box. Nineteen crayons were left in the box. How many crayons did Fran take out of the box?

3. William had some crayons when he came to school. Then Fran gave him 36 crayons. William now has 72 crayons. How many crayons did William have when he came to school?

4. The first eight counting numbers are 1, 2, 3, 4, 5, 6, 7, and 8. What is the sum of the first eight counting numbers?

5. Find the missing numbers in this sequence:

45, ___, ___, 72, 81, ___, ...

6. It is evening. What time was it 10 hours ago?

7. Write 703 in expanded form. Then use words to write this number.

8. Use digits and symbols to write that eight times zero equals nine times zero.

9. This metal object was found in a park. About how long is it?

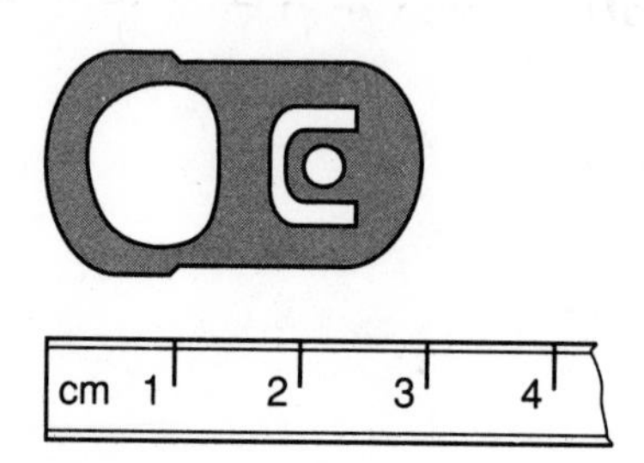

10. A 7 was in the hundreds' place. A 2 was in the ones' place. The number was between 742 and 762. What was the number?

11. Draw a pattern of X's to show the multiplication of 4×5.

12. 4×5 **13.** 2×5 **14.** 5×8

15. $\begin{array}{r} 93 \\ -\ 29 \\ \hline \end{array}$ **16.** $\begin{array}{r} 386 \\ +\ 275 \\ \hline \end{array}$ **17.** $\begin{array}{r} 61 \\ -\ 38 \\ \hline \end{array}$ **18.** $\begin{array}{r} 586 \\ +\ 376 \\ \hline \end{array}$

19. $\begin{array}{r} 61 \\ -\ 20 \\ \hline \end{array}$ **20.** $\begin{array}{r} 84 \\ -\ NA \\ \hline 37 \end{array}$ **21.** $\begin{array}{r} MC \\ -\ 76 \\ \hline 14 \end{array}$

22. $\begin{array}{r} KC \\ -\ 14 \\ \hline 37 \end{array}$ **23.** $\begin{array}{r} 36 \\ -\ MP \\ \hline 15 \end{array}$ **24.** $\begin{array}{r} KN \\ +\ 28 \\ \hline 63 \end{array}$

25. $486 + 294 + 8$

26. $46 + 37 + 28 + 19 + 55 + 62$

27. $86 - 49$

28. $91 - 55$

29. Change this addition problem to a multiplication problem:

$$7 + 7 + 7 + 7 + 7 + 7$$

30. $2 + 3 + 5 + 7 + 4 + 8 + 5 + 3 + 7 + 6 + 5 + 4$

LESSON 36 Subtracting Three-Digit Numbers

Speed Test: Multiplication Facts 0, 1, 2, 5 (Test C in Test Booklet)

When we subtract three-digit numbers, we subtract in the ones' column first. Then we subtract in the tens' column. Then we subtract in the hundreds' column.

Subtract ones	Subtract tens	Subtract hundreds
↓	↓	↓
365	365	365
− 123	− 123	− 123
2	42	242

If we find a column in which we cannot subtract, it will be necessary to regroup. To regroup, we borrow from the column to the left as we did when subtracting two-digit numbers.

Example 1 Find the difference: $365 - 87$

Solution We write the first number on top. We line up the last digits.

$$\begin{array}{r} 365 \\ -\ 87 \\ \hline (?) \end{array}$$

To subtract in the ones' column, we must borrow.

$$\begin{array}{r} 3\overset{5}{\not{6}}\,{}^{1}5 \\ -\ 8\ 7 \\ \hline 8 \end{array}$$

Now to subtract in the tens' column, we must borrow again.

$$\begin{array}{r} \overset{2}{\not{3}}\,{}^{1}\overset{5}{\not{6}}\,{}^{1}5 \\ -\ \ 8\ 7 \\ \hline 7\ 8 \end{array}$$

There is no digit in front of the 8. This is the same as a 0. We bring down the 2 to finish. The answer is **278**.

$$\begin{array}{r} \overset{2}{\not{3}}\,{}^{1}\overset{5}{\not{6}}\,{}^{1}5 \\ -\ \ 8\ 7 \\ \hline 2\ 7\ 8 \end{array}$$

Example 2

$$\begin{array}{r} 410 \\ -\ 112 \\ \hline \end{array}$$

Solution We borrow as is necessary. First we subtract in the ones' column. Then we subtract in the tens' column and then in the hundreds' column.

$$\begin{array}{r} 4\not{1}\,{}^{1}0 \\ -\ 11\ 2 \\ \hline 8 \end{array} \qquad \begin{array}{r} \overset{3}{\not{4}}\,{}^{1}\overset{0}{\not{1}}\,{}^{1}0 \\ -1\ 1\ 2 \\ \hline 9\ 8 \end{array} \qquad \begin{array}{r} \overset{3}{\not{4}}\,{}^{1}\overset{0}{\not{1}}\,{}^{1}0 \\ -1\ 1\ 2 \\ \hline \mathbf{2\ 9\ 8} \end{array}$$

Practice

a. $\begin{array}{r} 365 \\ -\ 287 \\ \hline \end{array}$ b. $\begin{array}{r} 430 \\ -\ 118 \\ \hline \end{array}$ c. $\begin{array}{r} 563 \\ -\ 356 \\ \hline \end{array}$

d. 240 − 65 e. 459 − 176 f. 157 − 98

Problem set 36

1. The room was full of students when the bell rang. Then forty-seven students ran away. Twenty-two students remained. How many students were there when the bell rang?

2. Fifty-six children peered through the window of the pet shop. Then they brought the puppies out. Now there were seventy-three children peering through the window. How many children came to the window when they brought the puppies out?

3. A nickel is worth 5¢. Gilbert had an even number of nickels in this pocket. Which of the following **could not be** the value of his nickels?
   (a) 45¢ (b) 70¢ (c) 20¢

4. It is morning. What time will it be in 1 hour and 15 minutes?

5. Draw a rectangle and shade one fifth of it.

6. To what number is the arrow pointing?

7. Write a multiplication problem for this pattern.

8. Write 843 in expanded form. Then use words to write 843.

9. $431 - 115$

10. $624 - 152$

11. $322 - 158$

12. 4×2

13. 4×5

14. 6×5

15. $\begin{array}{r} 74 \\ -\ 29 \\ \hline \end{array}$

16. $\begin{array}{r} 386 \\ +\ 278 \\ \hline \end{array}$

17. $\begin{array}{r} 61 \\ -\ 48 \\ \hline \end{array}$

18. $\begin{array}{r} 486 \\ +\ 275 \\ \hline \end{array}$

19. $\begin{array}{r} 51 \\ -\ 22 \\ \hline \end{array}$

20. $\begin{array}{r} 86 \\ -\ NA \\ \hline 43 \end{array}$

21. $\begin{array}{r} 25 \\ +\ XY \\ \hline 36 \end{array}$

22. $\begin{array}{r} NQ \\ -\ 24 \\ \hline 37 \end{array}$

23. $\begin{array}{r} 45 \\ -\ PR \\ \hline 16 \end{array}$

24. $\begin{array}{r} 57 \\ -\ N \\ \hline 49 \end{array}$

25. 483 + 296

26. 55 + 78 + 93 + 76 + 48

27. 86 − 59

28. 92 − 55

29. Change this addition problem to a multiplication problem:

$$1 + 1 + 1 + 1 + 1 + 1 + 1$$

30. 4 + 3 + 8 + 4 + 2 + 5 + 7 + 6 + 3 + 7 + 2 + 5

LESSON 37 Larger-Smaller-Difference Word Problems

Speed Test: 100 Addition Facts (Test A in Test Booklet)

There are 43 apples in the large basket.

There are 19 apples in the small basket.

When we compare the number of apples in the two baskets, we see that 43 is **greater than** 19. To find **how much greater** 43 is than 19, we can subtract.

$$\begin{array}{r} 43 \\ -\ 19 \\ \hline 24 \end{array}$$

As we think about this problem we realize that it is not a "some went away" problem because nothing went away. This is a different kind of problem. In this problem we are **comparing** two numbers. The thinking pattern we use to help us compare numbers is called the **larger-smaller-difference** pattern.

The larger-smaller-difference pattern is a subtraction pattern with three numbers. It is the same as the subtraction pattern for some went away problems. We write the two numbers we are comparing on the first two lines with the larger number on top. The smaller number is written on the second line. **The smaller number has a minus sign.** The bottom number tells us **how much larger** the top number is than the middle number. The bottom number also tells us **how much smaller** the middle number is than the top number.

Below we have drawn a larger-smaller-difference pattern box. Inside we have written the number of apples in the two baskets.

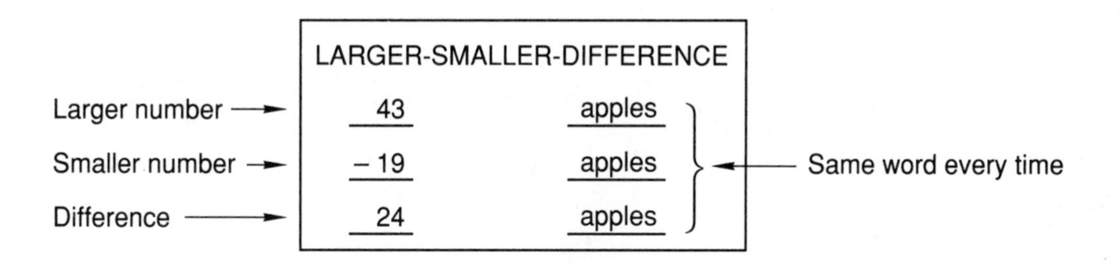

The difference tells us how many more. It also tells us how many fewer. There are **24 more** apples in the large basket than there are in the small basket. We can say this comparison another way. There are **24 fewer** apples in the small basket than there are in the large basket.

Example 1 Forty-two apples is how many more than 13 apples?

Solution To find out **how many more**, we use the larger-smaller-difference thinking pattern. The two numbers we are comparing are 42 and 13. We write these on the first two lines of the box with the larger number on top. The minus sign goes with the smaller number.

	LARGER-SMALLER-DIFFERENCE	
Larger number	42	apples
− Smaller number	− 13	apples
Difference	____	apples

To find the bottom number in this pattern, we subtract.

$$\begin{array}{r} 42 \text{ apples} \\ -\ 13 \text{ apples} \\ \hline 29 \text{ apples} \end{array} \qquad \begin{array}{r} 29 \\ +\ 13 \\ \hline 42 \end{array} \text{ check}$$

Forty-two apples is **29 apples** more than 13 apples.

Example 2 Seventeen apples is how many fewer than 63 apples?

Solution Whether we are finding how many more or how many fewer, we still use the larger-smaller-difference thinking pattern. We are comparing 17 and 63. We write 63 on top.

	LARGER-SMALLER-DIFFERENCE	
Larger number	63	apples
− Smaller number	− 17	apples
Difference	____	apples

To find the bottom number, we subtract.

$$\begin{array}{r} 63 \text{ apples} \\ -\ 17 \text{ apples} \\ \hline 46 \text{ apples} \end{array} \qquad \begin{array}{r} 46 \\ +\ 17 \\ \hline 63 \end{array} \text{ check}$$

Example 3 Seventeen is how much less than 42?

Solution Problems about numbers that ask **how much less** or **how much greater** also have a larger-smaller-difference pattern. Since we are just comparing numbers, there are no words to write in the box.

	LARGER-SMALLER-DIFFERENCE	
Larger number	42	___
– Smaller number	– 17	___
Difference	___	___

We find the bottom number by subtracting.

$$\begin{array}{r} 42 \\ -\ 17 \\ \hline 25 \end{array} \qquad \begin{array}{r} 25 \\ +\ 17 \\ \hline 42 \end{array} \text{ check}$$

Seventeen is **25 less than** 42.

Practice Answer each question. Draw a larger-smaller-difference pattern for each problem.

a. Forty-three is how much greater than twenty-seven?

b. Mary has 42 peanuts. Frank has 22 peanuts. How many fewer peanuts does Frank have?

c. Roger had 53 shells. Juanita had 95 shells. How many more shells did Juanita have?

Problem set 37

1. There were 43 parrots in the tree. Some flew away. Then there were 27 parrots in the tree. How many flew away?

2. One hundred fifty is how much greater than twenty-three?

3. Twenty-three is how much less than seventy-five?

4. It is evening. What time will it be 3 hours and 40 minutes from now?

5. Write 412 in expanded form. Then use words to write 412.

6. What fraction of this figure is shaded?

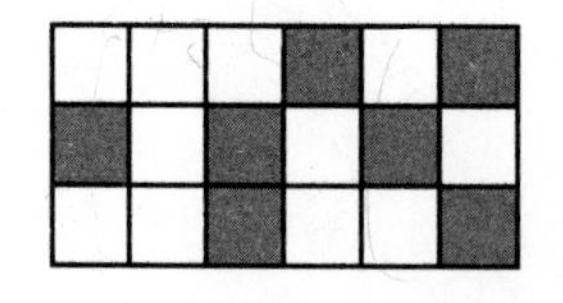

7. Matt drew a triangle that had two sides each 3 cm long. These sides made a right angle. Draw this triangle.

8. 2×5 **9.** 5×7 **10.** 2×7

11. To what number is the arrow pointing?

75 100 125 150

12. 5×8 **13.** 2×8 **14.** 5×9

15. $\begin{array}{r} 422 \\ -\ 248 \\ \hline \end{array}$ **16.** $\begin{array}{r} 907 \\ -\ \ 29 \\ \hline \end{array}$ **17.** $\begin{array}{r} 422 \\ -\ 144 \\ \hline \end{array}$ **18.** $\begin{array}{r} 701 \\ -\ 473 \\ \hline \end{array}$

19. $\begin{array}{r} 486 \\ +\ 295 \\ \hline \end{array}$ **20.** $\begin{array}{r} 307 \\ +\ 209 \\ \hline \end{array}$ **21.** $\begin{array}{r} 868 \\ XYZ \\ \hline 446 \end{array}$ **22.** $\begin{array}{r} 76 \\ -\ BC \\ \hline 28 \end{array}$

23. $\begin{array}{r} 22 \\ +\ AN \\ \hline 37 \end{array}$ **24.** $432 - 185$

25. $\begin{array}{r} 47 \\ -\ BC \\ \hline 21 \end{array}$ **26.** $\begin{array}{r} 77 \\ -\ FG \\ \hline 39 \end{array}$

27. $24 + 48 + 65 + 93 + 47 + 28 + 32$

28. Change this addition problem to a multiplication problem:

$$2 + 2 + 2 + 2 + 2$$

29. Draw a pattern of X's to show 3×5.

30. $2 + 5 + 7 + 8 + 6 + 4 + 9 + 3 + 7 + 8 + 4 + 6 + 5$

LESSON 38 Multiplication Facts (Squares)

Speed Test: Multiplication Facts 0, 1, 2, 5 (Test C in Test Booklet)

Squares are the multiplication facts that have the same factor twice: 1×1, 2×2, 3×3, and so on. They are called squares because squares can be used to illustrate these facts.

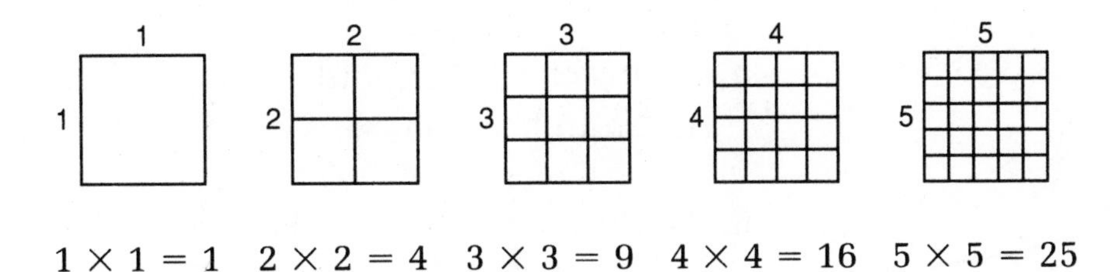

$1 \times 1 = 1$ $2 \times 2 = 4$ $3 \times 3 = 9$ $4 \times 4 = 16$ $5 \times 5 = 25$

These are squares that we will memorize:

$0 \times 0 = 0$	$6 \times 6 = 36$
$1 \times 1 = 1$	$7 \times 7 = 49$
$2 \times 2 = 4$	$8 \times 8 = 64$
$3 \times 3 = 9$	$9 \times 9 = 81$
$4 \times 4 = 16$	$10 \times 10 = 100$
$5 \times 5 = 25$	$11 \times 11 = 121$
	$12 \times 12 = 144$

To memorize facts, it is helpful to practice saying the facts aloud. Say these facts aloud two or three times a day. Get used to the sounds. The sounds help us remember the facts.

Example Which multiplication fact is illustrated by this square?

Solution The length is 6 and the width is 6. Counting, we find there are 36 squares. This is a picture of $\mathbf{6 \times 6 = 36}$.

Practice

a. 3×3 b. 6×6 c. 9×9

d. 4×4 e. 8×8 f. 7×7

g. Draw a square to show 3×3.

Problem set 38

1. Forty-two is how much greater than twenty-seven?

2. Forty-six is how much less than seventy?

3. Seven hundred fifteen cheerful hikers began the long march. Three hundred seventeen cheerful hikers finished the long march. How many cheerful hikers did not finish the long march?

4. Write 483 in expanded form. Then use words to write the number 483.

5. Draw a rectangle whose length is 5 cm and whose width is 3 cm.

6. It is evening. What time will it be 13 hours and 40 minutes from now?

7. What fraction of this shape is shaded?

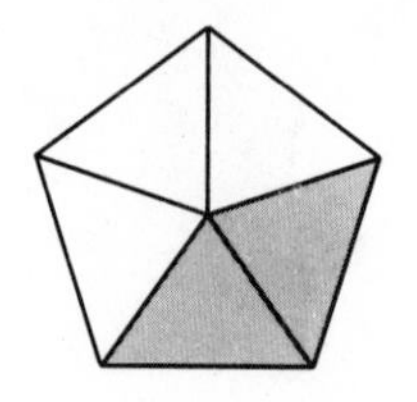

8. The addition problem $5 + 5 + 5$ can be written as a multiplication problem by writing 3×5. Write a multiplication problem for $7 + 7 + 7 + 7$.

9. Round 38 to the nearest ten. Round 44 to the nearest ten. Use digits and symbols to compare the rounded numbers.

10. The digit in the tens' place was 4. The digit in the ones' place was 2. The number was greater than 500 and less than 600. What was the number?

11. To what number is the arrow pointing?

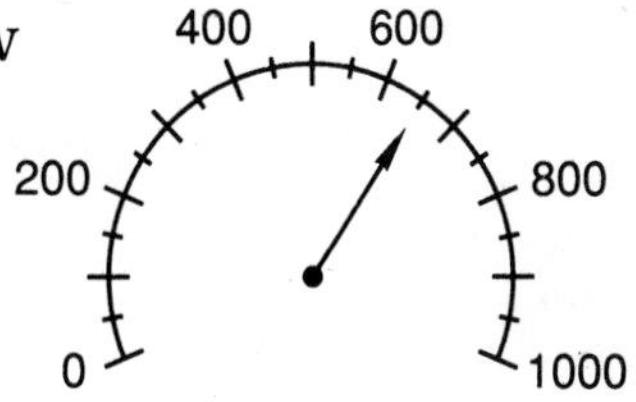

12. 4×4

13. 6×6

14. 7×7

15. 9×9

16. $\begin{array}{r} 433 \\ -\ 268 \\ \hline \end{array}$

17. $\begin{array}{r} 805 \\ -\ 47 \\ \hline \end{array}$

18. $\begin{array}{r} 401 \\ -\ 246 \\ \hline \end{array}$

19. $\begin{array}{r} 722 \\ -\ 444 \\ \hline \end{array}$

20. $\begin{array}{r} ABC \\ +\ 248 \\ \hline 486 \end{array}$

21. $\begin{array}{r} 327 \\ +\ 468 \\ \hline \end{array}$

22. $\begin{array}{r} 747 \\ -\ NAB \\ \hline 414 \end{array}$

23. $\begin{array}{r} 85 \\ -\ AB \\ \hline 27 \end{array}$

24. $24 + 89 + 64 + 28 + 37 + 445$

25. $437 - 288$

26. $742 - 258$

27. Draw a square to show 4×4.

28. $\begin{array}{r} 21 \\ 5 \\ 7 \\ 43 \\ 56 \\ 4 \\ +\ 17 \\ \hline \end{array}$

29. $\begin{array}{r} 19 \\ 25 \\ 46 \\ 54 \\ 36 \\ 94 \\ +\ 55 \\ \hline \end{array}$

30. $\begin{array}{r} 13 \\ 25 \\ 14 \\ 7 \\ 8 \\ 12 \\ +\ 6 \\ \hline \end{array}$

LESSON 39 Place Value through 999,999

Speed Test: Multiplication Facts 0, 1, 2, 5, and Squares.

We remember that the value of a digit in a number is the digit times its place value. The places in a three-digit number are the hundreds' place, the tens' place, and the ones' place. When we write 634, we mean 6 hundreds plus 3 tens plus 4 ones.

634 means 600 + 30 + 4

The places to the left of the hundreds' place are the thousands' place (1000s), the ten-thousands' place (10,000s), and the hundred-thousands' place (100,000s). We notice that as we move to the left, each new place has one more zero.

hundred thousands' place 100,000	ten thousands' place 10,000	thousands' place 1,000	hundreds' place 100	tens' place 10	ones' place 1
___	___	___ ,	___	___	___

Example 1 Write 5634 in expanded form.

Solution The 5 is in the thousands' place and has a value of 5000. So we write

5000 + 600 + 30 + 4

Example 2 Write 75,634 in expanded form.

Solution The 7 is in the ten-thousands' place and has a value of 70,000. So we write

70,000 + 5000 + 600 + 30 + 4

Example 3 Write 275,634 in expanded form.

Solution The 2 is in the hundred-thousands' place and has a value of 200,000. So we write

200,000 + 70,000 + 5000 + 600 + 30 + 4

Practice Write each of these numbers in expanded form.

a. 1284

b. 26,416

c. 462,184

Problem set 39

1. Two hundred thirty-three is how much greater than seventy-six?

2. One hundred twenty-three is how much less than five hundred seventy-two?

3. The first shipment contained 223 beds. The second shipment contained 243 beds. The third shipment contained only 175 beds. What was the total number of beds in all three shipments?

4. Write the number 78,062 in expanded form.

5. Draw two perpendicular lines. Then draw two parallel lines.

6. It is early in the morning. What time will it be in 4 hours and 45 minutes?

7. What fraction of this shape is shaded?

8. Change this addition problem to a multiplication problem:

$$7 + 7 + 7 + 7 + 7 + 7 + 7$$

9. Round 82 to the nearest ten. Round 88 to the nearest ten. Use digits and symbols to compare the rounded numbers.

10. The digit in the hundreds' place was 4. The digit in the ones' place was 2. The number was between 450 and 460. What was the number?

11. To what number is the arrow pointing?

100 200 300

12. 7×7 **13.** 9×9 **14.** 6×6 **15.** 5×6

16. $\begin{array}{r} 444 \\ -\ 268 \\ \hline \end{array}$ **17.** $\begin{array}{r} 927 \\ -\ 48 \\ \hline \end{array}$ **18.** $\begin{array}{r} 345 \\ +\ ABC \\ \hline 421 \end{array}$ **19.** $\begin{array}{r} 725 \\ -\ 566 \\ \hline \end{array}$

20. $\begin{array}{r} 478 \\ -\ 259 \\ \hline \end{array}$ **21.** $\begin{array}{r} 329 \\ +\ 428 \\ \hline \end{array}$ **22.** $\begin{array}{r} AB \\ -\ 48 \\ \hline 37 \end{array}$ **23.** $\begin{array}{r} XYZ \\ -\ 328 \\ \hline 404 \end{array}$

24. $88 + 24 + 35 + 66 + 58 + 229$

25. $582 - 428$ **26.** $923 - 744$

27. Draw a rectangle to show 3×4.

28. $\begin{array}{r} 2 \\ 5 \\ 17 \\ 28 \\ 6 \\ 4 \\ +\ 8 \\ \hline \end{array}$ **29.** $\begin{array}{r} 5 \\ 14 \\ 21 \\ 46 \\ 5 \\ 8 \\ +\ 24 \\ \hline \end{array}$ **30.** $\begin{array}{r} 19 \\ 26 \\ 5 \\ 8 \\ 4 \\ 7 \\ +\ 25 \\ \hline \end{array}$

LESSON 40 Multiplication Facts (9's)

Speed Test: Multiplication Facts 2, 5, and Squares (Test D in Test Booklet)

The 9's multiplication facts have patterns that can help us learn these facts.

Below we have listed some 9's multiplication facts. **Notice that the first digit of each answer is 1 less than the number that is multiplied by 9.**

$9 \times 2 = 18$	$(1 + 8 = 9)$
$9 \times 3 = 27$	$(2 + 7 = 9)$
$9 \times 4 = 36$	$(3 + 6 = 9)$
$9 \times 5 = 45$	$(4 + 5 = 9)$
$9 \times 6 = 54$	$(5 + 4 = 9)$
$9 \times 7 = 63$	$(6 + 3 = 9)$
$9 \times 8 = 72$	$(7 + 2 = 9)$
$9 \times 9 = 81$	$(8 + 1 = 9)$

Notice also that the two digits of the answer add up to 9.

These two patterns can help us quickly find the answer to a 9's multiplication. The first digit of the answer is 1 less than the number multiplied. The two digits add up to 9.

Example 1 What is the **first digit** of each answer?

(a) $\begin{array}{r} 9 \\ \times\ 6 \\ \hline ?\ \_ \end{array}$ (b) $\begin{array}{r} 3 \\ \times\ 9 \\ \hline ?\ \_ \end{array}$ (c) $\begin{array}{r} 9 \\ \times\ 7 \\ \hline ?\ \_ \end{array}$ (d) $\begin{array}{r} 4 \\ \times\ 9 \\ \hline ?\ \_ \end{array}$ (e) $\begin{array}{r} 9 \\ \times\ 8 \\ \hline ?\ \_ \end{array}$

Solution The first digit is 1 less than the number multiplied by 9.

(a) $\begin{array}{r} 9 \\ \times\ 6 \\ \hline \mathbf{5}\ \_ \end{array}$ (b) $\begin{array}{r} 3 \\ \times\ 9 \\ \hline \mathbf{2}\ \_ \end{array}$ (c) $\begin{array}{r} 9 \\ \times\ 7 \\ \hline \mathbf{6}\ \_ \end{array}$ (d) $\begin{array}{r} 4 \\ \times\ 9 \\ \hline \mathbf{3}\ \_ \end{array}$ (e) $\begin{array}{r} 9 \\ \times\ 8 \\ \hline \mathbf{7}\ \_ \end{array}$

Example 2 Complete these two-digit numbers so that the sum of the digits is 9.

(a) 3 __ (b) 6 __ (c) 4 __ (d) 5 __

(e) 8 __ (f) 1 __ (g) 2 __ (h) 7 __

Solution We find the second digit so that the two digits add up to 9.

(a) **36** (b) **63** (c) **45** (d) **54**

(e) **81** (f) **18** (g) **27** (h) **72**

Practice Find the answer to each multiplication fact. Remember, the first digit is 1 less than the number multiplied, and the two digits add up to 9.

a. 9×6 **b.** 5×9 **c.** 9×8 **d.** 3×9

e. 9×4 **f.** 7×9 **g.** 9×2 **h.** 9×9

Problem set 40

1. There are two hundred fifteen pages in the book. Ted has read eighty-six pages. How many more pages are left to read?

2. Use the digits 7, 8, and 9 once each to make an even number greater than 800.

3. Use digits and a comparison symbol to show that four hundred eighty-five is less than six hundred ninety.

4. What numbers are missing in this sequence?

   8, 16, __, __, 40, 48, __, . . .

5. If it is morning, what time is shown on this clock?

6. Use words to write 697.

7. Write 729 in expanded form.

8. Round 66 to the nearest ten.

9. Each side of this square is how long?

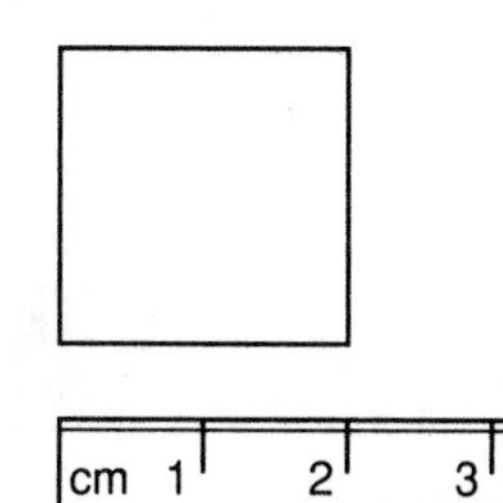

10. Find a letter of the alphabet that has just two perpendicular lines.

11. $12 - W = 7$

12. What fraction of this rectangle is shaded?

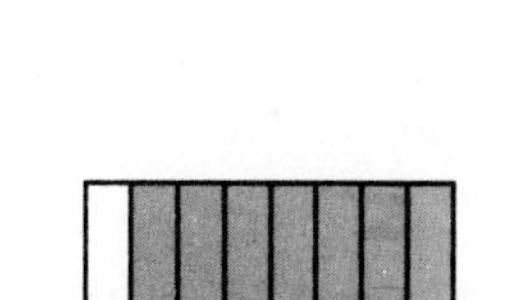

13. Draw a pattern of X's that shows the multiplication 5×5.

14. Is the value of 3 nickels and 2 dimes an even number of cents or an odd number of cents?

15. The arrow is pointing to what number on this number line?

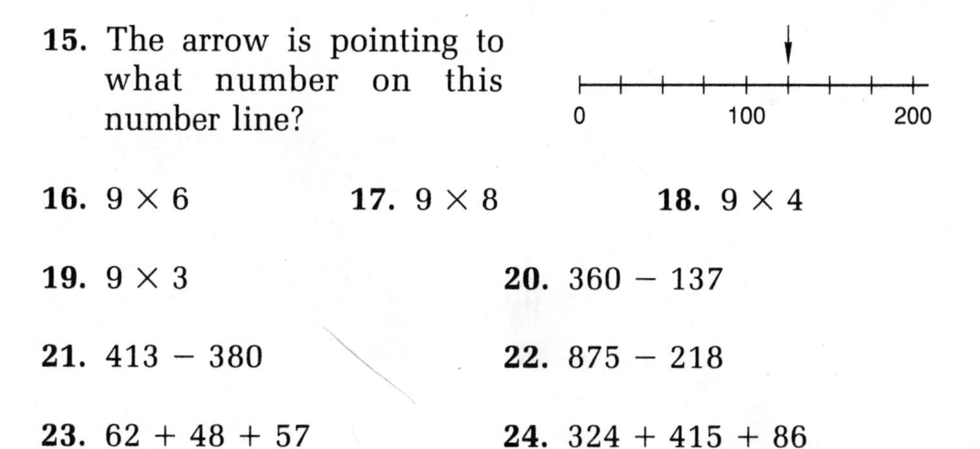

16. 9×6

17. 9×8

18. 9×4

19. 9×3

20. $360 - 137$

21. $413 - 380$

22. $875 - 218$

23. $62 + 48 + 57$

24. $324 + 415 + 86$

25. Compare: $47 + 36 \bigcirc 57 + 26$

26. Ninety-seven is how much less than two hundred?

27. Eighty-four is how much greater than forty-eight?

28. Change this addition problem to a multiplication problem:

$$6 + 6 + 6 + 6 + 6 + 6 + 6$$

29. $416 + 23 + 46 + 84$

30. $2 + 3 + 7 + 8 + 5 + 4 + 3 + 7 + 6 + 5 + 1 + 9$

LESSON 41 Writing Numbers through 999,999: Part 1

> **Speed Test:** Multiplication Facts 0, 1, 2, 3, 5, Squares, and 9s (Test D in Test Booklet)

We can use a comma to help us read numbers that have four, five, or six digits. To read

4507   34507   234507

we place a comma three digits from the right end of the numbers.

4,507   34,507   234,507

The comma is the thousands' comma. **The digits to the left of this comma tell the number of thousands.** First we read the part of the number to the left of the comma. Then we read the thousands' comma by saying "thousand." Then we read the part of the number to the right of the comma.

4,507	is read	four **thousand**, five hundred seven
34,507	is read	thirty-four **thousand**, five hundred seven
234,507	is read	two hundred thirty-four **thousand**, five hundred seven

When we write the number in words, we must remember

to place a comma after the word "thousand," as we did in these examples.

Example Use words to write 123456.

Solution First we place a comma three places from the right end of the number.

123,456

Now we read the part of the number to the left of the comma,

one hundred twenty-three

Next we write **thousand** and **put a comma after this word.**

one hundred twenty-three thousand,

Finally, we write the part of the number to the right of the comma.

one hundred twenty-three thousand, four hundred fifty-six

Practice Use words to write:

a. 907,411

b. 223,512

c. 14,518

Problem set 41

1. Wilbur saw that there were one hundred forty-two goldfish in the first tank. There were three hundred fifteen goldfish in the second tank. How many more goldfish were in the second tank?

2. Two hundred five is how much less than four hundred seventeen?

3. When Linda looked the first time, she saw 211 elves frolicking in the glen. When she looked the second time, there were 272 elves frolicking in the glen. How many more elves did she see the second time?

4. Write the number 863,425 in expanded form. Then write this number by using words.

5. Draw two parallel lines. Then draw a line that makes right angles when it crosses the parallel lines.

6. It is evening. What time will it be 2 hours and 25 minutes from now?

7. What fraction of this figure is shaded?

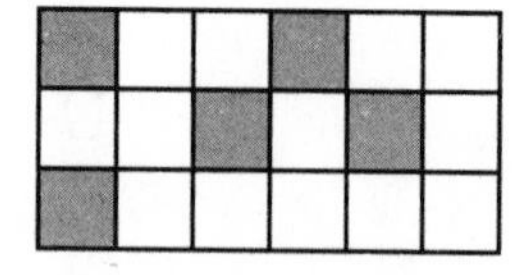

8. Change this addition problem to a multiplication problem:

$$9 + 9 + 9 + 9 + 9$$

9. Round 114 to the nearest ten. Then round 107 to the nearest ten. Then use digits and comparison symbols to compare the rounded numbers.

10. The digit in the thousands' place was 4. The digit in the ones' place was 6. The digit in the tens' place was 7. The number was between 4200 and 4300. What was the number?

11. To which number is the arrow pointing?

2000 2100 2200 2300

12. 5×8 **13.** 8×8 **14.** 9×3 **15.** 9×4

16. $\begin{array}{r} 737 \\ -\ 268 \\ \hline \end{array}$ **17.** $\begin{array}{r} 921 \\ -\ \ 58 \\ \hline \end{array}$ **18.** $\begin{array}{r} 464 \\ +\ 247 \\ \hline \end{array}$

19. $\begin{array}{r} 329 \\ +\ XYZ \\ \hline 547 \end{array}$ **20.** $\begin{array}{r} 488 \\ +\ 269 \\ \hline \end{array}$ **21.** $\begin{array}{r} 555 \\ -\ ABC \\ \hline 222 \end{array}$

22. $\begin{array}{r} 529 \\ -\ 438 \\ \hline \end{array}$

23. $\begin{array}{r} 82 \\ -\ AB \\ \hline 34 \end{array}$

24. 14 + 26 + 35 + 91 + 84 + 36 + 328

25. 472 − 188

26. 703 + 246

Use words to write each number.

27. 416,803

28. 607,051

29. $\begin{array}{r} 4 \\ 8 \\ 12 \\ 16 \\ 14 \\ 28 \\ +\ 37 \\ \hline \end{array}$

30. $\begin{array}{r} 5 \\ 8 \\ 7 \\ 14 \\ 6 \\ 21 \\ +\ 15 \\ \hline \end{array}$

LESSON 42 Writing Numbers through 999,999: Part 2

> **Speed Test:** Test 0, 1, 2, 5, Squares, and 9s (Test D in Test Booklet)

In the preceding lesson we practiced using words to write numbers through 999,999. In this lesson we will use digits to write numbers through 999,999.

We use a comma to help us write thousands. We write the number of thousands in front of the comma and the rest of the number behind the comma. The form looks like this:

_____ , _____

Example 1 Use digits to write eight hundred ninety-five thousand, two hundred seventy.

Solution It is a good idea to read the entire number before we begin writing it. We see the word "thousand," so we know it will have a **thousands' comma** after the digits that tell how many thousands.

______, ______

Then we go back and read the part of the number in front of the word "thousand" and write this number in front of the comma. For "eight hundred ninety-five thousand" we write:

895, ______

Now, to the right of the comma, we write the last part of the number.

895,270

When writing thousands, we must be sure there are **three digits after** the thousands' comma. Sometimes it may be necessary to use one or more zeros to be sure there are three digits after the comma.

Example 2 Use digits to write thirty-five thousand.

Solution We see the word "thousand," so it will have this form:

______, ______

In front of the word "thousand," we read "thirty-five," so for thirty-five thousand we write

35, ______

There is nothing written after the word thousand—no hundreds, no tens, no ones. However, we need to have three digits, so we write three zeros.

35,000

Example 3 Use digits to write seven thousand, twenty-five.

Solution For seven thousand, we write:

7, ____

After the word "thousand," we read "twenty-five." It would not be correct to write

7,25 NOT CORRECT

We need to write three digits after the thousands' comma because there are three whole number places after the comma: hundreds, tens, and ones. Since there are no hundreds, we put a zero in the hundreds' place.

7,025 CORRECT

Practice Use digits to write each number.

a. One hundred twenty-one thousand, three hundred forty

b. Twelve thousand, five hundred seven

c. Five thousand, seventy-five

d. Eighty-two thousand, five hundred

e. Seven hundred fifty thousand

f. Two thousand, one

Problem set 42

1. Four hundred sixty-five is how much greater than twenty-four?

2. Marcie had four hundred twenty green ones. Robert had one hundred twenty-three green ones. How many fewer green ones did Robert have?

3. Terry has forty-two red ones. Mary has one hundred thirty red ones. How many more red ones does Mary have?

4. Write the number 725,463 in expanded form.

5. Draw a circle whose diameter is 3 centimeters.

6. It is afternoon. What time will it be 4 hours and 40 minutes from now?

7. What fraction of the circles is shaded?

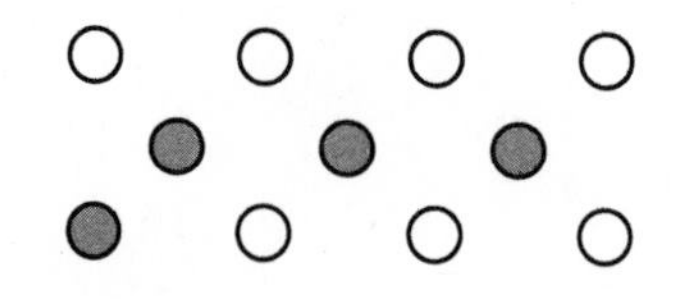

8. Change this addition problem to a multiplication problem:

$$8 + 8 + 8 + 8 + 8$$

9. Round 76 to the nearest ten. Round 59 to the nearest ten. Use digits and symbols to compare the rounded numbers.

10. The digit in the ones' place was 5. The digit in the tens' place was 6. The number was between 166 and 364. What was the number?

11. To what number is the arrow pointing?

2000 2100 2200 2300

12. 9×9 **13.** 5×7 **14.** 6×6 **15.** 4×4

16. $\begin{array}{r} 535 \\ -\ 268 \\ \hline \end{array}$ **17.** $\begin{array}{r} 903 \\ -\ 48 \\ \hline \end{array}$ **18.** $\begin{array}{r} 471 \\ -\ 346 \\ \hline \end{array}$

19. $\begin{array}{r} ABC \\ +\ 329 \\ \hline 715 \end{array}$ **20.** $\begin{array}{r} 478 \\ -\ 269 \\ \hline \end{array}$ **21.** $\begin{array}{r} 349 \\ +\ 428 \\ \hline \end{array}$

22. $\begin{array}{r} XY \\ -\ 42 \\ \hline 12 \end{array}$ **23.** $\begin{array}{r} ABC \\ -\ 127 \\ \hline 398 \end{array}$

24. 582 − 397

25. 903 − 476

Use words to write each number.

26. 417,526

27. 90,521

28. 365,123

29. Use digits to write nine hundred forty thousand, ten.

LESSON 43 Naming Mixed Numbers

Facts Practice: Multiplication Facts 2, 5, Squares (Test D in Test Booklet)

A **mixed number** is a **whole number** followed by a **fraction**. The mixed number $3\frac{1}{2}$ is read "three and one half."

Example How many circles are shaded?

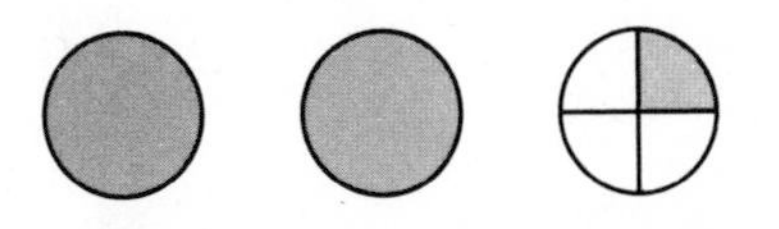

Solution Two whole circles are shaded and one fourth of another circle is shaded. The total number of shaded circles is two and one fourth, which we write as

$$2\frac{1}{4}$$

Practice What mixed numbers are pictured here?

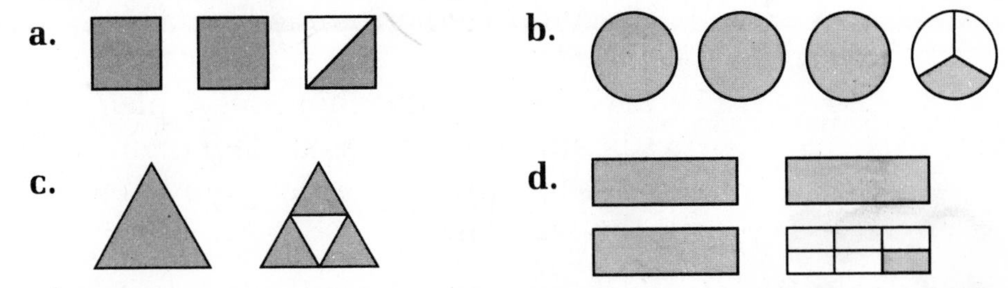

Problem set 43

1. There were seventy-one birds in the first tree. There were forty-two birds in the second tree. How many more birds were in the first tree?

2. At first there were four hundred ten screaming students. Then some quieted down, and only two hundred eighty-seven students were screaming. How many students quieted down?

3.. Ninety-seven were in the first bunch, fifty-seven were in the second bunch, and forty-eight were in the third bunch. How many were in all three bunches?

4. What mixed number is pictured in this figure?

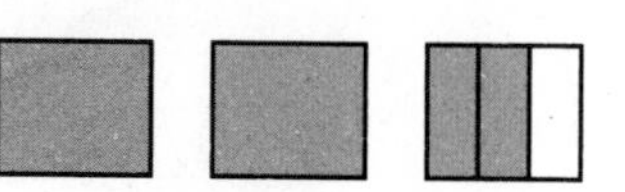

5. Write the number 506,148 in expanded form. Then write the number in words.

6. Which letter below has no right angles?

T H E N

7. It is morning. What time will it be 5 hours and 20 minutes from now?

8. Change this addition problem to a multiplication problem:

$$4 + 4 + 4 + 4 + 4 + 4 + 4 + 4$$

9. Round 176 to the nearest ten. Then round 187 to the nearest ten. Then use digits and a comparison symbol to compare the rounded numbers.

10. The digit in the thousands' place was 6. The digit in the ones' place was 3. The digit in the tens' place was 7. The number was between 6570 and 6670. What was the number?

11. To what number is the arrow pointing?

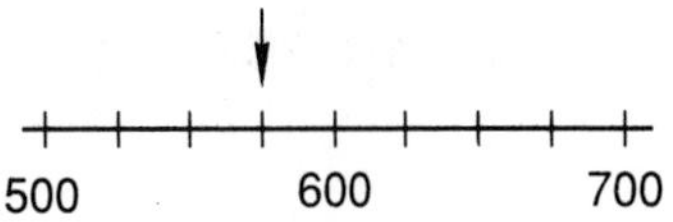

12. 9×7 **13.** 9×6 **14.** 6×7 **15.** 6×6

16. $\begin{array}{r} 732 \\ -\ 345 \\ \hline \end{array}$ **17.** $\begin{array}{r} 489 \\ +\ 257 \\ \hline \end{array}$ **18.** $\begin{array}{r} 464 \\ -\ 238 \\ \hline \end{array}$ **19.** $\begin{array}{r} 548 \\ +\ 999 \\ \hline \end{array}$

20. $\begin{array}{r} 487 \\ +\ XYZ \\ \hline 721 \end{array}$ **21.** $\begin{array}{r} 250 \\ -\ ABC \\ \hline 122 \end{array}$

22. $\begin{array}{r} ABC \\ -\ 338 \\ \hline 238 \end{array}$ **23.** $\begin{array}{r} 87 \\ -\ AB \\ \hline 54 \end{array}$

24. $22 + 46 + 84 + 97 + 96 + 47$

25. $403 - 22$ **26.** $703 - 444$ **27.** $\begin{array}{r} 14 \\ 21 \\ 18 \\ 5 \\ 7 \\ 6 \\ 14 \\ +\ 6 \\ \hline \end{array}$

Use words to write each number.

28. 403,705

29. 21,007

30. Use digits to write nine hundred seven thousand, seventy.

LESSON 44 Two Forms of Money

Speed Test: Multiplication Facts 2, 5, Squares (Test D in Test Booklet)

We can tell how much money by using a number and a cent sign ¢. If we put the cent sign behind the number, we are telling how many cents.

(a) 324¢ (b) 20¢ (c) 4¢

We can also use a dollar sign **$** to tell how much money. We always put the dollar sign in front of the number. Then we put a dot called a **decimal point** two places from the end of the number. The number to the left of the decimal point tells how many dollars. The number to the right of the decimal point tells how many cents.

(a) $3.24 (b) $0.20 (c) $0.04

The first amount is 3 dollars and 24 cents. The second amount is 20 cents. Since there were no dollars, we wrote a zero for dollars. We do not read the zero. The third amount is 4 cents.

The decimal point must be two digits from the end. If there is just one digit, we put in another zero in front of that digit.

$0.04 means the same as 4¢

We do not read the zeros in $0.04. We just say 4 cents.

Example 1 Write three hundred fifteen dollars and twenty-five cents using a dollar sign and a decimal point.

Solution When we use a dollar sign, we put a decimal point two places from the end.

$315.25

Example 2 Use words to write $30.76.

Solution We write the number of dollars, write "and," and then write the number of cents.

thirty dollars and seventy-six cents

Example 3 Use a dollar sign and a decimal point to write 7 cents.

Solution **A dollar sign requires a decimal point.** The decimal point goes two places from the end. If we write the 7,

7

we have only one digit, so we put a zero in front.

$.07

We may put another zero in front of the decimal point because there are no dollars.

$0.07

Example 4 Tommy has one quarter, one dime, and one nickel. Write how much money he has using the cent sign. Then write the amount again using the dollar sign and a decimal point.

Solution First we find how many cents. A quarter is 25 cents, a dime is 10 cents, and a nickel is 5 cents.

25¢ + 10¢ + 5¢ = **40¢**

Now we write 40 cents using the dollar sign and a decimal point.

$0.40

Practice Write these amounts with a cent sign instead of a dollar sign.

a. $0.17 **b.** $0.05 **c.** $4.60

Write these amounts with a dollar sign instead of a cent sign.

d. 195¢ **e.** 8¢ **f.** 30¢

g. Write the value of two quarters, two dimes, and one nickel with a dollar sign. Then use a cent sign to write this amount again.

Use words to write each amount.

h. $0.77

i. $12.25

j. $20.05

Problem set 44

1. The king saw one thousand, two hundred seventy. The queen saw two thousand, one hundred fifty-five. How many more did the queen see?

2. The king saw five hundred sixty-seven the first time he looked. Then he saw eight hundred forty-two the second time he looked. How many fewer did he see the first time he looked?

3. Jimbo had four dollars and sixty-five cents. Use a dollar sign and a decimal point to write this amount. Then write this amount again using a cent sign.

4. What temperature is shown on this thermometer?

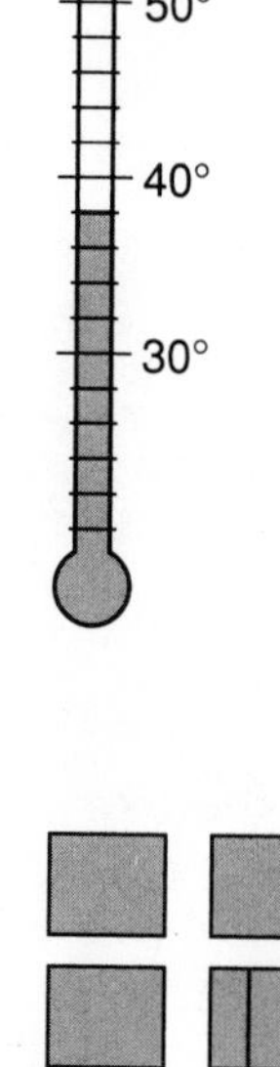

5. Which one of these angles does not look like a right angle?

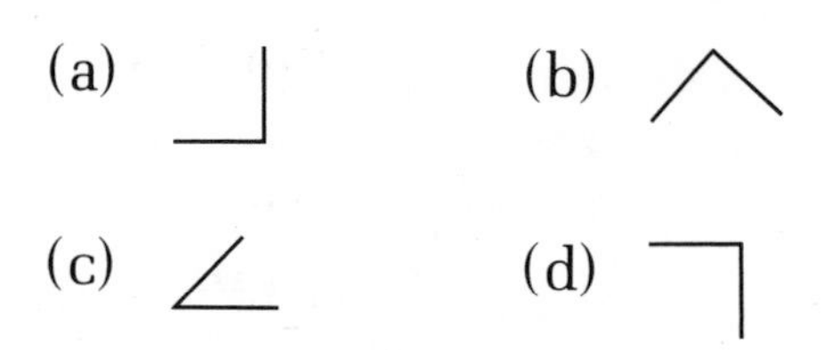

6. What mixed number is pictured?

7. Round 162 and 154 to the nearest 10. Then use a comparison symbol to compare the rounded numbers.

8. Write 250,516 in expanded form. Then use words to write this number.

9. It is evening. What time will it be 2 hours and 20 minutes from now?

10. The digit in the ones' place was 5. The digit in the tens' place was 6. The digit in the thousands' place was 8. The number was between 8390 and 8490. What was the number?

11. To what number is the arrow pointing?

3000 4000 5000 6000

12. Use words to write \$1.43. Then use a cent sign to write the same amount.

13. 6×9

14. 4×9

15. 8×9

16. 3×9

17. $\begin{array}{r} 603 \\ -\ 255 \\ \hline \end{array}$

18. $\begin{array}{r} 489 \\ +\ XYZ \\ \hline 766 \end{array}$

19. $\begin{array}{r} 532 \\ -\ 344 \\ \hline \end{array}$

20. $\begin{array}{r} ABC \\ +\ 294 \\ \hline 870 \end{array}$

21. $\begin{array}{r} 423 \\ -\ 245 \\ \hline \end{array}$

22. $\begin{array}{r} 670 \\ -\ XYZ \\ \hline 352 \end{array}$

23. $\begin{array}{r} 327 \\ -\ 158 \\ \hline \end{array}$

24. $\begin{array}{r} 93 \\ -\ BX \\ \hline 22 \end{array}$

25. $46 + 24 + 87 + 96$

26. $427 - 209$

27. Draw a rectangle that shows 2 × 5. Shade three tenths of the rectangle.

28. Use digits to write four hundred twenty-seven thousand, five hundred two.

29.
$$\begin{array}{r} 5 \\ 14 \\ 27 \\ 16 \\ 42 \\ 81 \\ +\ 6 \\ \hline \end{array}$$

30.
$$\begin{array}{r} 58 \\ 43 \\ 7 \\ 5 \\ 21 \\ 15 \\ +\ 84 \\ \hline \end{array}$$

LESSON 45 Reading Fractions and Mixed Numbers from a Number Line

Speed Test: Multiplication Facts 2, 5, 9, and Squares

To find mixed numbers on a number line, we must find the number of **spaces** between consecutive whole numbers. If there are 4 spaces between the whole numbers, each space equals $\frac{1}{4}$. If there are 6 spaces between the whole numbers, each space equals $\frac{1}{6}$.

Example 1 The arrow points to what number?

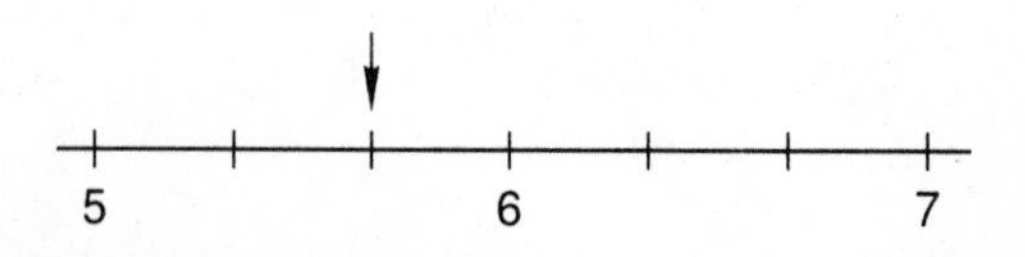

Solution There are three **spaces** between 5 and 6. Each space equals one third. The arrow points to $\mathbf{5\frac{2}{3}}$.

Example 2 The arrow points to what number on each number line?

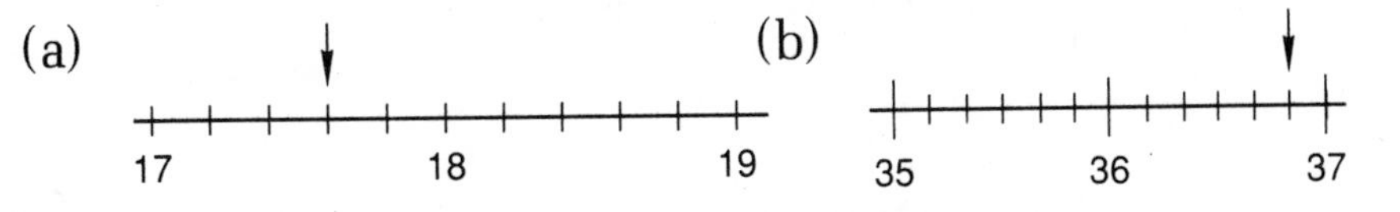

Solution (a) There are 5 spaces between the whole numbers. Each space equals $\frac{1}{5}$. The arrow points to $\mathbf{17\frac{3}{5}}$.

(b) There are 6 spaces between whole numbers. Each space equals $\frac{1}{6}$. The arrow points to $\mathbf{36\frac{5}{6}}$.

Practice Name the fraction or mixed number marked by the arrows on these number lines.

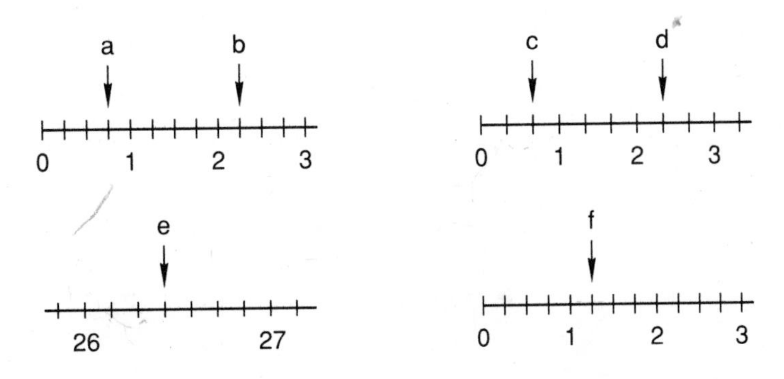

Problem set 45

1. On the way to school Ted saw four hundred twenty-seven petunia blossoms. Karen saw seven hundred fifteen petunia blossoms. How many more petunia blossoms did Karen see?

2. Circe had two hundred seventy-five pigs. After Odysseus came, Circe had four hundred sixty-two pigs. How many pigs did she get from Odysseus?

3. Four hundred seventy-five thousand, three hundred forty-two is a big number. Use digits to write this number.

4. The money in the piggy bank was worth seven dollars and sixty-five cents. Use a dollar sign to write this amount. Then use a cent sign to write this amount.

5. The arrow is pointing to what number?

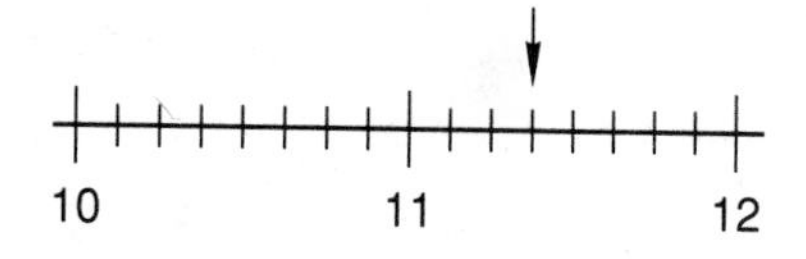

6. Draw a rectangle whose length is 5 cm and whose width is 3 cm.

7. What mixed number is shown by the shaded rectangles?

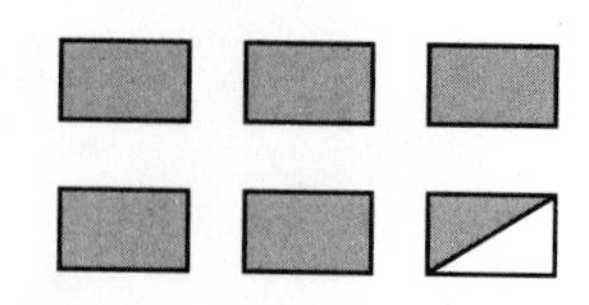

8. Round 516 to the nearest ten. Then round 524 to the nearest ten. Then use a comparison symbol to compare the rounded numbers.

9. Write 257,026 in expanded form. Then use words to write this number.

10. It is early morning. What time will it be 2 hours and 40 minutes from now?

11. Write each amount using a dollar sign. Then write each amount again using a cent sign.
(a) One dollar and forty-seven cents
(b) Seven cents

12. Use equals parts of a rectangle to show 4×6.

13. 7×9

14. 6×9

15. 4×9

16. 9×9

17. $603 - 355$

18. $499 + 288$

19. $501 - 322$

20. $XYZ + 296 = 531$

21. $ABC - 247 = 321$

22. $523 - XYZ = 145$ **23.** $327 - 269$

24. $28 + 46 + 48 + 64 + 32 + 344$

25. Use digits to write four hundred two thousand, seven hundred seventeen.

26.
$$\begin{array}{r} XY \\ -\ 14 \\ \hline 37 \end{array}$$

27.
$$\begin{array}{r} 74 \\ -\ AB \\ \hline 48 \end{array}$$

28.
$$\begin{array}{r} 67 \\ +\ XY \\ \hline 83 \end{array}$$

29.
$$\begin{array}{r} 426 \\ -\ 285 \\ \hline \end{array}$$

30.
$$\begin{array}{r} MPX \\ +\ 248 \\ \hline 716 \end{array}$$

LESSON 46 Multiplication Facts (Memory Group)

Speed Test: See Practice section of this lesson.

There are only 10 multiplication facts we have not practiced. We call these facts the **memory group.**

$3 \times 4 = 12$	$4 \times 7 = 28$
$3 \times 6 = 18$	$4 \times 8 = 32$
$3 \times 7 = 21$	$6 \times 7 = 42$
$3 \times 8 = 24$	$6 \times 8 = 48$
$4 \times 6 = 24$	$7 \times 8 = 56$

These facts must be memorized. Sounds are often helpful when we memorize things. If we say these facts aloud in a sing-song way, the sounds become familiar. When you practice facts, you should say the facts softly to yourself.

The multiplication facts should be practiced by doing **timed written tests** on a daily basis. A suggested goal is to complete a 100-facts written test in 4 minutes with no more than three errors. Then you should continue to practice often so you do not forget.

Practice Multiplication Facts: Memory Group (Test F in Test Booklet)

Problem set 46

1. There were two hundred twenty toys in the first pile. There were four hundred five toys in the second pile. How many more toys were in the second pile?

2. Five hundred seventy-five thousand, five hundred forty-two people lived in the city. Use digits to write that number.

3. Write 472,503 in expanded form. Then use words to write this number.

4. There were four hundred thirty-two boys. There were nine hundred eighteen boys and girls. How many girls were there?

5. The arrow is pointing to what number?

6. Which street is parallel to Broad Street?

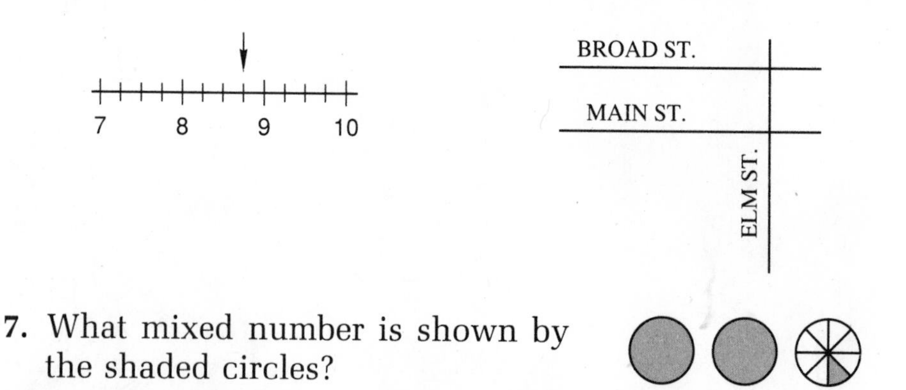

7. What mixed number is shown by the shaded circles?

8. Round 624 to the nearest ten. Then round 634 to the nearest ten. Use a comparison symbol to compare the rounded numbers.

9. It is morning. What time will it be 5 hours and 15 minutes from now?

10. Use a dollar sign and a decimal point to write each of these amounts.
(a) 1256¢ (b) 4¢ (c) 48¢

11. Draw shaded circles to show $2\frac{3}{4}$.

12. Draw a pattern of X's to show 4×7.

13. 4×7 **14.** 4×8 **15.** 4×6 **16.** 7×8

17. $\begin{array}{r} 723 \\ -\ 254 \\ \hline \end{array}$ **18.** $\begin{array}{r} 542 \\ +\ 269 \\ \hline \end{array}$ **19.** $\begin{array}{r} 943 \\ -\ 276 \\ \hline \end{array}$ **20.** $\begin{array}{r} 478 \\ +\ 249 \\ \hline \end{array}$

21. $\begin{array}{r} 849 \\ -\ 97 \\ \hline \end{array}$ **22.** $\begin{array}{r} 246 \\ +\ 97 \\ \hline \end{array}$ **23.** $\begin{array}{r} XYZ \\ -\ 581 \\ \hline 222 \end{array}$ **24.** $\begin{array}{r} ABC \\ +\ 843 \\ \hline 960 \end{array}$

25. $28 + 36 + 78 + 44 + 355$

26. $\begin{array}{r} 14 \\ 18 \\ 6 \\ 4 \\ 18 \\ +\ 15 \\ \hline \end{array}$ **27.** $\begin{array}{r} 29 \\ 5 \\ 13 \\ 27 \\ 63 \\ +\ 76 \\ \hline \end{array}$

28. Use digits to write five hundred forty thousand, seven hundred ten.

29. Use digits to write forty-two thousand, seven.

LESSON 47 Reading an Inch Scale to the Nearest Fourth

Speed Test: Multiplication (Test F in Test Booklet.)

An **inch** is a unit used for measuring length. An inch is this long.

———————

To measure lengths in inches, we use an inch scale. Inch scales are found on rulers and on tape measures. An inch scale has marks between the inch marks. These marks let us read the inch scale to the nearest half inch, nearest quarter inch, or nearest eighth inch. One quarter inch is the same as one fourth inch. In this lesson we will practice reading to the nearest quarter inch.

When reading an inch scale we should remember that $\frac{2}{4}$ is equal to $\frac{1}{2}$. The two circles below show that equal parts of the circles are shaded.

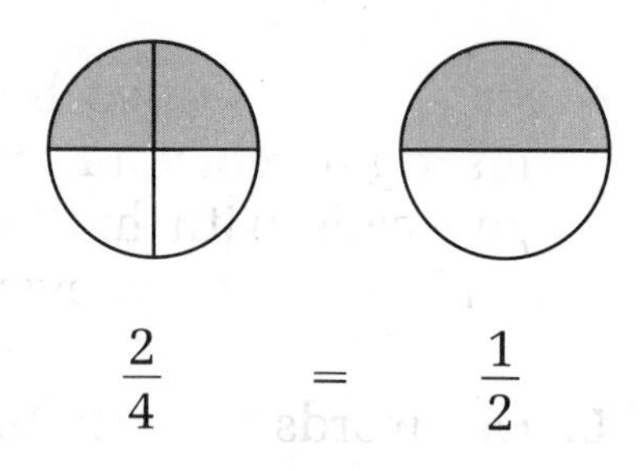

The fraction $\frac{1}{2}$ is the reduced form of $\frac{2}{4}$. Instead of naming a fraction $\frac{2}{4}$ we will use the fraction $\frac{1}{2}$. Half inch marks are usually drawn slightly longer than quarter inch marks.

Example How long is the nail to the nearest quarter inch?

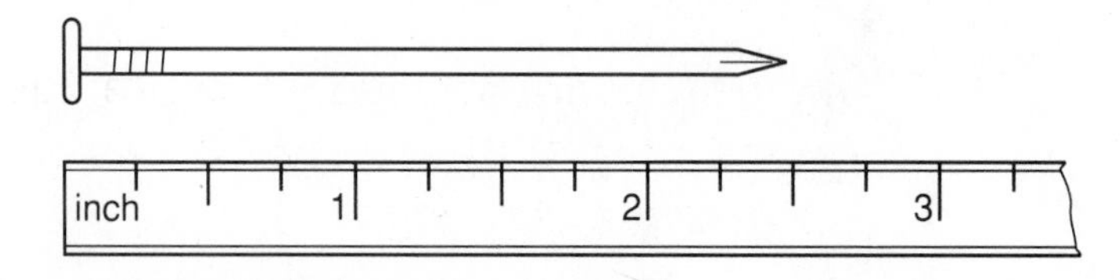

Solution The nail is 2 inches plus a fraction. It is closest to $2\frac{2}{4}$ inches. Instead of writing $\frac{2}{4}$, we use $\frac{1}{2}$. The nail is $2\frac{1}{2}$ inches long. We abbreviate this by writing **$2\frac{1}{2}$ in.**

Practice **a.** Draw a picture that shows that $\frac{2}{4}$ equals $\frac{1}{2}$.

Name each point marked by an arrow on this inch scale.

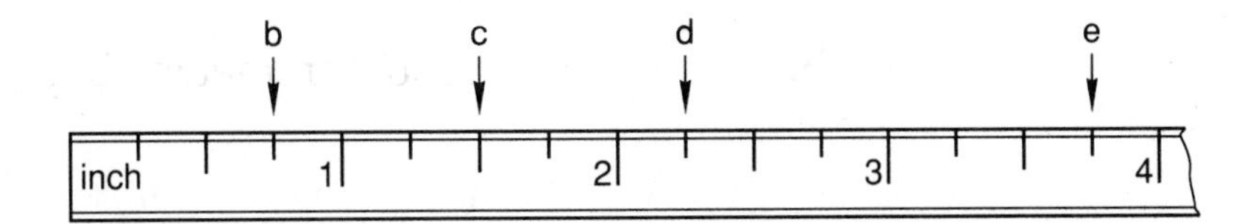

Problem set 47

1. Ann is twelve years old. Ann's mother is thirty-five years old. Ann's mother is how many years older than Ann?

2. Four hundred sixty-eight thousand, five hundred two boxes were in the warehouse. Use digits to write this number.

3. Write the number 483,905 in expanded form. Then use words to write this number.

4. Dan smashed two hundred forty-three pop cans with his right foot and smashed three hundred sixty-four pop cans with his left foot. Was the total number of smashed cans an even number or an odd number?

5. Use words to write the mixed number $102\frac{1}{2}$.

6. The zero holds what place in 407,918?

7. Use digits and a comparison symbol to show that seven hundred eighty-six is greater than seven hundred sixty-eight.

8. Use a dollar sign to write the value of 2 dollars, 1 quarter, 2 dimes, and 3 nickels. Write the value again using a cent sign.

9. It is morning. What time will it be 2 hours and 20 minutes from now?

10. One kilometer is how many meters?

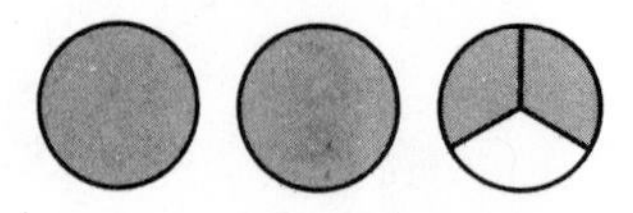

11. How many of these circles are shaded?

12. Find the length of this screw to the nearest quarter inch.

13. $\begin{array}{r} 9 \\ \times\ 8 \\ \hline \end{array}$ **14.** $\begin{array}{r} 6 \\ \times\ 7 \\ \hline \end{array}$ **15.** $\begin{array}{r} 8 \\ \times\ 6 \\ \hline \end{array}$ **16.** $\begin{array}{r} 6 \\ \times\ 4 \\ \hline \end{array}$ **17.** $\begin{array}{r} 7 \\ \times\ 3 \\ \hline \end{array}$

18. $\begin{array}{r} 486 \\ +\ 247 \\ \hline \end{array}$ **19.** $\begin{array}{r} 486 \\ -\ 247 \\ \hline \end{array}$ **20.** $\begin{array}{r} 293 \\ +\ 678 \\ \hline \end{array}$ **21.** $\begin{array}{r} 893 \\ -\ 678 \\ \hline \end{array}$

22. $\begin{array}{r} 463 \\ -\ XY \\ \hline 411 \end{array}$ **23.** $\begin{array}{r} 463 \\ +\ PQ \\ \hline 527 \end{array}$ **24.** $\begin{array}{r} 418 \\ -\ RST \\ \hline 216 \end{array}$ **25.** $\begin{array}{r} 24 \\ 36 \\ 84 \\ 125 \\ +\ 127 \\ \hline \end{array}$

26. $46 + 42 + 48 + MN = 198$
(*Hint:* First add $46 + 42 + 48$. Then find how much more is needed to total 198.)

27. Draw a rectangle that shows 3×5. Then shade four fifteenths of it.

28. $\begin{array}{r} 7 \\ 14 \\ 28 \\ 56 \\ 49 \\ 14 \\ +\ 17 \\ \hline \end{array}$

29. Use words to write 702,502.

30. Use digits to write five hundred forty thousand, seven hundred fifty-five.

LESSON 48 U.S. Units of Length

Speed Test: Multiplication (Test F in Test Booklet)

In Lesson 22 we described some units of length in the metric system. In the United States we also use other units of length. Some of these are inches, feet, yards, and miles. In the preceding lesson we said that an inch is this long.

———————

Twelve inches equal 1 **foot.** You may have a 1-foot ruler for use at your desk. A **yard** is 3 feet. One *big* step is about 1 yard. So a yard is about the same length as a meter. Actually, a yard is about $3\frac{1}{2}$ inches less than a meter. A **mile** is 1760 yards, which is 5280 feet.

We often use the abbreviations in., ft, yd, and mi to stand for inches, feet, yards, and miles.

Example 1 How many 1-foot rulers laid end to end would it take to equal 2 yards?

Solution A yard is 3 feet long. So it would take 3 one-foot rulers to reach 1 yard. To reach another yard would take 3 more one-foot rulers. Therefore, to reach 2 yards would take **6 one-foot rulers.**

Example 2 How many *big* steps would it take to walk a mile?

Solution A mile is 1760 yards, which is about **1760 *big* steps.**

Practice

a. How many 1-foot rulers laid end to end would it take to reach 3 yards?

b. A mile is how many feet?

c. One yard is how many inches?

Problem set 48

1. Use the digits 4, 5, and 6 once each to make an even number greater than 600.

2. Myrtle the Turtle laid one hundred fifty eggs in the sand. Her friend Gertle laid one hundred seventy-five eggs. Together, how many eggs did they lay?

3. Jill is 6 years younger than her brother. If her brother is 15 years old, how old is Jill?

4. What numbers are missing in this sequence?

$$\_\_, 40, \_\_, \_\_, 64, 72, \ldots$$

5. Use words to write the mixed number $23\frac{2}{3}$.

6. We have used letters to stand for missing digits in numbers. We can also use a letter to stand for a number that has more than one digit. If we write

$$N + 100 = 142$$

we see that the letter N must stand for the number 42. Find the value of N in this expression:

$$219 = N + 10 + 9$$

7. Write 809,742 in expanded form. Then use words to write this number.

8. Use digits and a comparison symbol to show that one hundred eighteen is less than one hundred eighty.

9. Round 128 to the nearest ten.

10. Write the value of 3 quarters and 3 dimes with a dollar sign. Then use words to write this amount of money.

11. Find the length of this rectangle to the nearest quarter inch.

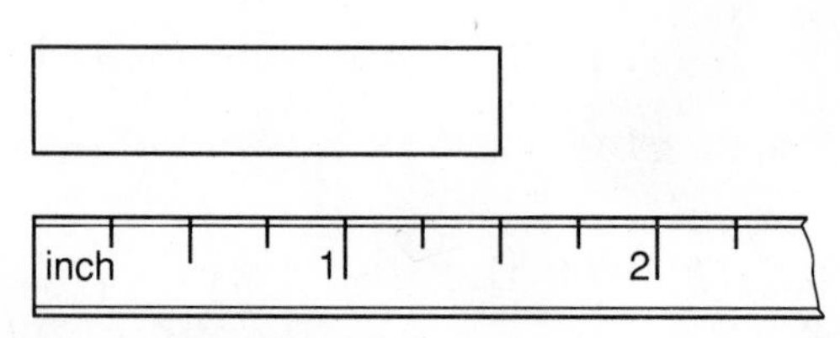

12. Draw a square and shade one ninth of it. (*Hint*: Draw a square. Then draw an evenly spaced tic-tac-toe pattern inside the square.)

13. Three one-foot rulers laid end to end reach how many inches?

14. Compare: One mile ◯ 1000 yards

15. $\begin{array}{r} 3 \\ \times\ 7 \\ \hline \end{array}$ **16.** $\begin{array}{r} 4 \\ \times\ 6 \\ \hline \end{array}$ **17.** $\begin{array}{r} 6 \\ \times\ 7 \\ \hline \end{array}$ **18.** $\begin{array}{r} 7 \\ \times\ 8 \\ \hline \end{array}$ **19.** $\begin{array}{r} 8 \\ \times\ 9 \\ \hline \end{array}$

20. $\begin{array}{r} 720 \\ -\ 425 \\ \hline \end{array}$ **21.** $\begin{array}{r} 426 \\ +\ 157 \\ \hline \end{array}$ **22.** $\begin{array}{r} 463 \\ -\ 286 \\ \hline \end{array}$ **23.** $\begin{array}{r} 436 \\ -\ 147 \\ \hline \end{array}$

24. $675 + 275$

25. $36 + 42 + 8 + 95 + N = 205$

26. Draw a pattern of X's to show 6×7.

27. $\begin{array}{r} 723 \\ -\ 256 \\ \hline \end{array}$ **28.** $\begin{array}{r} CDE \\ +\ 419 \\ \hline 784 \end{array}$

29. $\begin{array}{r} 235 \\ +\ FJK \\ \hline 764 \end{array}$ **30.** $\begin{array}{r} 235 \\ -\ 147 \\ \hline \end{array}$

LESSON 49

Lines and Segments

Speed Test: Multiplication (Test F in Test Booklet)

A pencil line has two ends. We say that a mathematical line has no ends. A mathematical line just goes on and on in both directions. Sometimes we draw arrowheads on a pencil line to show this idea.

A line never ends.

Part of a mathematical line is a **line segment** or just a **segment**. A segment has two endpoints.

A segment is part of a line. It has two ends.

Sometimes the endpoints of a segment are labeled with letters, and the segment is named using the letters. When we write the name of a segment, either letter can come first.

This is segment AB. This is also segment BA.

When there is more than one segment in a picture, we use the letters to make it clear what segment we are talking about.

Example The length of segment AB is 3 cm. The length of segment BC is 4 cm. What is the length of segment AC?

Solution Two short segments can form a longer segment. From A to B is one segment. From B to C is a second segment. Together they form the third segment, segment AC. We are told the lengths of segments AB and BC. If we add these lengths together, the sum will equal the length of segment AC.

$$3 \text{ cm} + 4 \text{ cm} = 7 \text{ cm}$$

The length of segment AC is **7 cm**.

Practice **a.** The length of segment AB is 5 cm. The length of segment BC is 4 cm. What is the length of segment AC?

b. The length of segment RS is 4 cm. The length of segment RT is 10 cm. What is the length of segment ST? (*Hint*: This time you will need to subtract.)

R S T

Problem set 49

1. A group of quail is called a covey. A group of cows is called a herd. A group of fish is called a school. There were twenty-five fish in the small school. There were one hundred twelve fish in the big school. How many fewer fish were in the small school?

2. A 36-inch yardstick was divided into two pieces. One piece was 12 inches long. How many inches long was the other piece?

3. Mrs. Green mailed forty-seven postcards from Paris. Her husband mailed sixty-two postcards from Paris. Her son mailed seventy-five postcards from Paris. In all, how many postcards did the Greens mail from Paris?

4. Write the number 416,528 in expanded form. Then use words to write this number.

5. Use digits and a comparison symbol to write that one thousand, four hundred sixty is less than one thousand, six hundred forty.

6. Use a dollar sign and a decimal point to write the value of 3 dollars, 2 quarters, 1 dime, and 2 nickels. Then write this amount of money using words.

7. Write the value of 2 quarters, 1 dime, and 2 nickels with a dollar sign. Write the value again using a cents sign.

8. How many squares are shaded?

9. Find the length of the segment to the nearest quarter inch.

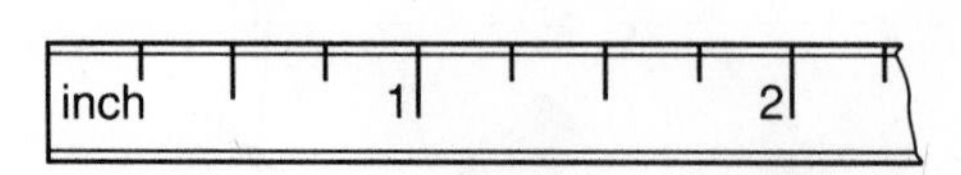

10. The length of segment AB is 3 cm. The length of segment BC is 4 cm. What is the length of segment AC?

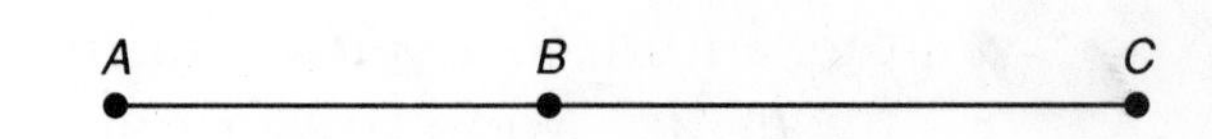

11. It is evening. What time will it be 1 hour and 50 minutes from now?

12. Use a comparison symbol to compare 294 and 302.

13. Draw a circle that is 2 centimeters in diameter.

14. $\begin{array}{r} 9 \\ \times\ 7 \\ \hline \end{array}$ **15.** $\begin{array}{r} 7 \\ \times\ 3 \\ \hline \end{array}$ **16.** $\begin{array}{r} 6 \\ \times\ 8 \\ \hline \end{array}$ **17.** $\begin{array}{r} 6 \\ \times\ 6 \\ \hline \end{array}$ **18.** $\begin{array}{r} 4 \\ \times\ 8 \\ \hline \end{array}$

19. $\begin{array}{r} 416 \\ -\ XYZ \\ \hline 179 \end{array}$ **20.** $\begin{array}{r} 498 \\ +\ 765 \\ \hline \end{array}$ **21.** $\begin{array}{r} 741 \\ -\ ABC \\ \hline 423 \end{array}$

22. $\begin{array}{r} 536 \\ +\ XYZ \\ \hline 721 \end{array}$ **23.** $\begin{array}{r} 713 \\ -\ 427 \\ \hline \end{array}$

24. $48 + 64 + 28 + 32 + N = 189$

25. Draw a pattern of X's to show 3×7.

26. Use digits to write four hundred fifteen thousand, seven.

27. $\begin{array}{r} 4 \\ 5 \\ 8 \\ 9 \\ 6 \\ 5 \\ +\ 4 \\ \hline \end{array}$ **28.** $\begin{array}{r} 9 \\ 3 \\ 14 \\ 7 \\ 8 \\ 5 \\ +\ 24 \\ \hline \end{array}$ **29.** $\begin{array}{r} 8 \\ 5 \\ 4 \\ 23 \\ 46 \\ 84 \\ +\ 21 \\ \hline \end{array}$ **30.** $\begin{array}{r} 7 \\ 2 \\ 8 \\ 9 \\ 4 \\ 7 \\ +\ 6 \\ \hline \end{array}$

LESSON 50 Subtracting Across Zero

> **Speed Test:** Multiplication Facts—Memory Group (Test F in Test Booklet)

In the problem below we must regroup twice before we can subtract the first time.

$$\begin{array}{r} 405 \\ -\ 126 \\ \hline \end{array}$$

We cannot borrow from the tens' column because the tens' column has a zero. The first step is to borrow 1 from the hundreds' column.

$$\begin{array}{r} \overset{3}{\cancel{4}}\,{}^{1}0\;5 \\ -\;1\;2\;6 \\ \hline \end{array}$$

Now we have a 10 in the tens' column. The next step is to borrow 1 from the tens' column.

$$\begin{array}{r} \overset{3}{\cancel{4}}\,\overset{9}{{}^{\cancel{1}}\cancel{0}}\,{}^{1}5 \\ -\;1\;2\;6 \\ \hline \end{array}$$

Now we subtract.

$$\begin{array}{r} \overset{3}{\cancel{4}}\,\overset{9}{{}^{\cancel{1}}\cancel{0}}\,{}^{1}5 \\ -\;1\;2\;6 \\ \hline \mathbf{2\;7\;9} \end{array}$$

We check by adding.

$$\begin{array}{r} 126 \\ +\,279 \\ \hline 405 \end{array} \quad \text{check}$$

Example Find the difference: 300 − 123

Solution First we change 3 hundreds to 2 hundreds and 10 tens. Next we change 10 tens to 9 tens and 10 ones. Then we subtract.

$$\begin{array}{r} 300 \\ -\,123 \\ \hline \end{array} \qquad \begin{array}{r} \overset{2}{\cancel{3}}\,{}^{1}0\;0 \\ -\;1\;2\;3 \\ \hline \end{array} \qquad \begin{array}{r} \overset{2}{\cancel{3}}\,\overset{9}{{}^{\cancel{1}}\cancel{0}}\,{}^{1}0 \\ -\;1\;2\;3 \\ \hline \mathbf{1\;7\;7} \end{array}$$

Now we check by adding.

$$\begin{array}{r} 123 \\ +\,177 \\ \hline 300 \end{array} \quad \text{check}$$

Practice

a. $\begin{array}{r} 300 \\ -\ 132 \\ \hline \end{array}$ b. $\begin{array}{r} 405 \\ -\ 156 \\ \hline \end{array}$ c. $\begin{array}{r} 201 \\ -\ 102 \\ \hline \end{array}$

d. 400 − 86 e. 308 − 124 f. 703 − 198

Problem set 50

1. There were one hundred forty-seven birds in the first flock. There were six hundred forty-three birds in the second flock. How many more birds were in the second flock?

2. Kevin had a dime, a quarter, and a penny. Write this amount with a dollar sign and a decimal point.

3. San Francisco is 400 miles north of Los Angeles, as shown. Santa Barbara is 110 miles north of Los Angeles. How far is it from Santa Barbara to San Francisco?

San Francisco

Santa Barbara

Los Angeles

4. Use the digits 5, 6, and 7 once each to make an odd number greater than 700.

5. Use digits and a comparison symbol to show that five hundred ninety-four is greater than five hundred forty-nine.

6. It is dark now. What time will it be 7 hours and 20 minutes from now?

7. Write the number 427,948 in expanded form. Then use words to write the number.

8. Draw a rectangle and shade one seventh of it.

9. There was a 6 in the ones' place and a 7 in the tens' place. The digit in the thousands' place was 4. The number was between 4000 and 4100. What was the number?

10. Which letter below has no right angles?

F E Z L

11. Find the length of segment BC.

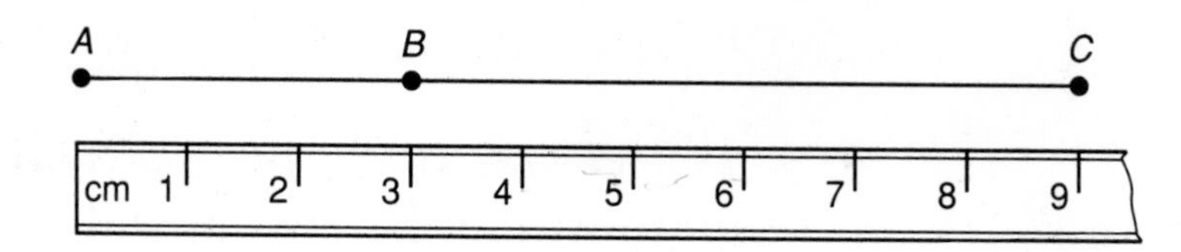

12. The arrow points to what number on this inch scale?

13. $\begin{array}{r} 7 \\ \times\ 8 \\ \hline \end{array}$ **14.** $\begin{array}{r} 9 \\ \times\ 6 \\ \hline \end{array}$ **15.** $\begin{array}{r} 8 \\ \times\ 8 \\ \hline \end{array}$ **16.** $\begin{array}{r} 6 \\ \times\ 7 \\ \hline \end{array}$ **17.** $\begin{array}{r} 9 \\ \times\ 8 \\ \hline \end{array}$

18. $\begin{array}{r} XYZ \\ +\ 179 \\ \hline 406 \end{array}$ **19.** $\begin{array}{r} 651 \\ -\ ABC \\ \hline 426 \end{array}$

20. $\begin{array}{r} 405 \\ -\ MNP \\ \hline 123 \end{array}$ **21.** $\begin{array}{r} 486 \\ -\ 297 \\ \hline \end{array}$

22. $44 + 444 + 378$ **23.** $193 + 58 + 572$

24. $26 + 28 + 44 + 78 + N = 289$

25. Use digits to write seven hundred thousand, seventeen.

26.		**27.**		**28.**		**29.**		**30.**	
	27		19		31		8		26
	4		81		14		15		14
	16		5		28		93		81
	28		7		15		47		9
	7		14		6		86		5
	+ 15		+ 21		+ 7		+ 19		+ 6

LESSON 51 Rounding Numbers to the Nearest Hundred · Using One Letter

Speed Test: Multiplication Facts—Memory Group (Test F in Test Booklet)

Rounding numbers We have practiced rounding numbers to the nearest ten. Now we will learn to round numbers to the nearest hundred. To round numbers to the nearest hundred, we find the hundred number to which the number is nearest. The hundred numbers are the numbers in this sequence.

100, 200, 300, 400, ...

A number line can help us understand rounding to the nearest hundred.

Example 1 Round 472 to the nearest hundred.

Solution The number 472 is between the hundred numbers 400 and 500. Halfway between 400 and 500 is 450. Since 472 is greater than 450, it is nearer 500. We see this on the number line below.

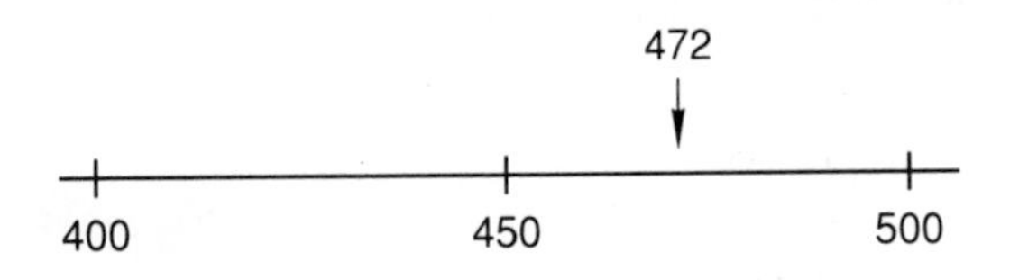

Rounding 472 to the nearest hundred gives us **500.**

Example 2 Round 362 and 385 to the nearest 100. Then use a comparison symbol to compare the rounded numbers.

Solution The number 362 is closer to 400 than it is to 300. The number 385 is closer to 400 than it is to 300. Both 362 and 385 round to 400. If we compare the rounded numbers, we get

$$400 = 400$$

Using one letter These two problems ask the same question.

$$\begin{array}{r} N \\ -\ 324 \\ \hline 467 \end{array} \qquad \begin{array}{r} XYZ \\ -\ 324 \\ \hline 467 \end{array}$$

In the problem on the left, the missing number is represented by N. In the problem on the right, the missing number is represented by XYZ. **In mathematics we usually use a single letter to represent the missing number.** Since the sum of the two bottom numbers in a subtraction pattern equals the top number, we add to find N.

$$\begin{array}{r} 324 \\ +\ 467 \\ \hline 791 \end{array}$$

Now we check.

$$\begin{array}{r} 791 \\ -\ 324 \\ \hline 467 \end{array}$$

Thus N stands for the number **791**.

Practice Round each number to the nearest hundred.

a. 813 **b.** 685 **c.** 427 **d.** 573

Find the missing number.

e. $\begin{array}{r} 641 \\ -\ N \\ \hline 276 \end{array}$ **f.** $\begin{array}{r} 641 \\ +\ N \\ \hline 743 \end{array}$ **g.** $\begin{array}{r} N \\ -\ 432 \\ \hline 748 \end{array}$

Problem set 51

1. Two hundred fifty-two came when the bell rang once. Then some more came when the bell rang twice. Four hundred two came in all. How many came when the bell rang twice?

2. Wilbur had sixty-seven grapes. He ate some grapes. Then he had thirty-eight grapes. How many grapes did he eat?

3. The distance from Whery to Radical is 42 km. The distance from Whery to Appletown is 267 km. How far is it from Radical to Appletown?

4. It is afternoon. What time will it be $2\frac{1}{2}$ hours from now?

5. Write the number 982,407 in expanded form. Then use words to write the number.

6. Round 673 and 637 to the nearest 100. Then use a comparison symbol to compare the rounded numbers.

7. How many squares are shaded?

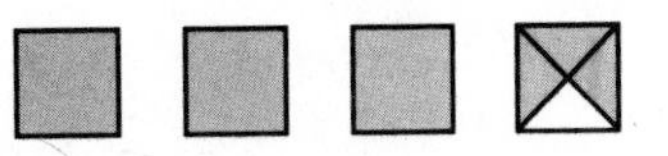

8. Find the length of the pin to the nearest quarter inch.

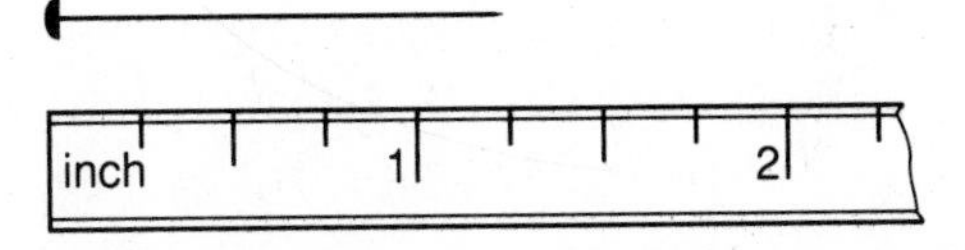

9. The length of segment AB is 7 cm. The length of segment AC is 12 cm. How long is segment BC?

A B C

10. One way to remember the difference between the words "parallel" and "perpendicular" is to look at the three l's in parallel. The l's are parallel. Now here's the question. Are the line segments in a plus sign parallel or perpendicular?

11. To what number is the arrow pointing?

1300 1500 1700

12. Use the digits 4, 7, and 8 once each to write an odd number greater than 500.

13. 6×8 **14.** 7×7 **15.** 9×8 **16.** 7×6

17. $\begin{array}{r} XYZ \\ +\ 338 \\ \hline 507 \end{array}$ **18.** $\begin{array}{r} 406 \\ -\ 228 \\ \hline \end{array}$ **19.** $\begin{array}{r} 705 \\ -\ N \\ \hline 416 \end{array}$ **20.** $\begin{array}{r} N \\ -\ 422 \\ \hline 305 \end{array}$

21. $55 + 555 + 378$ **22.** $293 + 58 + 681$

23. $8 + 56 + 78 + 24 + 55 + 387 + N = 609$

24. Draw a pattern of X's to show 6×8.

25. Write each amount of money with a dollar sign and a decimal point.
(a) 77¢ (b) 8¢ (c) 430¢

26. Use words to write 703,742.

27. Use digits to write nine hundred two thousand, five hundred seven.

28.
$$\begin{array}{r} 14 \\ 38 \\ 56 \\ 41 \\ 8 \\ 6 \\ +\ 14 \\ \hline \end{array}$$

29.
$$\begin{array}{r} 56 \\ 23 \\ 41 \\ 25 \\ 83 \\ 64 \\ +\ \ 5 \\ \hline \end{array}$$

30.
$$\begin{array}{r} 5 \\ 7 \\ 28 \\ 9 \\ 6 \\ 14 \\ +\ 26 \\ \hline \end{array}$$

LESSON 52 Adding and Subtracting Money

Speed Test: Multiplication Facts—Memory Group (Test F in Test Booklet)

To add or subtract money written with a dollar sign, we first line up the decimal points. Then we add if the sign is plus and subtract if the sign is minus.

Example 1 (a) \$3.45 + \$0.75 (b) \$5.35 − \$2

Solution (a) First we line up the decimal points and put a dollar sign and a decimal point in the answer.

$$\begin{array}{r} \$3.45 \\ +\ \$0.75 \\ \hline \$\ \ .\ \ \ \ \end{array}$$

Now we add.

$$\begin{array}{r} \$3.45 \\ +\ \$0.75 \\ \hline \mathbf{\$4.20} \end{array}$$

(b) First we use a decimal point to write \$2. To do this we put a decimal point and two zeros behind the \$2.

\$2 means \$2.00

Now we line up the decimal points. We put a dollar sign and a decimal point in the answer. Then we subtract.

$$\begin{array}{r} \$5.35 \\ -\ \$2.00 \\ \hline \$\ \ . \quad \end{array} \qquad \begin{array}{r} \$5.35 \\ -\ \$2.00 \\ \hline \mathbf{\$3.35} \end{array}$$

Example 2 (a) 67¢ − 59¢ (b) 45¢ + 80¢

Solution To add or subtract money in cent form, we line up the last digits and add or subtract as the sign shows. If the answer equals $1 or more, we usually change from cent form to dollar form.

(a)
$$\begin{array}{r} \overset{5}{\cancel{6}}\,{}^{1}7¢ \\ -\ 5\ 9¢ \\ \hline \mathbf{8¢} \end{array}$$

(b)
$$\begin{array}{r} 45¢ \\ +\,80¢ \\ \hline 125¢ \end{array} = \mathbf{\$1.25}$$

Example 3 $3.75 + $4 + 15¢

Solution If both forms of money are in the problem, we change to one form first. Most people change the cent form to the dollar form. We make these changes.

$4	means	$4.00
15¢	means	$0.15

Next we line up the decimal points. We put a decimal point and a dollar sign in the answer. Then we add.

$$\begin{array}{r} \$3.75 \\ \$4.00 \\ +\ \$0.15 \\ \hline \$\ \ . \quad \end{array} \qquad \begin{array}{r} \$3.75 \\ \$4.00 \\ +\ \$0.15 \\ \hline \mathbf{\$7.90} \end{array}$$

Practice

a. $6.32 + $5

b. $3.25 − $1.75

c. 46¢ + 64¢

d. 98¢ − 89¢

e. $1.46 + 87¢

f. 76¢ − $0.05

Problem set 52

1. One hundred pennies are separated into two piles. In one pile there are thirty-five pennies. How many pennies are in the other pile?

2. Maria skied three hundred forty-two kilometers the first month. She skied seven hundred fifteen kilometers the second month. How many more kilometers did she ski the second month?

3. The apple cost 35¢. Harry gave the vendor a dollar bill. How much change did he get back?

4. Purcell is a town between Chickasha and Konawa. It is 61 miles from Chickasha to Konawa. It is 24 miles from Chickasha to Purcell. How far is it from Purcell to Konawa? Draw a sketch of this problem.

5. Round 752 to the nearest hundred. Then round 572 to the nearest hundred. Use a comparison symbol to compare the rounded numbers.

6. Write the number 347,284 in expanded form. Then use words to write the number.

7. Are railroad tracks parallel or perpendicular?

8. Draw a square to show 3 × 3. Then shade two ninths of the square.

9. It is morning. What time was it 2 hours ago?

10. To what number is the arrow pointing?

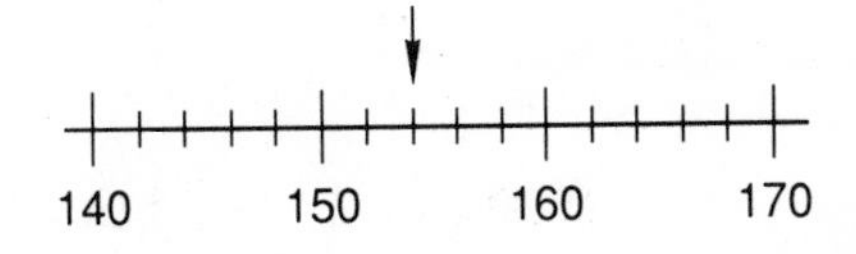

11. \$2.45 + \$3

12. \$3.25 − \$2.47

13. \$5.86 − \$2.20

14. \$2.15 + \$3 + 7¢

15. 4¢ + 23¢ + 7¢

16. $\begin{array}{r} 507 \\ -\ N \\ \hline 456 \end{array}$

17. $\begin{array}{r} N \\ -\ 207 \\ \hline 423 \end{array}$

18. $\begin{array}{r} 486 \\ +\ 957 \\ \hline \end{array}$

19. $\begin{array}{r} 405 \\ -\ 228 \\ \hline \end{array}$

20. $\begin{array}{r} N \\ +\ 215 \\ \hline 703 \end{array}$

21. 6×8

22. 4×3

23. 7×9

24. 4×6

25. 5×4

26. $24 + 28 + 36 + 48 + 97 + 355 + N = 699$

27. Use digits to write four hundred two thousand, five hundred two.

28. $\begin{array}{r} 5 \\ 7 \\ 5 \\ 14 \\ 17 \\ 17 \\ 21 \\ 8 \\ +\ 6 \\ \hline \end{array}$

29. $\begin{array}{r} 19 \\ 42 \\ 17 \\ 5 \\ 2 \\ 8 \\ 9 \\ 5 \\ +\ 6 \\ \hline \end{array}$

30. $\begin{array}{r} 5 \\ 14 \\ 7 \\ 28 \\ 46 \\ 9 \\ 46 \\ 85 \\ +\ 14 \\ \hline \end{array}$

LESSON 53 Multiplying Two-Digit Numbers, Part 1

Speed Test: 64 Multiplication Facts (Test G in Test Booklet)

To multiply 32 by 3, we write the 32 on top. We write the 3 under the 2 in 32. Then we multiply from right to left.

Multiply the ones ↓

$$\begin{array}{r} 32 \\ \times\ \ 3 \\ \hline 6 \end{array}$$

Multiply the tens ↓

$$\begin{array}{r} 32 \\ \times\ \ 3 \\ \hline 96 \end{array}$$

Example 1 Multiply: 20×3

Solution We will write the numbers one above the other. We place the number with more digits on top. We multiply by the 0, then by the 2. The product is **60**.

$$\begin{array}{r} 20 \\ \times 3 \\ \hline 60 \end{array}$$

Example 2 Multiply: 42×3

Solution First we multiply 3 by 2 and get 6. Then we multiply 3 by 4 and get 12.

$$\begin{array}{r} 42 \\ \times\ \ 3 \\ \hline \mathbf{126} \end{array}$$

Practice

a. $\begin{array}{r} 31 \\ \times\ \ 2 \\ \hline \end{array}$

b. $\begin{array}{r} 31 \\ \times\ \ 4 \\ \hline \end{array}$

c. $\begin{array}{r} 42 \\ \times\ \ 4 \\ \hline \end{array}$

d. $\begin{array}{r} 30 \\ \times\ \ 2 \\ \hline \end{array}$

e. $\begin{array}{r} 30 \\ \times\ \ 4 \\ \hline \end{array}$

f. $\begin{array}{r} 24 \\ \times\ \ 0 \\ \hline \end{array}$

Problem set 53

1. Juan compared two numbers. The first number was forty-two thousand, three hundred seventy-six. The second number was forty-two thousand, eleven. Use digits and a comparison symbol to show the comparison.

2. Sally found two hundred forty-three pines cones. Jane found four hundred ninety-six pine cones. How many more pine cones did Jane find?

3. The ticket cost $3.25. The man paid for the ticket with a $5 bill. How much change did he get?

4. Use a comparison symbol to compare 1780 yards and 1 mile.

5. Write the number 415,708 in expanded form. Then use words to write the number.

6. The length of segment RT is 9 cm. The length of segment ST is 5 cm. What is the length of segment RS?

R S T

7. How many circles are shaded?

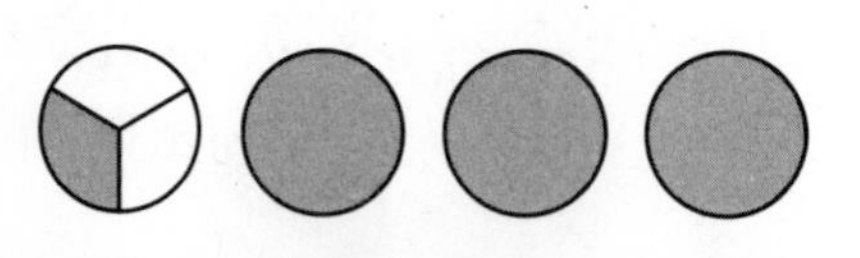

8. Find the length of this nail to the nearest quarter inch.

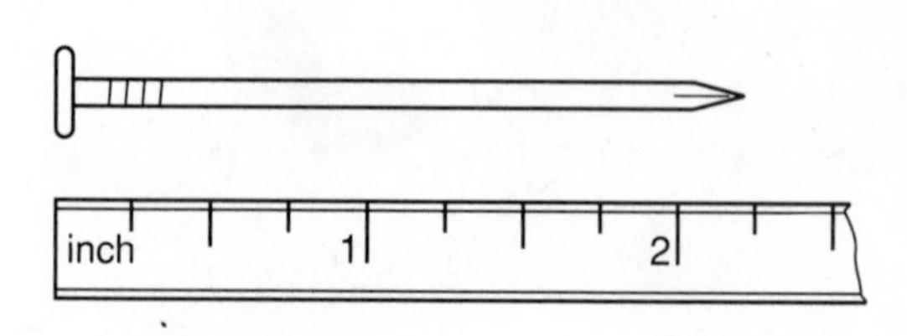

9. Compare: $4 \times 6 \bigcirc 3 \times 8$

10. $\begin{array}{r} \$4.03 \\ -\ \$1.68 \\ \hline \end{array}$ **11.** $\begin{array}{r} \$4.33 \\ +\ \$5.28 \\ \hline \end{array}$ **12.** $\begin{array}{r} \$5.22 \\ -\ \$2.46 \\ \hline \end{array}$ **13.** $\begin{array}{r} \$7.08 \\ -\ \$0.59 \\ \hline \end{array}$

14. $\begin{array}{r} 21 \\ \times\ 6 \\ \hline \end{array}$ **15.** $\begin{array}{r} 40 \\ \times\ 7 \\ \hline \end{array}$ **16.** $\begin{array}{r} 73 \\ \times\ 2 \\ \hline \end{array}$ **17.** $\begin{array}{r} 51 \\ \times\ 6 \\ \hline \end{array}$

18. $2 + 47¢ + 21¢

19. $3.42 + 75¢ + 68¢

20. $\begin{array}{r} 418 \\ +\ 285 \\ \hline \end{array}$

21. $\begin{array}{r} 387 \\ +\ N \\ \hline 496 \end{array}$

22. $\begin{array}{r} 462 \\ -\ N \\ \hline 143 \end{array}$

23. $\begin{array}{r} N \\ -\ 472 \\ \hline 246 \end{array}$

24. $24 + 14 + 86 + 75 + 97 + 431 + N = 889$

25. Write this addition problem as a multiplication problem:

$$21 + 21 + 21 + 21 + 21 + 21$$

Use words to write each number:

26. 903,742

27. 815,002

28. Use digits to write seven hundred forty-two thousand, nine hundred fifteen.

29. $\begin{array}{r} 426 \\ -\ N \\ \hline 21 \end{array}$

30. $\begin{array}{r} N \\ +\ 628 \\ \hline 793 \end{array}$

LESSON 54

Parentheses

> **Speed Test:** 64 Multiplication Facts (Test G in Test Booklet)

Some problems seem to have two answers.

$$4 + 3 \times 2$$

If we add $4 + 3$ first, we get one answer. If we multiply 3×2 first, we get a different answer.

$4 + 3 \times 2$	problem	$4 + 3 \times 2$	problem
7×2	added	$4 + 6$	multiplied
14	multiplied	10	added

The person who makes up the problem can use parentheses to tell us exactly what to do. **We always do the work inside the parentheses first.**

Example 1 $(3 \times 4) + 5$

Solution The person who made up this problem wants us to multiply first. If we multiply first, we get

$$12 + 5 = \mathbf{17}$$

Example 2 $3 \times (4 + 5)$

Solution The person who made up this problem wants us to add first. If we add first, we get

$$3 \times 9 = \mathbf{27}$$

Practice

a. $8 - (4 + 2)$

b. $(8 - 4) + 2$

c. $9 - (6 - 3)$

d. $(9 - 6) - 3$

e. $10 + (2 \times 3)$

f. $(10 + 2) \times 3$

Problem set 54

1. There were four hundred twenty-four in the first bunch. There were seven hundred forty-two in the first two bunches. How many were in the second bunch?

2. Samantha saw fifty-four. Roger saw three hundred twenty. How many more did Roger see?

3. The space shuttle orbited 155 miles above the earth. The weather balloon floated 25 miles above the earth. The space shuttle was how much higher than the weather balloon?

4. How much change should you get back if you give the clerk \$5.00 for a box of cereal that costs \$2.85?

5. Use the digits 4, 5, 6, and 7 to write an odd number between 4700 and 4790.

6. Use digits and a comparison symbol to show that five hundred sixteen is less than five hundred sixty.

7. It is morning. What time was it 40 minutes ago?

8. Write 472,856 in expanded form. Then use words to write the number.

9. How many circles are shaded?

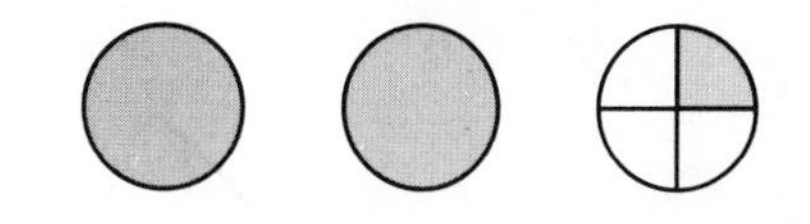

10. Use shaded circles as in Problem 9 to represent the mixed number $4\frac{3}{4}$.

11. Use a ruler to measure this line segment to the nearest quarter inch.

12. Use a centimeter scale to measure the segment in Problem 11 to the nearest centimeter.

13. $8 \times (4 + 3)$

14. $(8 \times 4) + 3$

15. $\begin{array}{r} \$4.07 \\ -\ \$2.26 \\ \hline \end{array}$

16. $\begin{array}{r} \$5.02 \\ -\ \$2.47 \\ \hline \end{array}$

17. $\begin{array}{r} \$5.83 \\ -\ \$2.97 \\ \hline \end{array}$

18. $\begin{array}{r} \$3.92 \\ +\ \$5.14 \\ \hline \end{array}$

19. $\begin{array}{r} 42 \\ \times\ 3 \\ \hline \end{array}$

20. $\begin{array}{r} 83 \\ \times\ 2 \\ \hline \end{array}$

21. $\begin{array}{r} 40 \\ \times\ 4 \\ \hline \end{array}$

22. $\begin{array}{r} 41 \\ \times\ 6 \\ \hline \end{array}$

23. \$2 + 42¢ + 7¢

24. \$5.24 + 23¢ + 2¢

25. $36 + 18 + 26 + 84 + 78 + 204 + N = 499$

26. Write this addition problem as a multiplication problem:

$$40 + 40 + 40 + 40$$

27. Eight times four is how much less than 35?

28. Use words to write the number 906,413.

29. Use digits to write five hundred fourteen thousand, nine hundred eighty-seven.

30. Use digits to write seven hundred five thousand, two hundred two.

LESSON 55

Division

Speed Test: 100 Multiplication Facts (Test C in Test Booklet)

To review addition and subtraction, we remember that addition problems have three numbers. These numbers form a pattern. The largest number is on the bottom.

$$\begin{array}{r} 4 \\ +\ 3 \\ \hline 7 \end{array} \rightarrow \text{largest number}$$

Subtraction problems also have three numbers. These three numbers form a pattern. The largest number is on top. **Subtraction undoes addition.** We can subtract 4 from 7 and get 3. We can subtract 3 from 7 and get 4.

$$\begin{array}{rcccr} 7 & \leftarrow & \text{larger number} & \rightarrow & 7 \\ -\ 4 & \leftarrow & \text{smaller number} & \rightarrow & -\ 3 \\ \hline 3 & \leftarrow & \text{difference} & \rightarrow & 4 \end{array}$$

If we know two of the numbers in an addition or subtraction pattern, we can find the other number.

Multiplication problems also have three numbers. The multiplied numbers are **factors**. The answer is the **product**.

$$\text{factor} \times \text{factor} = \text{product}$$

If we know the two factors, we multiply to find the product. If the factors are 4 and 3, the product is 12.

$$4 \times 3 = 12$$

If we know one factor and the product, we can find the other factor.

$$4 \times ? = 12 \qquad ? \times 3 = 12$$

The process of finding the missing factor is called *division*.

We know how to use a multiplication table to find the product of 3 and 4.

$$\begin{array}{r} 4 \\ \downarrow \\ 3 \rightarrow \overline{)12} \end{array}$$

If we know the product is 12 and one factor is 3, we could write

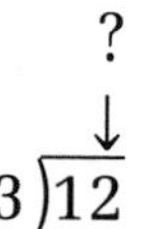

To solve this problem, we need to find the missing factor. We think, "3 times what number equals 12?" Since we know the multiplication facts, we know the missing factor is 4. We write our answer this way.

$$\begin{array}{r} 4 \\ 3\overline{)12} \end{array}$$

A division problem is like a miniature multiplication table. The product is inside the division box. The two factors are outside the division box. One factor is in front and the other is on top. We can find division facts on a multiplication table. We show how to find $4\overline{)32}$ on the multiplication table below. First we find 4 in the left-hand column. Then we follow this row across until we see 32. Then we go up this column and find that the answer is 8.

Multiplication Table

	0	1	2	3	4	5	6	7	**8**	9	10	11	12
0	0	0	0	0	0	0	0	0	0	0	0	0	0
1	0	1	2	3	4	5	6	7	8	9	10	11	12
2	0	2	4	6	8	10	12	14	16	18	20	22	24
3	0	3	6	9	12	15	18	21	24	27	30	33	36
→ **4**	0	4	8	12	16	20	24	28	32	36	40	44	48
5	0	5	10	15	20	25	30	35	40	45	50	55	60
6	0	6	12	18	24	30	36	42	48	54	60	66	72
7	0	7	14	21	28	35	42	49	56	63	70	77	84
8	0	8	16	24	32	40	48	56	64	72	80	88	96
9	0	9	18	27	36	45	54	63	72	81	90	99	108
10	0	10	20	30	40	50	60	70	80	90	100	110	120
11	0	11	22	33	44	55	66	77	88	99	110	121	132
12	0	12	24	36	48	60	72	84	96	108	120	132	144

Example 1 $\begin{array}{r} ? \\ 3\overline{)15} \end{array}$

Solution To find the missing number, we think, "3 times what number is 15?" We remember that $3 \times 5 = 15$, so we know that the missing number must be **5**.

Example 2 $2\overline{)18}$

Solution This is the way division problems are often written. We search for the number that goes above the box. We think, "2 times what number is 18?" We remember that $2 \times 9 = 18$, so the answer is **9**. We write 9 above the 18.

$$\begin{array}{r} 9 \\ 2\overline{)18} \end{array}$$

Practice Find the answer to each division.

a. $2\overline{)12}$ **b.** $3\overline{)21}$ **c.** $4\overline{)20}$ **d.** $5\overline{)30}$

e. $6\overline{)42}$ **f.** $7\overline{)28}$ **g.** $8\overline{)48}$ **h.** $9\overline{)36}$

Problem set 55

1. Four hundred ninety-five were on the first train. Seven hundred sixty-two were on the first two trains. How many were on the second train?

2. There were one hundred twenty-five fish in the first tank. There were three hundred forty-two fish in the second tank. How many fewer fish were in the first tank?

3. Cyrus baled 82 bales of hay the first day. He baled 92 bales of hay on the second day. He baled 78 bales of hay on the third day. How many bales of hay did he bale in the three days?

4. Use the digits 1, 3, 7, and 8 once each to write an odd number greater than 7825.

5. Draw shaded rectangles to picture the mixed number $4\frac{1}{3}$.

6. Write 205,782 in expanded form. Then use words to write this number.

7. It is morning. What time was it 12 hours and 42 minutes ago?

8. To the nearest quarter inch, how long is this rectangle?

inch 1 2

9. Martin took two dozen *big* steps. About how many meters did he walk?

10. To what number is the arrow pointing?

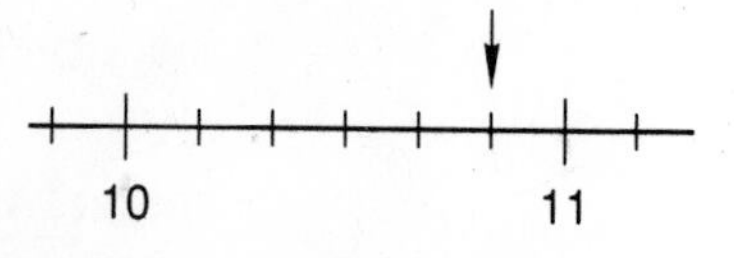

11. $64 + (9 \times 4)$

12. $6.25 + 39¢ + $3

13. $\begin{array}{r} \$4.02 \\ - \$2.47 \\ \hline \end{array}$

14. $\begin{array}{r} \$5.07 \\ - \$2.48 \\ \hline \end{array}$

15. $\begin{array}{r} N \\ + 223 \\ \hline 715 \end{array}$

16. $\begin{array}{r} 938 \\ - ABC \\ \hline 223 \end{array}$

17. $\begin{array}{r} 42 \\ \times \ 3 \\ \hline \end{array}$

18. $\begin{array}{r} 81 \\ \times \ 5 \\ \hline \end{array}$

19. $\begin{array}{r} 32 \\ \times \ 4 \\ \hline \end{array}$

20. $6\overline{)30}$

21. $7\overline{)21}$

22. $8\overline{)56}$

23. $9\overline{)81}$

24. $7\overline{)28}$

25. $3\overline{)15}$

26. $4\overline{)24}$

27. $8\overline{)24}$

28. $5\overline{)40}$

29. $7\overline{)63}$

30. $6\overline{)42}$

LESSON 56 Another Way to Show Division

Speed Test: 100 Multiplication Facts (Test H in Test Booklet)

We have different ways to show division. Here we show two different ways to show "fifteen divided by three."

$$3\overline{)15} \qquad 15 \div 3$$

The first way uses a division box. The second way uses a division sign. We should be able to solve problems in either form and to change from one form to the other form.

Example 1 (a) Use digits and a division box to show "twenty-four divided by six."

(b) Use digits and a division sign to show "twenty-four divided by eight."

Solution (a) $6\overline{)24}$ (b) $\mathbf{24 \div 8}$

Example 2 Solve: $28 \div 4$

Solution We read this, "28 divided by 4." It means the same as $4\overline{)28}$. So 28 divided by 4 means "4 times what number equals 28?" We remember that $4 \times 7 = 28$, so

$$28 \div 4 = \mathbf{7}$$

Multiplication facts have three numbers. Division facts have the same three numbers. Multiplication facts and division facts should be practiced together. Frequent timed written tests on multiplication and division facts will help you memorize these facts.

Practice **a.** $49 \div 7$ **b.** $45 \div 9$ **c.** $40 \div 8$

d. $36 \div 6$ **e.** $32 \div 8$ **f.** $27 \div 3$

Use digits and a division box to show:

g. twenty-seven divided by nine

h. twenty-eight divided by seven

Problem set 56

1. The number was twenty-seven thousand, four hundred thirty-three. Use digits to write this number. Then write the number in expanded form.

2. Frank hit the target two hundred forty-three times. Vanessa hit the target five hundred seven times. How many more times did Vanessa hit the target than Frank hit the target?

3. Four hundred forty-two came. When the bell rang, some more came. Six hundred three came in all. How many came when the bell rang?

4. Use the digits 1, 5, 6, and 8 once each to write an even number greater than 8420.

5. Draw shaded circles to picture the mixed number $4\frac{2}{3}$.

6. To the nearest quarter inch, how long is the pin?

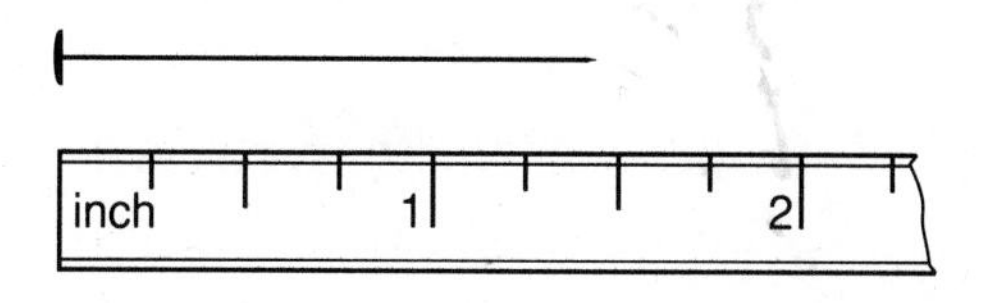

7. It is evening. What time was it 2 hours and 25 minutes ago?

8. One sixth of this rectangle is shaded. Draw another rectangle and shade five sixths of it.

9. The length of segment PQ is 2 cm. The length of segment PR is 11 cm. How long is segment QR?

P Q R

10. $95 - (7 \times 8)$

11. $2.53 + 45¢ + $3

12. $\begin{array}{r} \$3.04 \\ -\ \$1.22 \\ \hline \end{array}$

13. $\begin{array}{r} \$8.46 \\ -\ \$3.58 \\ \hline \end{array}$

14. $\begin{array}{r} 842 \\ -\ XYZ \\ \hline 215 \end{array}$

15. $\begin{array}{r} N \\ -\ 516 \\ \hline 293 \end{array}$

16. $\begin{array}{r} 43 \\ \times\ 3 \\ \hline \end{array}$

17. $\begin{array}{r} 51 \\ \times\ 5 \\ \hline \end{array}$

18. $\begin{array}{r} 42 \\ \times\ 4 \\ \hline \end{array}$

19. $\begin{array}{r} 51 \\ \times\ 0 \\ \hline \end{array}$

20. $28 \div 7$ **21.** $81 \div 9$ **22.** $35 \div 7$ **23.** $16 \div 4$

24. $28 \div 4$ **25.** $42 \div 7$ **26.** $48 \div 8$ **27.** $45 \div 5$

28. Five times five is how much less than 28?

29. Use words to write the number 912,052.

30. Use digits to write the number six hundred fifteen thousand, nine hundred three.

LESSON 57 Multiplying Two-Digit Numbers, Part 2 · Missing Numbers in Multiplication

Speed Test: 90 Division Facts, Form A (Test I in Test Booklet)

Multiplying two-digit numbers

We have practiced multiplying two-digit numbers in two steps. First we multiply the digit in the ones' place. Then we multiply the digit in the tens' place.

Step 1	Step 2
$\begin{array}{r} 12 \\ \times 4 \\ \hline 8 \end{array}$	$\begin{array}{r} 12 \\ \times 4 \\ \hline 48 \end{array}$

The product of the first step is often a two-digit number. We do not write both digits below the line. Instead we write the last digit below the line in the ones' column and carry the first digit to the tens' column.

Step 1

$$\begin{array}{r} {}^{1} \\ 12 \\ \times\ \ 8 \\ \hline 6 \end{array}$$

For step 2, we multiply the second digit and then **add** the digit carried to this column.

Step 2

$(8 \times 1) + 1 = 9$

$$\begin{array}{r} {}^{1} \\ 12 \\ \times\ \ 8 \\ \hline 96 \end{array}$$

Example 1 Find the product: 64×8

Solution We line up the last digits. Then in (1) we multiply 8×4 and get 32. We write the 2 below the line and the 3 above the 6. Then in (2) we multiply 8×6 and add 3 to get 51. The product of 64 and 8 is **512**.

(1)
$$\begin{array}{r} {}^{3} \\ 64 \\ \times\ \ 8 \\ \hline 2 \end{array}$$

(2)
$$\begin{array}{r} {}^{3} \\ 64 \\ \times\ \ 8 \\ \hline 512 \end{array}$$

Missing numbers in multiplication

Recall that in a multiplication problem there are two factors and a product.

$$\text{factor} \times \text{factor} = \text{product}$$

When one of the factors of a multiplication problem is missing, we can find the missing factor by dividing the product by the known factor.

Example 2 $4 \times N = 20$

Solution The problem asks, "4 times what number equals 20?" This is like the division $4\overline{)20}$. The missing factor is **5**.

Example 3 $A \times 7 = 56$

Solution The problem asks, "What number times 7 equals 56?" This is like the division $7\overline{)56}$. The missing factor is **8**.

Practice

a. $\begin{array}{r} 16 \\ \times\ 4 \\ \hline \end{array}$

b. $\begin{array}{r} 24 \\ \times\ 3 \\ \hline \end{array}$

c. $\begin{array}{r} 645 \\ \times\ 6 \\ \hline \end{array}$

d. 53×7

e. 35×8

f. 64×9

g. $4 \times N = 24$

h. $A \times 8 = 24$

i. $12 = 2 \times R$

Problem set 57

1. When Louie counted them, he counted five hundred two. Diana's count was two hundred sixty-five. By how much was Diana's count less than Louie's?

2. There were four hundred seventy-two birds in the first flock. There were one hundred forty-seven birds in the second flock. How many fewer birds were in the second flock?

3. Rae saw them coming. First she saw forty-two of them. Then she saw seventy-five more. Then she saw eighty-six more. How many of them did she see in all?

4. Use the digits 1, 3, 6, and 8 once each to write an odd number that is between 8000 and 8350.

5. Write 306,020 in expanded form. Then use words to write the number.

6. Draw shaded circles to picture the number $6\frac{1}{8}$.

7. How many feet are in 1 mile? How many yards are in 1 mile?

8. To what number is the arrow pointing on this inch scale?

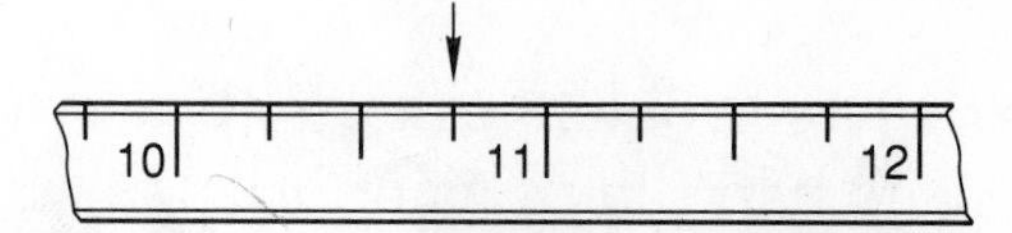

9. How many centimeters are in a meter? How many meters are in a kilometer?

10. To what number is the arrow pointing?

11. $61 + (4 \times 5)$

12. $3.25 + 37¢ + $3

13. 27×7

14. $\begin{array}{r} 33 \\ \times\ \ 6 \\ \hline \end{array}$

15. $\begin{array}{r} 24 \\ \times\ \ 5 \\ \hline \end{array}$

16. $\begin{array}{r} 19 \\ \times\ \ 6 \\ \hline \end{array}$

17. $\begin{array}{r} 42 \\ \times\ \ 7 \\ \hline \end{array}$

18. $\begin{array}{r} \$5.06 \\ -\ \$2.28 \\ \hline \end{array}$

19. $\begin{array}{r} \$3.28 \\ +\ \$4.33 \\ \hline \end{array}$

20. $\begin{array}{r} \$5.27 \\ -\ \$3.68 \\ \hline \end{array}$

21. $\begin{array}{r} 14 \\ 28 \\ 45 \\ 36 \\ 92 \\ +\ 47 \\ \hline \end{array}$

22. $\begin{array}{r} 329 \\ +\ \ N \\ \hline 568 \end{array}$

23. $\begin{array}{r} XYZ \\ -\ 222 \\ \hline 407 \end{array}$

24. $\begin{array}{r} 525 \\ -\ N \\ \hline 207 \end{array}$

25. $28 \div 7$

26. $7\overline{)35}$

27. $6\overline{)54}$

28. $63 \div 7$

29. $5 \times N = 20$

30. $N \times 4 = 28$

LESSON **58**

Equal Groups Word Problems

> **Speed Test:** 90 Division Facts, Form A (Test I in Test Booklet)

We have found that some word problems have an addition pattern. The addition pattern has 3 numbers. If we know two of the numbers, we can find the third number.

$$\begin{array}{r} 5 \text{ marbles} \\ + \ 7 \text{ marbles} \\ \hline 12 \text{ marbles} \end{array}$$

Some word problems have a subtraction pattern. The subtraction pattern has three numbers. If we know two of the numbers, we can find the third number.

$$\begin{array}{r} 12 \text{ marbles} \\ - \ 7 \text{ marbles} \\ \hline 5 \text{ marbles} \end{array}$$

Some word problems have a multiplication pattern. The multiplication pattern has three numbers. If we know two of the numbers, we can find the third number.

$$\begin{array}{ll} \ \ 5 \text{ marbles in each bag} & \\ \times \ 7 & \text{bags} \\ \hline 35 \text{ marbles} & \end{array}$$

We call the multiplication pattern the **equal groups** pattern. Look at this pattern carefully. It looks different from the other patterns we have practiced. The top number of the pattern is the number in each equal group. The second number of the pattern is the number of groups. The bottom number is the total number in all the groups.

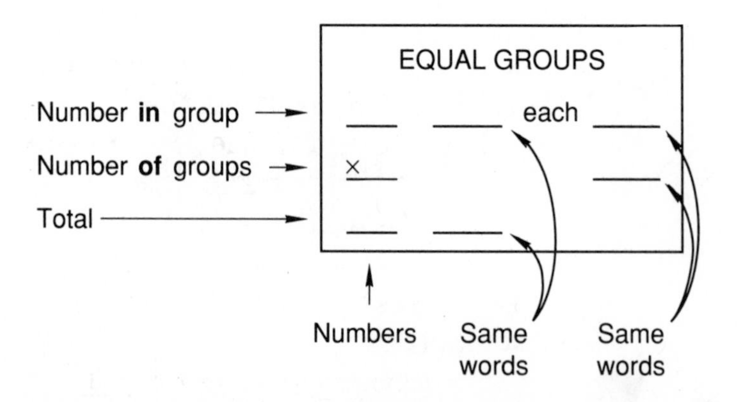

To find the bottom number in the equals groups pattern, we multiply. To find the first or second number, we divide.

Example 1 Ted has 5 cans of tennis balls. There are 3 balls in each can. How many tennis balls does he have?

Solution The words "in each" are a clue to this problem. The words "in each" usually mean that the problem is an equal groups problem. We write the number and words that go with "in each" on the first line. This is the number in each group.

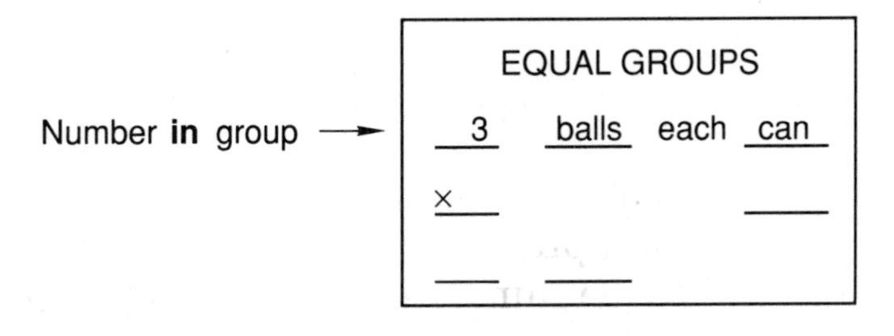

The other number and word in the problem is 5 cans. This is the number of groups. The word "cans" lines up with "can" in the first line of the box.

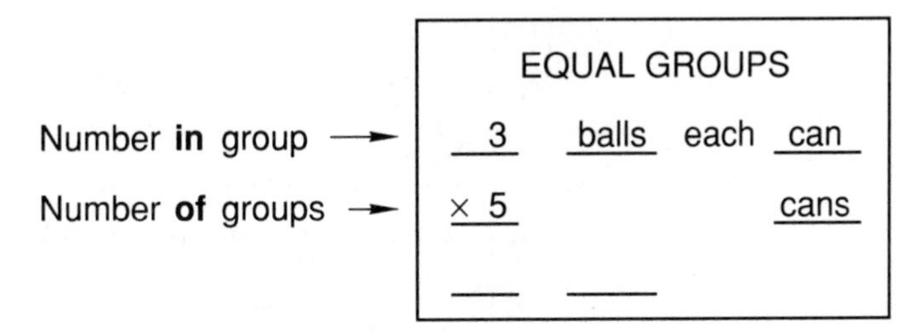

To find the bottom number, we multiply the first two numbers. The word on the bottom line is the same as the word above it on the first line. We complete the equal groups pattern box and answer the question.
Ted has **15 tennis balls.**

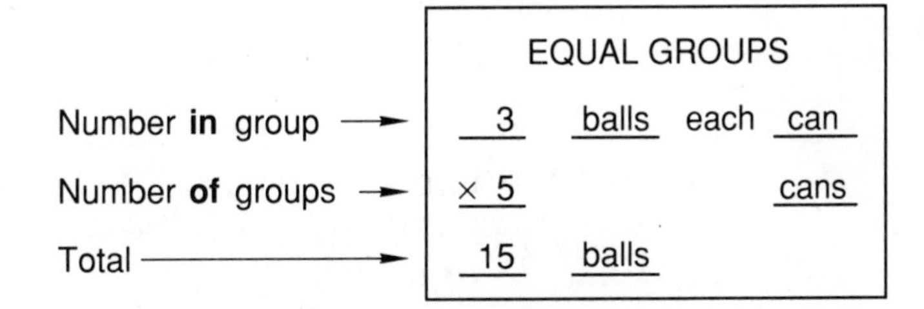

Example 2 Ted has 21 tennis balls in cans. There are 3 balls in each can. How many cans does he have?

Solution There are two numbers in this problem. The words "in each" are a clue. They show us the number and words to write on the first line.

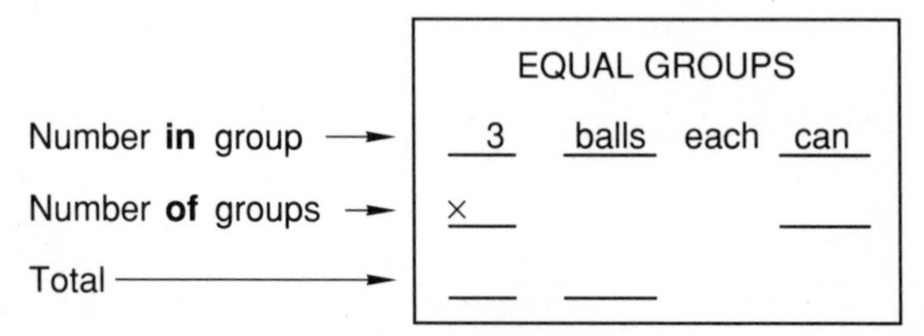

The other number is 21 and the other word is "balls." We need to decide if this is the number of groups or the total. The pattern box shows us how the words should line up. Altogether Ted had 21 tennis balls. This is the total.

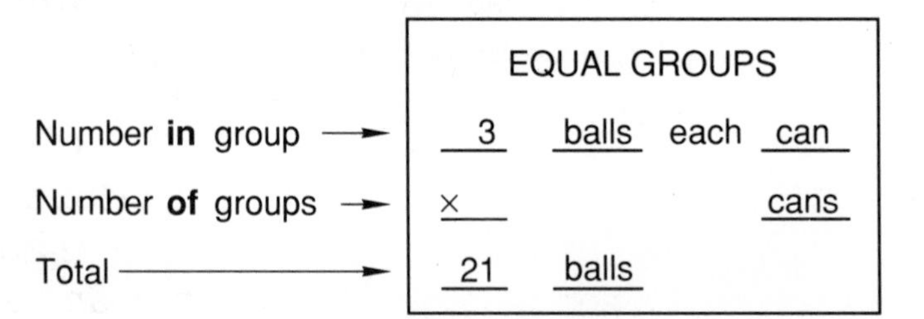

Now we need to find the missing number. **To find the first or second number of an equal groups pattern, we divide.**

$$3\overline{)21}\quad 7$$

We complete the pattern box, matching the words, and then answer the question.

	EQUAL GROUPS
Number **in** group →	3 balls each can
Number **of** groups →	× 7 cans
Total →	21 balls

Seven times 3 balls is 21 balls, so our answer is correct. Ted has **7 cans**.

Example 3 Ted has 5 cans of rubber balls. He has 40 rubber balls in all. How many balls are in each can if the same number of balls is in each can?

Solution The words "in each" show us that this is an equal groups problem. We are not given an "in each" number, but we are given the other words that go with "in each," which we can write on the first line.

	EQUAL GROUPS			
Number **in** group →	N	balls	each	can
Number **of** groups →	× 5			cans
Total →	40	balls		

Now we need to find the missing number on the first line. **To find the first or second number in an equal groups pattern, we divide.**

$$5\overline{)40}\quad 8$$

We complete the pattern box and answer the question.

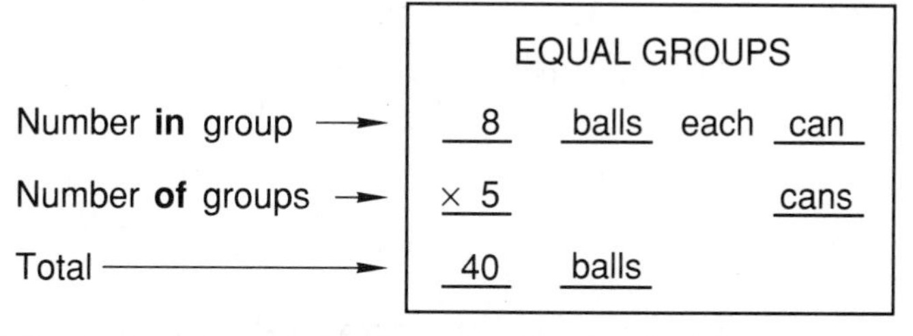

We see that 5 times 8 balls equals 40 balls, so our answer is correct. There are **8 tennis balls in each can.**

Equal groups problems are easy after we find out where to put the numbers in the pattern.

Practice **a.** There were 8 birds in each flock. There were 6 flocks in all. How many birds were there in all?

b. There are 6 people in each car. There are 9 cars. How many people are there in all?

c. There were 35 people. There were 7 cars. The number of people in each car was the same. How many people were in each car?

Problem set 58

1. There were 8 boys in each row. There were 4 rows. How many boys were there?

2. The same number of girls sat in each row. There were 9 rows. There were 63 girls. How many girls were in each row?

3. Four hundred twenty-three came in the building. At noon some more came. Now there were seven hundred five in the building. How many came at noon?

4. The big animal weighed four hundred seventy-five pounds. The small animal weighed one hundred eleven pounds. How much more did the big animal weigh?

5. Shade circles to picture the number $6\frac{3}{4}$.

6. To what number is the arrow pointing?

7. It is evening. What time will it be 7 hours and 22 minutes from now?

8. Draw a rectangle and shade $\frac{7}{8}$ of it.

9. Use digits to write three hundred forty-nine thousand, eight hundred eleven.

10. Seven times eight is how much less than 60?

11. \$4.63 + 23¢ + \$6 + 6¢

12. $\begin{array}{r} \$3.07 \\ -\ \$2.28 \\ \hline \end{array}$ **13.** $\begin{array}{r} \$4.78 \\ +\ \$3.96 \\ \hline \end{array}$ **14.** $\begin{array}{r} 707 \\ -\ \ N \\ \hline 485 \end{array}$ **15.** $\begin{array}{r} ABC \\ -\ 423 \\ \hline 548 \end{array}$

16. $303 - (5 \times 8)$

17. $(4 + 3) \times 8$

18. $N \times 6 = 30$

19. $8 \times A = 24$

20. $(432 + 376) - 148$

21. $(587 - 238) + 415$

22. $\begin{array}{r} 45 \\ \times\ 6 \\ \hline \end{array}$ **23.** $\begin{array}{r} 23 \\ \times\ 7 \\ \hline \end{array}$ **24.** $\begin{array}{r} 34 \\ \times\ 8 \\ \hline \end{array}$ **25.** $\begin{array}{r} 83 \\ \times\ 5 \\ \hline \end{array}$ **26.** $\begin{array}{r} 64 \\ \times\ 9 \\ \hline \end{array}$

27. $56 \div 7$

28. $64 \div 8$

29. $6\overline{)30}$

30. $9\overline{)45}$

LESSON 59 Perimeter

Speed Test: 90 Division Facts, Form A (Test I in Test Booklet)

The Greek word *peri* means around. The Greek word *metron* means measure. We put these words together to form the word **perimeter**. Perimeter means the measure around. The distance around a shape is its perimeter. To find the perimeter of a shape, we add the lengths of all of its sides.

Example 1 Claudia ran around the perimeter of the block. How far did Claudia run?

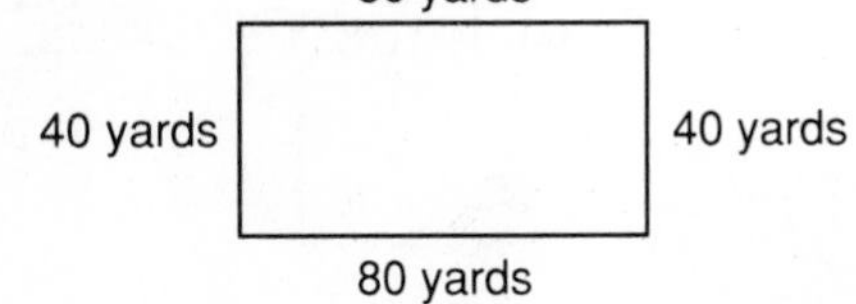

Solution The perimeter of the block is the distance around the block. This block is a rectangle. The two longer sides are both 80 yards, and the two shorter sides are both 40 yards. We are told that Claudia ran around the perimeter, so we know that she ran all the way around. We find that distance by adding all four sides.

$$\begin{array}{r} 80 \text{ yards} \\ 40 \text{ yards} \\ 80 \text{ yards} \\ +\ 40 \text{ yards} \\ \hline 240 \text{ yards} \end{array}$$

Claudia ran **240 yards**.

Example 2 What is the perimeter of this square?

Solution We are told the shape is a square. That means all four sides are the same length. We see that each side is 2 centimeters long. The perimeter is the distance all the way around. We add all four sides.

Perimeter = 2 cm + 2 cm + 2 cm + 2 cm = **8 cm**

Practice Find the perimeter of each shape.

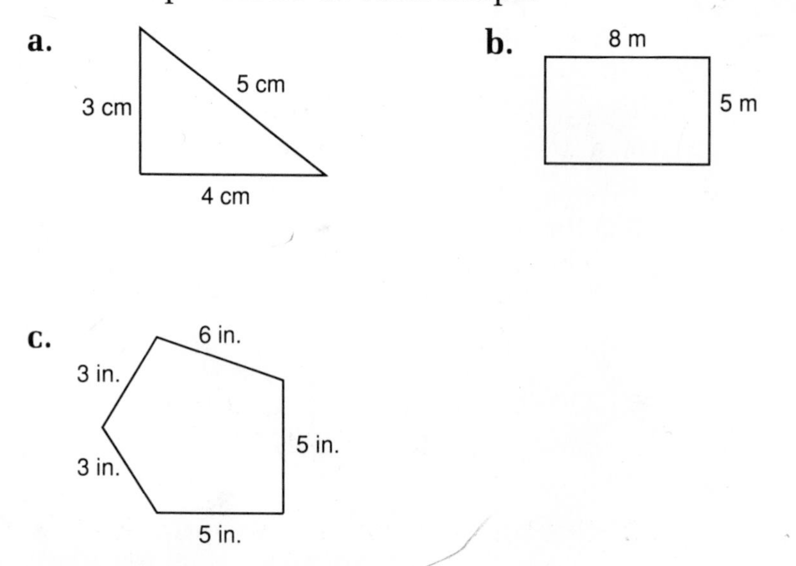

d. What is the perimeter of this square?

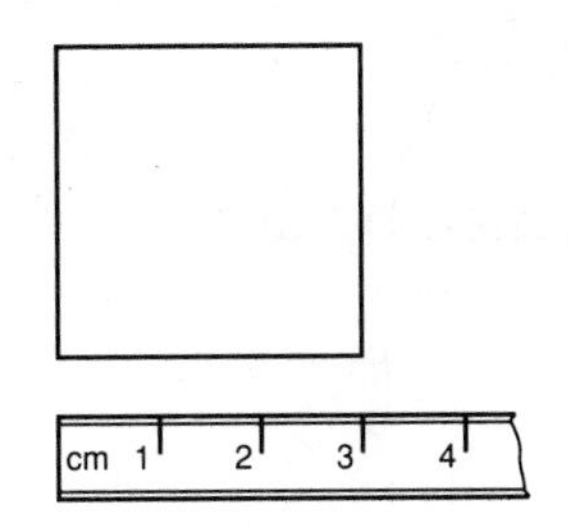

e. What does the word "perimeter" mean?

Problem set 59

1. Each of the 3 lifeboats carried 12 people. How many people were in the 3 lifeboats?

2. The tape cost $6.98. The tax was 42¢. What was the total price?

3. Sarah did six hundred twenty sit-ups. Syd did four hundred seventeen sit-ups. Sarah did how many more sit-ups than Syd?

4. The coach separated 28 players into 4 equal teams. How many players were on each team?

5. Justin ran around the perimeter of the block. How far did Justin run?

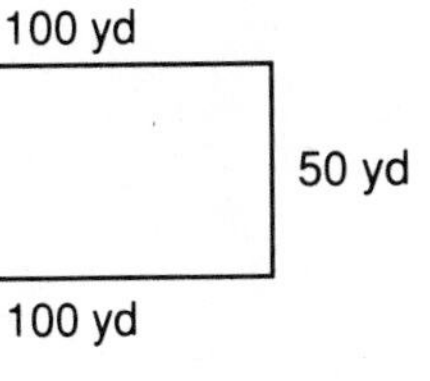

6. Shade circles to draw a picture of $3\frac{3}{8}$.

7. Use the digits 1, 2, 3, and 4 to write an odd number between 4205 and 4223.

8. Six times seven is how much less than 45?

9. It is evening. What time will it be 9 hours and 30 minutes from now?

10. To what number is the arrow pointing?

72 73 74

11. $(4 \times 5) + 203$

12. $4.63 + $2 + 47¢ + 65¢

13. $\begin{array}{r} 43 \\ \times\ 6 \\ \hline \end{array}$

14. $\begin{array}{r} 54 \\ \times\ 8 \\ \hline \end{array}$

15. $\begin{array}{r} 37 \\ \times\ 3 \\ \hline \end{array}$

16. $\begin{array}{r} 46 \\ \times\ 4 \\ \hline \end{array}$

17. $47 + 55 + 84 + 63 + 24 + 27$

18. $\begin{array}{r} 87 \\ \times\ 5 \\ \hline \end{array}$

19. $\begin{array}{r} 78 \\ \times\ 6 \\ \hline \end{array}$

20. $\begin{array}{r} N \\ \times\ 7 \\ \hline 63 \end{array}$

21. $\begin{array}{r} 8 \\ \times\ Q \\ \hline 32 \end{array}$

22. $6.08 − $4.29

23. $7.03 − $4.26

24. $\begin{array}{r} 325 \\ -\ 198 \\ \hline \end{array}$

25. $\begin{array}{r} 472 \\ +\ 348 \\ \hline \end{array}$

26. $\begin{array}{r} N \\ -\ 279 \\ \hline 484 \end{array}$

27. $\begin{array}{r} 462 \\ +\ ABC \\ \hline 529 \end{array}$

28. $3\overline{)24}$

29. $4\overline{)24}$

30. $36 \div 6$

31. $36 \div 9$

LESSON 60 Adding Numbers with More Than Three Digits · Checking One-Digit Division

Speed Test: 90 Division Facts, Form A (Test I in Test Booklet)

Adding numbers with more than three digits To add numbers that have more than three digits, we add in the ones' column first. Then we add in the tens' column, the hundreds' column, the thousands' column, the ten thousands' column, and so forth. When the sum of the digits in one column is a two-digit number, we record the second digit below the line. We carry the first digit one column to the left.

Example 1 Add:

$$\begin{array}{r} 43{,}287 \\ +\ 68{,}595 \\ \hline \end{array}$$

Solution We add the digits in the ones' column first. When the sum is a two-digit number, we write the last digit below the line and the first digit above the next column. The sum is **111,882**.

$$\begin{array}{r} {}^{1}\ \ \ {}^{1}{}^{1}\ \\ 43{,}287 \\ +\ 68{,}595 \\ \hline 111{,}882 \end{array}$$

Example 2 Add: 456 + 1327 + 52 + 3624

Solution When we write the numbers in a column, we are careful to line up the last digit in each number. We add the digits one column at a time, starting from the right. The sum is **5459**.

$$\begin{array}{r} {}^{111}\ \ \\ 456 \\ 1327 \\ 52 \\ +\ 3624 \\ \hline 5459 \end{array}$$

Checking one-digit division We can check division by multiplying. We multiply the numbers outside the division box.

Division: $3\overline{)12}$ with quotient 4

$$\begin{array}{r} 4 \\ \times\ 3 \\ \hline 12 \end{array}$$ check

We see that the multiplication answer matches the number inside the division box. We usually show this by writing the multiplication answer under the number in the division box.

$$\begin{array}{r} 4 \\ 3\overline{)12} \\ 12 \end{array}$$

12 ← **Step 1.** Divide 12 by 3 and write 4.
12 ← **Step 2.** Multiply 4 by 3 and write 12.

Example 3 Divide. Check the answer by multiplying.

(a) $3\overline{)18}$ (b) $4\overline{)32}$

Solution First we divide and write the answer above the box. Then we multiply and write the product below the box.

(a) $\begin{array}{r} 6 \\ 3\overline{)18} \\ 18 \end{array}$ (b) $\begin{array}{r} 8 \\ 4\overline{)32} \\ 32 \end{array}$

Practice

a. $\begin{array}{r} 4356 \\ +\ 5644 \\ \hline \end{array}$ **b.** $\begin{array}{r} 46{,}027 \\ +\ 39{,}682 \\ \hline \end{array}$ **c.** $\begin{array}{r} 360{,}147 \\ +\ 96{,}894 \\ \hline \end{array}$

d. 436 + 5714 + 88 **e.** 43,284 + 572 + 7635

Divide. Check the answer by multiplying.

f. $3\overline{)21}$ **g.** $7\overline{)42}$ **h.** $6\overline{)48}$

Problem set 60

1. The coach had forty-two players. The coach wanted to make six equal teams. How many players should be on each team?

2. There were 7 pancakes in each stack. If there were 42 pancakes in all, how many stacks were there?

3. Five hundred forty-two laughed out loud. Nine hundred thirty-three merely smiled. What was the total of the smilers and the laughers?

4. Marion found four hundred forty-three. Henry found six hundred two. How many more did Henry find?

5. Write the number 476,032 in expanded form. Then use words to write the number.

6. Use the digits 1, 2, 3, and 4 once each to write an odd number that is between 4225 and 4300.

7. How many circles are shaded?

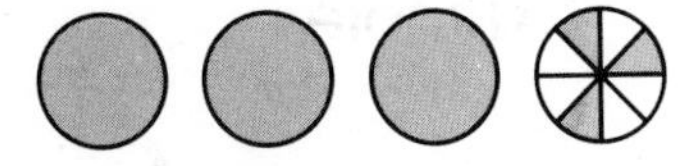

8. Which of these angles is not a right angle?

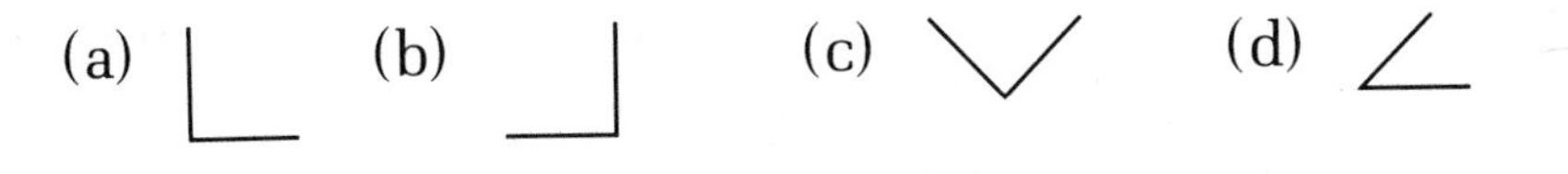

9. This is a rectangle.
(a) What is the length?
(b) What is the width?
(c) What is the perimeter?

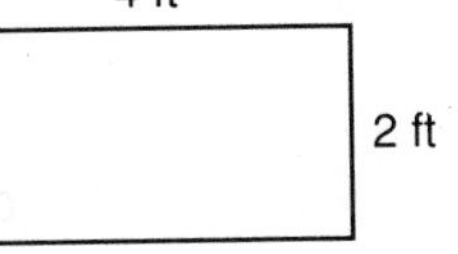

10. Thirty is how much more than 4 times 7?

11. 3864 + 598 + 4237 + 12

12. $\begin{array}{r} 63,285 \\ +\ 97,642 \\ \hline \end{array}$

13. $\begin{array}{r} \$5.00 \\ -\ \$4.81 \\ \hline \end{array}$

14. $\begin{array}{r} N \\ +\ 398 \\ \hline 614 \end{array}$

15. $\begin{array}{r} 46 \\ \times\ 6 \\ \hline \end{array}$

16. $\begin{array}{r} 85 \\ \times\ 5 \\ \hline \end{array}$

17. $\begin{array}{r} 37 \\ \times\ 7 \\ \hline \end{array}$

18. $\begin{array}{r} 48 \\ \times\ 8 \\ \hline \end{array}$

19. $\begin{array}{r} F \\ \times\ 8 \\ \hline 72 \end{array}$

20. $\begin{array}{r} 46,385 \\ +\ 5,296 \\ \hline \end{array}$

21. $\begin{array}{r} 462,586 \\ +\ 39,728 \\ \hline \end{array}$

22. $\begin{array}{r} XYZ \\ -\ 478 \\ \hline 263 \end{array}$

23. $\begin{array}{r} 478 \\ -\ ABC \\ \hline 203 \end{array}$

24. Compare: $7 \times 6 \bigcirc 44$

Divide. Check the answer by multiplying.

25. $2\overline{)18}$ **26.** $7\overline{)21}$ **27.** $8\overline{)56}$

28. Use words to write the number 416,007.

29. $\begin{array}{r} 47 \\ \times\ \ 6 \\ \hline \end{array}$ **30.** $\begin{array}{r} 28 \\ \times\ \ 5 \\ \hline \end{array}$

LESSON 61 Subtracting Numbers with More Than Three Digits

Speed Test: 90 Division Facts, Form A (Test I in Test Booklet)

To subtract numbers with more than three digits, we begin by subtracting in the ones' column. We borrow if it is necessary. Then we move one column to the left and subtract in the tens' column. We borrow if it is necessary. Then we move one column at a time to the hundreds' column, the thousands' column, the ten thousands' column, and so on. We subtract in each column and borrow when it is necessary. Sometimes we must subtract across several zeros.

Example 1 Subtract: 36,152 − 9,415

Solution We write the first number on top. We write the second number below. We line up digits with the same place value. We begin by subtracting in the ones' column. Then we subtract in the other columns.

$$\begin{array}{r} \overset{2}{\cancel{3}}\,\overset{15}{\cancel{6}},\,{}^{1}1\,\overset{4}{\cancel{5}}\,{}^{1}2 \\ -\quad 9,\,4\,1\,5 \\ \hline \mathbf{2\,6,\,7\,3\,7} \end{array}$$

Example 2 Subtract: $\begin{array}{r} 5000 \\ -\ 2386 \\ \hline \end{array}$

Solution We need to find some more ones for the ones' place before we can subtract. We can only get ones from the tens' place, but there are no tens. We can only get tens from the hundreds' place, but there are no hundreds. To get hundreds, we go to the thousands' place. We exchange 1 thousand for 10 hundreds, 1 hundred for 10 tens, and 1 ten for 10 ones. Then we can subtract.

$$\begin{array}{r} \scriptstyle 4\ 9\ 9\ \\ \not{5}\,{}^{1}\not{0}\,{}^{1}\not{0}\,{}^{1}0 \\ -\ 2\ 3\ 8\ 6 \\ \hline \mathbf{2\ 6\ 1\ 4} \end{array}$$

Practice a.
$$\begin{array}{r} 4783 \\ -\ 2497 \\ \hline \end{array}$$

b.
$$\begin{array}{r} 4000 \\ -\ 527 \\ \hline \end{array}$$

Problem set 61

1. There were 8 buses. Each bus could seat 60 students. How many students could ride in all the buses?

2. Each van could carry 9 students. There were 63 students. How many vans were needed?

3. Fifty-six Easter eggs were to be divided into 7 equal piles. How many eggs would be in each pile?

4. There are 10 swimmers in the race. Only 3 can win medals. How many swimmers will not win a medal?

5. There were two hundred sixty-seven apples in the first bin. There were four hundred sixty-five apples in the second bin. How many fewer apples were in the first bin?

6. Two hundred seventy-five came inside before noon. Then more came. There were four hundred sixty inside by 3 o'clock. How many came after noon?

7. Write the number 813,724 in expanded form. Then use words to write the number.

8. Use the digits 0, 2, 5, and 7 to write an even number that is greater than 1000 and less than 2750.

9. Use shaded circles to draw a picture of $3\frac{3}{4}$.

10. This shape has four sides, but it is not a rectangle. What is the perimeter of this shape in meters?

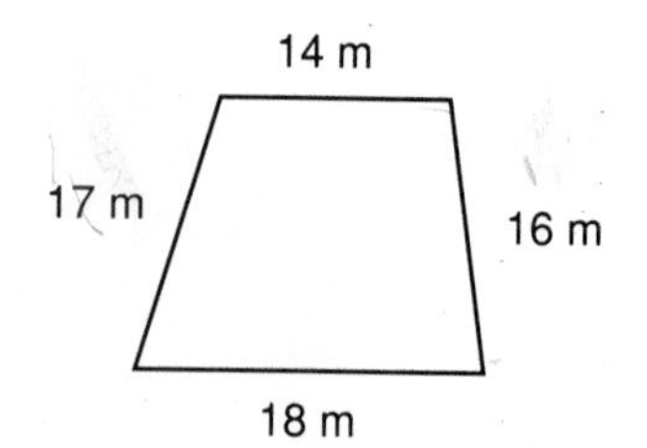

11. $849 + 73 + 615$

12. $38{,}295 + 467 + 9512$

13. $\begin{array}{r} 47{,}586 \\ +\ 23{,}491 \\ \hline \end{array}$

14. $\begin{array}{r} \$5.00 \\ -\ \$3.26 \\ \hline \end{array}$

15. $\begin{array}{r} N \\ +\ 258 \\ \hline 604 \end{array}$

16. $\begin{array}{r} 49 \\ \times\ 6 \\ \hline \end{array}$

17. $\begin{array}{r} 84 \\ \times\ 5 \\ \hline \end{array}$

18. $\begin{array}{r} 76 \\ \times\ 8 \\ \hline \end{array}$

19. $\begin{array}{r} 35 \\ \times\ 9 \\ \hline \end{array}$

20. $\begin{array}{r} 43 \\ \times\ 7 \\ \hline \end{array}$

21. $\begin{array}{r} 47{,}205 \\ -\ 23{,}426 \\ \hline \end{array}$

22. $\begin{array}{r} 400 \\ -\ \ N \\ \hline 256 \end{array}$

23. $\begin{array}{r} 4000 \\ -\ 2468 \\ \hline \end{array}$

24. Use a dollar sign and decimal point to show each amount of money.
(a) 54¢ (b) 6¢ (c) 340¢

25. Thirteen is how much more than 4 times 3?

26. Compare: $4 \times 8 \bigcirc 31$

Divide. Check the answer by multiplying.

27. $3\overline{)27}$ **28.** $7\overline{)28}$ **29.** $8\overline{)72}$ **30.** $6\overline{)54}$

LESSON 62 One-Digit Division with a Remainder

Speed Test: 90 Division Facts, Form B (Test J in Test Booklet)

We can divide 12 objects into equal groups of 4. Here we show 12 dots divided into equal groups of 4.

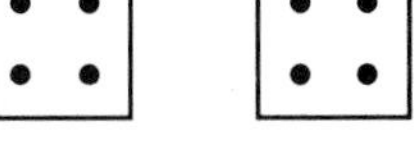

12 dots 3 equal groups

However, we cannot divide 13 dots into equal groups of 4 because there is 1 dot too many. We call the extra dot the **remainder**.

13 dots 3 equal groups Remainder

We can show that 13 is to be divided into groups of 4 by writing

$$4\overline{)13}$$

As we look at this problem we may wonder what to write for the answer. The answer is not exactly 3 because 3 × 4 is 12, which is less than 13. However, the answer is not 4 because 4 × 4 is 16, which is more than 13. Since we *can* make 3 groups of four, we write 3 for our answer. Then we multiply 3 × 4 and write 12 below the 13.

$$\begin{array}{r} 3 \\ 4\overline{)13} \\ 12 \end{array}$$ ← three groups

We see that 13 is more than 12. Now we find out how much is left over after making 3 groups of 4. To do this, we subtract 12 from 13.

$$\begin{array}{r} 3 \\ 4\overline{)13} \\ -12 \\ \hline 1 \end{array}$$

← three groups

subtract

← 1 left over (remainder)

There is 1 left over. The amount left over is the **remainder**. We use a small r to stand for remainder. We can write the answer as

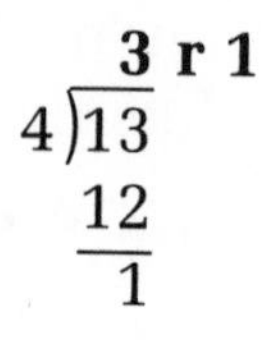

We cannot divide 14 dots into equal groups of 3 because there are 2 dots too many.

14 dots | 4 equal groups | Remainder

This is the reason that if we divide 14 by 3 we get 4 with a remainder of 2.

$$\begin{array}{r} 4 \\ 3\overline{)14} \\ 12 \\ \hline 2 \end{array}$$

← remainder

The answer is 4 with a remainder of 2. We can use a small r to stand for remainder and write

$$14 \div 3 = \mathbf{4\ r\ 2}$$

From these examples we see that if a division does not come out even, the number we get when we subtract is the remainder.

Example 1 Divide: $3\overline{)16}$

Solution Three will not divide 16 evenly. Three times 6 equals 18. This is too much. Three times 5 is 15. This is less than 16. We will try 5.

$$3\overline{)16}^{\;5}$$

Now we multiply and then we subtract.

$$\begin{array}{r} 5 \\ 3\overline{)16} \\ \underline{15} \\ 1 \end{array} \leftarrow \text{remainder}$$

The answer is **5 r 1**.

Example 2 Divide: 20 ÷ 6

Solution First we write the problem using a division box.

$$6\overline{)20}$$

Now we think: 6 times 4 equals 24. This is too much. We think: 6 times 3 equals 18. This is less than 20. We will use 3.

$$\begin{array}{r} 3 \\ 6\overline{)20} \end{array}$$

Next we multiply, and then we subtract.

$$\begin{array}{r} 3 \\ 6\overline{)20} \\ \underline{18} \\ 2 \end{array} \leftarrow \text{remainder}$$

We write the answer as **3 r 2**.

Practice Divide and write the answer with a remainder.

a. $3\overline{)17}$ **b.** $5\overline{)12}$ **c.** $4\overline{)23}$

d. 15 ÷ 2 **e.** 20 ÷ 6 **f.** 25 ÷ 3

Problem set 62

1. Harry had 56 washers. He wanted to put them into equal piles of 8 washers. How many piles would he have?

2. There were 42 children waiting for a ride. There were 7 cars available. If the same number rode in each car, how many would be in a car?

3. The first time, the king's guess was forty-nine thousand, six hundred two. The second time, the king's guess was one hundred twelve thousand, nine hundred forty-five. His second guess was how much greater than his first guess?

4. Four thousand, two were waiting in the school yard. Some went away at 10 o'clock. Then there were two thousand, thirteen waiting in the school yard. How many went away at 10 o'clock?

5. Write the number 416,020 in expanded form. Then use words to write the number.

6. Round 4725 to the nearest hundred. Then round 4575 to the nearest hundred. Then find the sum of the rounded numbers.

7. What number is three hundred seventeen less than five thousand, one hundred forty-three?

8. One side of a square is 4 feet long. What is the perimeter of the square?

9. How many circles are shaded?

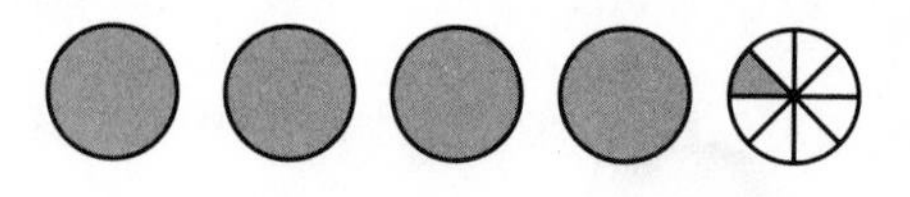

10. $54 + (42 \div 6)$

11. \$6.35 + \$12.49 + 42¢

12. \$100.00 − \$59.88

13. 51,438 − 47,495

14. $\begin{array}{r} 60 \\ \times\ 9 \\ \hline \end{array}$

15. $\begin{array}{r} 57 \\ \times\ 4 \\ \hline \end{array}$

Divide and write the answer with a remainder.

16. $25 \div 4$

17. $22 \div 5$

18. $\begin{array}{r} 38 \\ \times 7 \\ \hline \end{array}$

19. $7\overline{)30}$

20. $\begin{array}{r} 46 \\ \times\ 8 \\ \hline \end{array}$

21. $39 \div 6$

22. $\begin{array}{r} XYZ \\ -\ 165 \\ \hline 402 \end{array}$

23. $\begin{array}{r} 400 \\ -\ ABC \\ \hline 283 \end{array}$

24. $\begin{array}{r} 468{,}576 \\ -\ 329{,}684 \\ \hline \end{array}$

25. $\begin{array}{r} 4000 \\ -\ 2444 \\ \hline \end{array}$

Use words to write each number.

26. 913,052

27. 87,906

28. $\begin{array}{r} 48 \\ 76 \\ 25 \\ 336 \\ 98 \\ +\ 75 \\ \hline \end{array}$

29. $\begin{array}{r} 43 \\ 58 \\ 9 \\ 6 \\ 454 \\ +\ 28 \\ \hline \end{array}$

30. $\begin{array}{r} 95 \\ 7 \\ 87 \\ 423 \\ 76 \\ +\ 98 \\ \hline \end{array}$

LESSON 63 Months and Years

Speed Test: 90 Division Facts, Form B (Test J in Test Booklet)

A year is the number of days it takes for the earth to go around the sun. It takes the earth almost exactly $365\frac{1}{4}$ days to go around the sun. To make the days come out even, we have 3 years in a row that have 365 days each. Then we have a year that has 366 days. The year with 366 days is called a **leap year**. We put the extra day in February. The rest of the months have 30 days or 31 days.

Months of the year	Number of days
January	31
February	28 (29 in leap years)
March	31
April	30
May	31
June	30
July	31
August	31
September	30
October	31
November	30
December	31

February has 28 or 29 days. Four months have 30 days. All the rest have 31 days. If we say the four months that have 30 days, we can remember the number of days in the other months. The following jingle helps us remember which months have 30 days.

Thirty days hath September,
April, June, and November.
All the rest have 31 excepting February.
February has 28 or 29.

Example 1 How many days does December have?

Solution "Thirty days hath September, April, June, and November." This tell us that December does not have 30 days. December must have **31 days**.

Example 2 According to this calendar, May 10, 2014, is what day of the week?

MAY						2014
S	M	T	W	T	F	S
				1	2	3
4	5	6	7	8	9	10
11	12	13	14	15	16	17
18	19	20	21	22	23	24
25	26	27	28	29	30	31

Solution The letters across the top of the calendar stand for the days of the week. We see that May 10 is a **Saturday**, the second Saturday of the month.

Example 3 How many years were there from 1620 to 1776?

Solution To find the number of years from one date to another, we may subtract.* We subtract the earlier date from the later date. In this problem we subtract 1620 from 1776.

$$\begin{array}{r} 1776 \\ -\ 1620 \\ \hline 156 \end{array}$$

We find that there were **156 years** from 1620 to 1776.

Practice

a. How many days long is a leap year?

b. According to the calendar in Example 2, what is the date of the fourth Friday of the month?

c. How many years were there from 1918 to 1943?

---

*Years have been numbered forward (A.D.) and backward (B.C.) from the birth of Jesus of Nazareth. In this book, all year dates should be considered years A.D.

Problem set 63

1. There were 7 students in each row. If there were 56 students in all, how many rows were there?

2. There were 7 nails in each board. If there were 42 boards, how many nails were there?

3. Helen saw forty-two thousand, nine hundred in the first hour. She saw thirty-five thousand, four hundred forty-seven in the second hour. How many more did she see in the first hour?

4. Write the number 318,785 in expanded form. Then use words to write the number.

5. According to this calendar, what day of the week was December 25, 1957?

DECEMBER						1957
S	M	T	W	T	F	S
1	2	3	4	5	6	7
8	9	10	11	12	13	14
15	16	17	18	19	20	21
22	23	24	25	26	27	28
29	30	31				

6. Round 6523 to the nearest hundred. Round 5602 to the nearest hundred. Then use comparison symbols to compare the rounded numbers.

7. One side of a rectangle is 10 kilometers long. Another side is 20 kilometers long. Draw a picture of the rectangle and show the lengths of the sides. What is the perimeter of the rectangle?

8. How many circles are shaded?

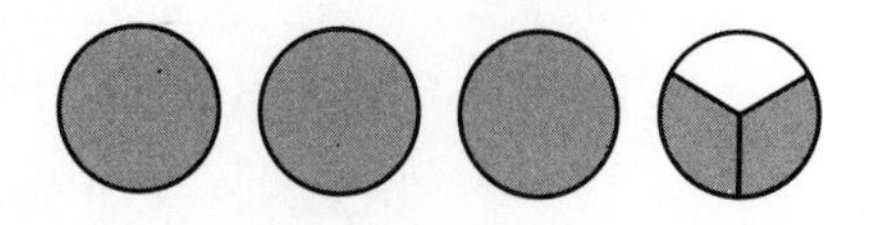

9. To what number is the arrow pointing?

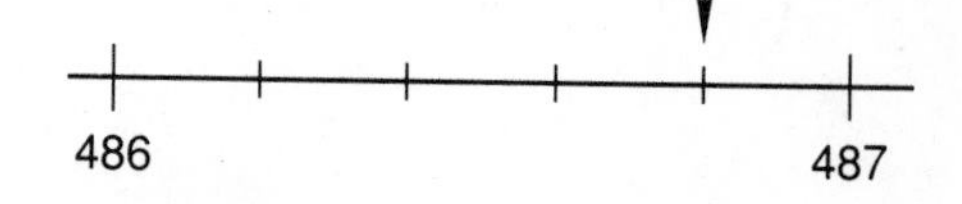

10. When Elmer emptied his bank, he found 17 pennies, 4 nickels, 5 dimes, and 2 quarters. What was the value of the coins in his bank?

11. $\begin{array}{r} 794{,}150 \\ +\ \ 9{,}863 \\ \hline \end{array}$

12. $\begin{array}{r} 51{,}786 \\ +\ 36{,}357 \\ \hline \end{array}$

13. $\begin{array}{r} 876 \\ 40 \\ 317 \\ 55 \\ 11 \\ +\ \ 5 \\ \hline \end{array}$

14. $\begin{array}{r} \$20.00 \\ -\ \$18.47 \\ \hline \end{array}$

15. $\begin{array}{r} 41{,}315 \\ -\ 29{,}418 \\ \hline \end{array}$

16. $\begin{array}{r} 46 \\ \times\ 7 \\ \hline \end{array}$ **17.** $\begin{array}{r} 54 \\ \times\ 8 \\ \hline \end{array}$ **18.** $\begin{array}{r} 39 \\ \times\ 9 \\ \hline \end{array}$ **19.** $\begin{array}{r} 58 \\ \times\ 6 \\ \hline \end{array}$ **20.** $\begin{array}{r} 30 \\ \times\ 4 \\ \hline \end{array}$

21. $5\overline{)36}$ **22.** $43 \div 7$ **23.** $9\overline{)64}$ **24.** $32 \div 6$

25. $Y \times 4 = 32$ **26.** $6 \times N = 42$

27. Use digits to write four hundred forty-two thousand, nine hundred seventy-six.

28. $\begin{array}{r} 7 \\ 8 \\ 6 \\ 5 \\ 4 \\ +\ 3 \\ \hline \end{array}$

29. $\begin{array}{r} 24 \\ 35 \\ 9 \\ 8 \\ 17 \\ +\ 26 \\ \hline \end{array}$

30. $\begin{array}{r} 25 \\ 43 \\ 5 \\ 2 \\ 7 \\ +\ 428 \\ \hline \end{array}$

LESSON 64 Multiples

Speed Test: 90 Division Facts, Form A (Test I in Test Booklet)

If we multiply 4 by the numbers 1, 2, 3, 4, 5, 6, . . . , we get

4, 8, 12, 16, 20, 24, . . .

These numbers are **multiples of 4**. The multiples of 4 are the numbers we get if we count by fours. The following numbers are the multiples of 6.

6, 12, 18, 24, 30, 36, . . .

The multiples of 6 are the numbers we get if we count by sixes.

The multiples of any number are the answers we get when we multiply the number by 1, 2, 3, 4, 5, 6, ...

Example 1 List the first four multiples of 7.

Solution To find the first four multiples of 7, we multiply 7 by 1, then by 2, then by 3, and then by 4.

$$\begin{array}{r} 7 \\ \times\ 1 \\ \hline 7 \end{array} \qquad \begin{array}{r} 7 \\ \times\ 2 \\ \hline 14 \end{array} \qquad \begin{array}{r} 7 \\ \times\ 3 \\ \hline 21 \end{array} \qquad \begin{array}{r} 7 \\ \times\ 4 \\ \hline 28 \end{array}$$

The first four multiples of 7 are **7, 14, 21, and 28**. Notice that the multiples of 7 are the numbers we say when we count by 7.

Example 2 (a) What is the fourth multiple of 6?

(b) What is the third multiple of 8?

Solution (a) To find the fourth multiple of 6, we multiply 6 by 4. The fourth multiple of 6 is **24**.

(b) To find the third multiple of 8, we multiply 8 by 3. The third multiple of 8 is **24**.

Practice **a.** List the first five multiples of 6.

b. List the third, fourth, and fifth multiples of 9.

c. What is the seventh multiple of 8?

Problem set 64

1. Kent bought a toy for $1.85 and sold it for 75¢ more. For what price did he sell the toy?

2. Two thousand people entered the contest. Only seven will win prizes. How many will not win prizes?

3. There were 37 people in each boxcar. If there were 8 boxcars, how many people were riding in boxcars?

4. Twenty-seven thousand people lived in the big town. Only eight thousand, four hundred seventy-two lived in the small town. How many more people lived in the big town?

5. Draw a rectangle that is 4 cm long and 3 cm wide. What is the perimeter of the rectangle?

6. Fiona found the third multiple of 4. Then she subtracted 2 from this number. What was her answer?

7. Write the number 220,500 in expanded form. Then use words to write the number.

8. If it is afternoon, what time was it 40 minutes ago?

9. How many years were there from 1776 to 1789?

10. What is the length of line segment ST?

R S T

cm 1 2 3 4 5 6 7 8 9 10

11. $\begin{array}{r} 400 \\ -\ 222 \\ \hline \end{array}$

12. $\begin{array}{r} 705 \\ -\ 423 \\ \hline \end{array}$

13. $\begin{array}{r} 4587 \\ -\ 2364 \\ \hline \end{array}$

14. $\begin{array}{r} \$25.42 \\ -\ \$\ 7.25 \\ \hline \end{array}$

15. $\begin{array}{r} 64 \\ \times\ 5 \\ \hline \end{array}$

16. $\begin{array}{r} 70 \\ \times\ 6 \\ \hline \end{array}$

17. $\begin{array}{r} 89 \\ \times\ 4 \\ \hline \end{array}$

18. $\begin{array}{r} 63 \\ \times\ 7 \\ \hline \end{array}$

19. $\begin{array}{r} 46 \\ 295 \\ 329 \\ 746 \\ 761 \\ +\ 356 \\ \hline \end{array}$

20. $8\overline{)15}$

21. $5\overline{)32}$

22. $63 \div 7$

23. $33 \div 6$

24. $44 \div 7$

25. $59 \div 8$

26. Use digits to write seventy thousand, seven hundred two.

27. $\begin{array}{r} 7 \\ 5 \\ 4 \\ 3 \\ 2 \\ 6 \\ 9 \\ 5 \\ +\ N \\ \hline 44 \end{array}$

28. $\begin{array}{r} 8 \\ 4 \\ 6 \\ 2 \\ 5 \\ 2 \\ 5 \\ 7 \\ +\ N \\ \hline 43 \end{array}$

29. $\begin{array}{r} 2 \\ 3 \\ 5 \\ 6 \\ 4 \\ 8 \\ 7 \\ N \\ +\ 2 \\ \hline 42 \end{array}$

30. $\begin{array}{r} 1 \\ 3 \\ 5 \\ 2 \\ 7 \\ 6 \\ 2 \\ 2 \\ +\ N \\ \hline 33 \end{array}$

LESSON 65 Using Pictures to Compare Fractions

Speed Test: 90 Division Facts, Form A (Test I in Test Booklet)

We have practiced comparing whole numbers using comparison symbols. In this lesson we will begin comparing fractions. To begin, we will compare one half and one third.

$$\frac{1}{2} \bigcirc \frac{1}{3}$$

One way to compare fractions is to draw pictures of the fractions and compare the pictures. We will draw two circles of the same size. We shade $\frac{1}{2}$ of one circle and $\frac{1}{3}$ of the other circle.

$\frac{1}{2}$ $\frac{1}{3}$

As we compare the pictures, we see that more of the circle is shaded when $\frac{1}{2}$ is shaded than when $\frac{1}{3}$ is shaded. So $\frac{1}{2}$ is greater than $\frac{1}{3}$.

$$\frac{1}{2} > \frac{1}{3}$$

Example Compare: $\frac{1}{4} \bigcirc \frac{1}{3}$
Draw two rectangles to show the comparison.

Solution We draw two rectangles of the same size. We shade $\frac{1}{4}$ of one rectangle and $\frac{1}{3}$ of the other rectangle. We see that $\frac{1}{4}$ is slightly less than $\frac{1}{3}$.

$$\frac{1}{4} < \frac{1}{3}$$

Practice Compare these fractions. Draw a picture for each comparison.

a. $\frac{1}{2} \bigcirc \frac{2}{3}$ **b.** $\frac{1}{2} \bigcirc \frac{1}{4}$ **c.** $\frac{2}{5} \bigcirc \frac{3}{4}$

Problem set 65

1. James had fifty-six pies. Seven pies would go on one tray. How many trays did he need?

2. Five hundred forty-seven birds were on the lake. At sunrise some flew away. Then only two hundred fourteen birds were left. How many flew away?

3. Three hundred forty-two rabbits munched grass in the meadow. That afternoon some more rabbits came. Now seven hundred fifty-two rabbits munched grass in the meadow. How many rabbits came in the afternoon?

4. Write the number 614,823 in expanded form. Then use words to write the number.

5. List the months of the year that have 31 days. How many days does February have?

6. Find the eighth multiple of 6. Then subtract 15. What is the answer?

7. Compare these fractions. Draw two rectangles to show the fractions.

$$\frac{1}{4} \bigcirc \frac{1}{6}$$

8. Round 497 to the nearest hundred. Round 325 to the nearest hundred. Then find the sum of the rounded numbers.

9. What is the perimeter of this rectangle?

10. $\begin{array}{r} 36{,}427 \\ +\ 9{,}615 \\ \hline \end{array}$

11. $\begin{array}{r} 43{,}675 \\ +\ 52{,}059 \\ \hline \end{array}$

12. $\begin{array}{r} 948 \\ 55 \\ 427 \\ 86 \\ 94 \\ +\ 5 \\ \hline \end{array}$

13. $\begin{array}{r} \$10.00 \\ -\ \$\ 5.46 \\ \hline \end{array}$

14. $\begin{array}{r} 36{,}024 \\ -\ 15{,}539 \\ \hline \end{array}$

15. $\begin{array}{r} 56 \\ \times\ 8 \\ \hline \end{array}$

16. $\begin{array}{r} 73 \\ \times\ 9 \\ \hline \end{array}$

17. $\begin{array}{r} 46 \\ \times\ 7 \\ \hline \end{array}$

18. $\begin{array}{r} 84 \\ \times\ 6 \\ \hline \end{array}$

19. $\begin{array}{r} 40 \\ \times\ 5 \\ \hline \end{array}$

20. $56 \div 5$

21. $7\overline{)48}$

22. $0 \div 7$

23. $7\overline{)25}$

24. $9\overline{)60}$

25. $N \times 6 = 54$

26. $4 \times W = 36$

27.
$$\begin{array}{r} XYZ \\ +\ 942 \\ \hline 1065 \end{array}$$

28.
$$\begin{array}{r} N \\ -\ 942 \\ \hline 517 \end{array}$$

29.
$$\begin{array}{r} 4 \\ 7 \\ 8 \\ 21 \\ +\ N \\ \hline 49 \end{array}$$

30.
$$\begin{array}{r} 5 \\ 8 \\ 13 \\ 14 \\ +\ N \\ \hline 47 \end{array}$$

LESSON 66 Rate Word Problems

Speed Test: 64 Multiplication Facts (Test G in Test Booklet)

A **rate** tells **how far** or **how many in one time group**.

The car went 30 miles in 1 hour.

This tells us that the car's rate is 30 miles per hour. The car would go 60 miles in 2 hours. The car would go 90 miles in 3 hours. The car would go 120 miles in 4 hours.

rate	×	time	=	how far
30	×	1	=	30
30	×	2	=	60
30	×	3	=	90
30	×	4	=	120

Jethro could read 7 books in 1 month.

This tells us that Jethro's rate is 7 books per month. Jethro could read 14 books in 2 months. Jethro could read 21 books in 3 months. Jethro could read 28 books in 4 months.

rate	×	time	=	how many
7	×	1	=	7
7	×	2	=	14
7	×	3	=	21
7	×	4	=	28

Rate problems have the same pattern that equal groups problems have. The rate tells how many in one time group.

Example 1 Vera could read 2 pages in 1 minute. How many pages can she read in 9 minutes?

Solution This is a rate problem. A rate problem is an equal groups problem. Let's draw the pattern.

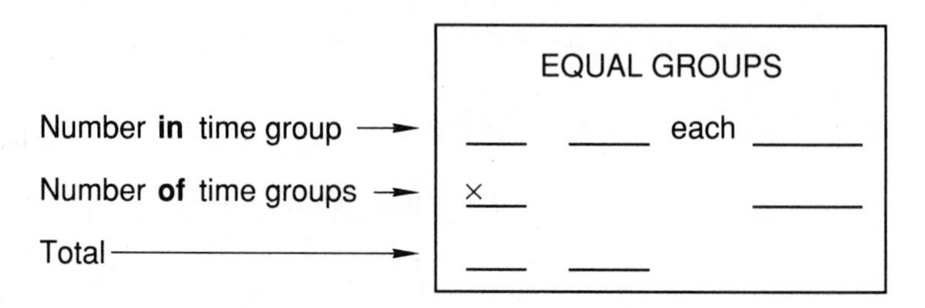

We do not see the words "in each" in a rate problem, but there are words that mean "in each." The words "in 1 minute" mean "in each minute." We write "2 pages each minute" on the first line and "9 minutes" on the second line.

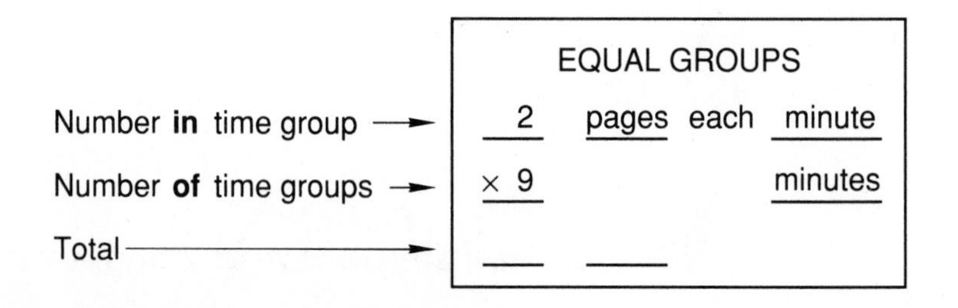

Now we find the missing number. To find the bottom number of an equal groups pattern, we multiply.

$$\begin{array}{r} 2 \\ \times\ 9 \\ \hline 18 \end{array}$$

We complete the pattern box and answer the question.

	EQUAL GROUPS			
Number **in** time group →	2	pages	each	minute
Number **of** time groups →	× 9			minutes
Total →	18	pages		

Vera could read **18 pages** in two minutes.

Example 2 Nalcomb earns 3 dollars a week for doing chores. How much money does he earn in 14 weeks?

Solution This is a rate problem. A rate problem is an equal groups problem. The words "3 dollars a week" mean "3 dollars **each** week."

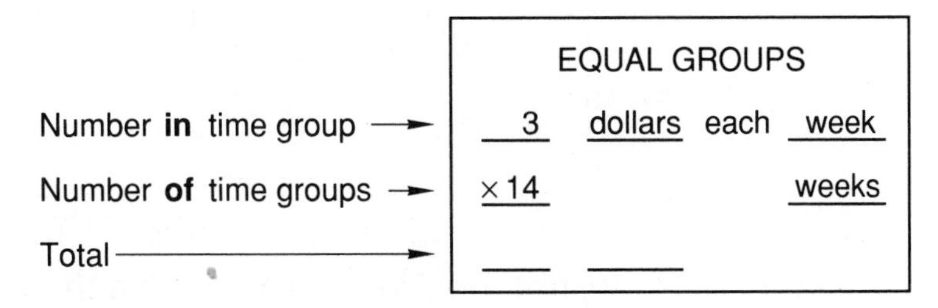

We need to find the missing number. We find the bottom number of an equal groups pattern by multiplying. We rearrange the factors so that the two-digit number is on top.

$$\begin{array}{r} 14 \\ \times\ \ 3 \\ \hline 42 \end{array}$$

We complete the pattern box and answer the question.

	EQUAL GROUPS
Number **in** time group →	3 dollars each week
Number **of** time groups →	× 14 weeks
Total →	42 dollars

Nalcomb earns **42 dollars** in 14 weeks.

Practice

a. Tammy drove 55 miles in one hour. At that rate, how far can she drive in 6 hours?

b. Jeff swims 20 laps every day. How many laps will he swim in 1 week?

Problem set 66

1. Marybeth could jump 42 times in 1 minute. How many times could she jump in 8 minutes?

2. Robo could run 7 miles in 1 hour. At that rate, how far could Robo run in 3 hours?

3. Quite a few were sitting in the stadium at sunrise. At 8 o'clock fourteen thousand, forty-two more came. Now there were seventeen thousand sitting in the stadium. How many were sitting in the stadium at sunrise?

4. Write the number 472,846 in expanded form. Then use words to write this number.

5. Round 7427 to the nearest hundred. Then round 6528 to the nearest hundred. Then add the rounded numbers.

6. Round 582 to the nearest hundred. Then subtract 18. What is the answer?

7. It is afternoon. What time was it 6 hours and 5 minutes ago?

8. Find the fourth multiple of 6. Then find the sixth multiple of 7. What is the sum of these two numbers?

9. How many years were there from 1492 until 1800?

10. A square has one side that is 7 inches long. What is the perimeter of the square?

11. $\begin{array}{r} 86{,}134 \\ +\ 76{,}487 \\ \hline \end{array}$

12. $\begin{array}{r} 498{,}258 \\ +\ 276{,}584 \\ \hline \end{array}$

13. $\begin{array}{r} 963 \\ 718 \\ 25 \\ 4 \\ 423 \\ +\ 75 \\ \hline \end{array}$

14. $\begin{array}{r} 70{,}003 \\ -\ 36{,}418 \\ \hline \end{array}$

15. $\begin{array}{r} N \\ -\ 432 \\ \hline 257 \end{array}$

16. $\begin{array}{r} 93 \\ \times\ 5 \\ \hline \end{array}$

17. $\begin{array}{r} 84 \\ \times\ 6 \\ \hline \end{array}$

18. $\begin{array}{r} 77 \\ \times\ 7 \\ \hline \end{array}$

19. $\begin{array}{r} 80 \\ \times\ 8 \\ \hline \end{array}$

20. $\begin{array}{r} 94 \\ \times\ 9 \\ \hline \end{array}$

21. $57 \div 8$

22. $7\overline{)65}$

23. $45 \div 6$

24. $7\overline{)51}$

25. $7 \times N = 42$

26. $R \times 8 = 48$

27. Compare these fractions. Draw a picture to show each fraction.

$$\frac{2}{3} \bigcirc \frac{3}{4}$$

28. $\begin{array}{r} 423 \\ 7 \\ 8 \\ 15 \\ +\ N \\ \hline 458 \end{array}$

29. $\begin{array}{r} 2 \\ 24 \\ 36 \\ N \\ +\ 14 \\ \hline 82 \end{array}$

30. $\begin{array}{r} 7 \\ 5 \\ 4 \\ 3 \\ +\ N \\ \hline 30 \end{array}$

LESSON 67 Multiplying Three-Digit Numbers

Speed Test: 90 Division Facts, Form A (Test I in Test Booklet)

When we multiply three-digit numbers, we multiply the ones' digit first. Then we multiply the tens' digit. Then we multiply the hundreds' digit.

Ones' digit	Tens' digit	Hundreds' digit
$\begin{array}{r} \downarrow \\ 123 \\ \times\quad 3 \\ \hline 9 \end{array}$	$\begin{array}{r} \downarrow\ \\ 123 \\ \times\quad 3 \\ \hline 69 \end{array}$	$\begin{array}{r} \downarrow\quad \\ 123 \\ \times\quad 3 \\ \hline 369 \end{array}$

In the next problem we get 18 when we multiply the ones' digit. We write the 8 in the ones' column and carry the 1 to the tens' column. Then we multiply the tens' digit.

Ones' digit	Tens' digit	Hundreds' digit
$\begin{array}{r} 1\downarrow \\ 456 \\ \times\quad 3 \\ \hline 8 \end{array}$	$\begin{array}{r} 1\downarrow\ \\ 456 \\ \times\quad 3 \\ \hline 68 \end{array}$	$\begin{array}{r} 1\downarrow 1 \\ 456 \\ \times\quad 3 \\ \hline 1368 \end{array}$

Three times 5 is 15 plus 1 is 16. We write the 6 and carry the 1 to the hundreds' column. Three times 4 is 12, plus 1 equals 13. The product is 1368.

Example Multiply: 654×7

Solution We write the 7 just below the 4. Then we multiply from left to right. We carry 2 after the first multiplication. We carry 3 after the second multiplication.

$$\begin{array}{r} {}^{32}\ \\ 654 \\ \times\quad 7 \\ \hline 4578 \end{array}$$

Practice

a. $\begin{array}{r} 234 \\ \times \quad 3 \\ \hline \end{array}$

b. $\begin{array}{r} 340 \\ \times \quad 4 \\ \hline \end{array}$

c. $\begin{array}{r} 406 \\ \times \quad 5 \\ \hline \end{array}$

d. 314×6

e. 261×8

Problem set 67

1. Elizabeth pays 9 dollars every week for karate lessons. How much does she pay for 4 weeks of karate lessons?

2. It takes 4 apples to make 1 pie. How many apples does it take to make 5 pies?

3. Elmer has to get up at 6 a.m. What time should he go to sleep in order to get 8 hours of sleep?

4. Nine hundred seventy-three were hiding in the forest. Hoover found three hundred eighty-eight of them. How many were still hiding in the forest?

5. Write the number 798,402 in expanded form. Then use words to write the number.

6. Find the fourth multiple of 7. Then find the third multiple of 6. Add these multiples. What is the answer?

7. According to this calendar, what is the date of the second Tuesday in September 2042?

SEPTEMBER						2042
S	M	T	W	T	F	S
	1	2	3	4	5	6
7	8	9	10	11	12	13
14	15	16	17	18	19	20
21	22	23	24	25	26	27
28	29	30				

8. If $5 + N = 23$, what number is N?

9. What is the perimeter of this figure? Dimensions are in feet.

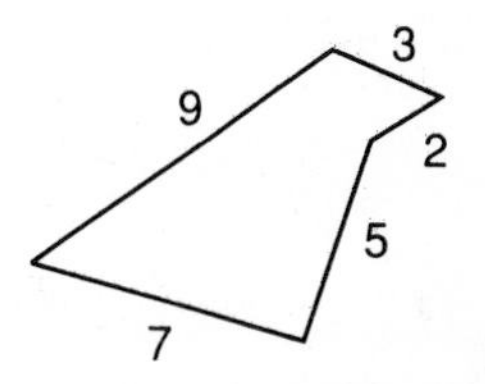

10. Compare these fractions. Shade two circles to show the fractions.

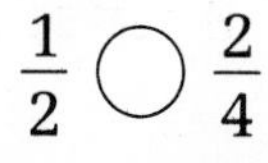

11. The arrow points to what number on this number line?

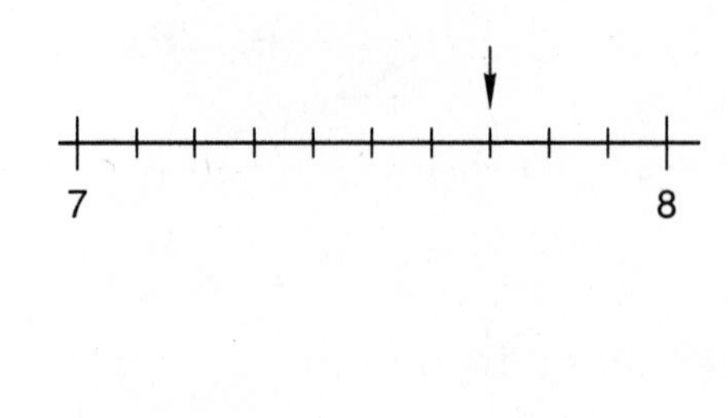

12. Joan started her walk in the morning. She finished her walk in the afternoon. How long did she walk?

13. \$3 + \$4.39 + \$12.62

14. $47 + 362 + 85 + 454$

15. \$20.00 − \$7.29

16. $41{,}059 - 36{,}275$

17. $\begin{array}{r} 768 \\ \times\ 3 \\ \hline \end{array}$

18. $\begin{array}{r} 280 \\ \times\ 4 \\ \hline \end{array}$

19. $\begin{array}{r} 436 \\ -\ XYZ \\ \hline 252 \end{array}$

20. $5\overline{)36}$

21. $7\overline{)45}$

22. $4\overline{)35}$

23. $17 \div 3$

24. $49 \div 6$

25. $57 \div 8$

26. $\begin{array}{r} 6 \\ 5 \\ N \\ 4 \\ 7 \\ +\ 3 \\ \hline 31 \end{array}$

27. $\begin{array}{r} N \\ 5 \\ 7 \\ 3 \\ 2 \\ +\ 12 \\ \hline 38 \end{array}$

28. $\begin{array}{r} 4 \\ 7 \\ 8 \\ N \\ 3 \\ +\ 16 \\ \hline 44 \end{array}$

29. $\begin{array}{r} 81 \\ 15 \\ 22 \\ 36 \\ 84 \\ +\ 22 \\ \hline \end{array}$

30. $\begin{array}{r} 7 \\ 15 \\ 423 \\ 16 \\ 8 \\ +\ 22 \\ \hline \end{array}$

LESSON 68 Multiples of 10 and 100 · Letters in Equations

Speed Test: 90 Division Facts, Form A (Test I in Test Booklet)

Multiples of 10 and 100

If we multiply 10 by 1, 2, 3, 4, 5, and so on, the answers we get are the multiples of 10.

Multiples of 10   10, 20, 30, 40, 50, . . .

The multiples of 10 are the numbers we use to count by tens. **Every multiple of 10 has a zero as its last digit**. If we multiply 100 by 1, 2, 3, 4, 5, and so on, the answers we get are the multiples of 100.

Multiples of 100   100, 200, 300, 400, 500, . . .

The multiples of 100 are the numbers we use to count by hundreds. **The last two digits of every multiple of 100 are zeros.**

Example 1 What are the fourth, sixth, and ninth multiples of 10?

Solution These multiples are the answers we get when we multiply 10 by 4, 6, and 9.

Fourth multiple of 10 is **40**
Sixth multiple of 10 is **60**
Ninth multiple of 10 is **90**

Example 2 What are the third, fifth, and seventh multiples of 100?

Solution These numbers are the answers we get when we multiply 100 by 3, 5, and 7.

Third multiple of 100 is **300**
Fifth multiple of 100 is **500**
Seventh multiple of 100 is **700**

Example 3 Find the sum of the fourth multiple of 10, the third multiple of 100, and the second multiple of 8.

Solution

$$\begin{array}{r} 4 \times 10 = 40 \\ 3 \times 100 = 300 \\ 2 \times 8 = 16 \\ \hline \mathbf{356} \end{array} \text{ sum}$$

Letters in equations We have used letters to represent individual digits. **We can also use letters to represent numbers that have any number of digits.**

Example 4 Find N: $3 \times 5 = 2 + N$

Solution It takes two steps to find N. First we multiply 3 and 5 to get 15. Now we have

$$15 = 2 + N$$

Since $2 + 13$ equals 15, we know that N equals **13**.

Practice

a. What is the sum of the fourth multiple of 100, the fifth multiple of 10, and the third multiple of 5?

b. Find N: $2 + 17 = N + 5$

c. Find W: $W + 3 = 5 \times 3$

Problem set 68

1. There were forty-two apples in each big basket. There were seven big baskets. How many apples were in the seven big baskets?

2. There were forty-eight pears in all. Six pears were in each box. How many boxes were there?

3. One thousand, nine hundred forty-two ducks were in the first flight. Two thousand, seven hundred five ducks were in the second flight. How many ducks were in both flights?

4. Two thousand, three hundred forty-seven geese landed on the lake the first month. Four thousand, three hundred two geese landed on the lake in the first two months. How many geese landed on the lake in the second month?

5. Two hundred forty-two thousand, nine hundred forty-two is a big number. Three hundred fifty-five thousand, eleven is also a big number. How much greater is the second number?

6. Find the sum of the fourth multiple of 10, the third multiple of 7, and the second multiple of 100.

7. How many years were there from 1492 to 1701?

8. It is dark outside. What time was it 3 hours and 20 minutes ago?

9. Compare these fractions. Shade circles to show the fractions.

$$\frac{2}{5} \bigcirc \frac{1}{4}$$

10. Janine could pack 40 packages in 1 hour. How many packages could she pack in 5 hours?

11. If $25 + N = 5 \times 7$, what number is N?

12. One side of a rectangle is 2 miles long. Another side is 3 miles long. Draw a picture of the rectangle and show the length of each side. What is the perimeter of the rectangle?

13. $\begin{array}{r} \$37.75 \\ + \$45.95 \\ \hline \end{array}$

14. $\begin{array}{r} 43{,}793 \\ + 76{,}860 \\ \hline \end{array}$

15. $\begin{array}{r} 480 \\ 97 \\ 126 \\ 53 \\ + 2362 \\ \hline \end{array}$

16. $\begin{array}{r} \$50.00 \\ - \$42.87 \\ \hline \end{array}$

17. $\begin{array}{r} 43{,}793 \\ - 26{,}860 \\ \hline \end{array}$

18. 483×4 **19.** 360×4 **20.** 207×8

21. $8\overline{)43}$ **22.** $5\overline{)43}$ **23.** $7\overline{)43}$

24. $29 \div 4$ **25.** $32 \div 6$ **26.** $55 \div 9$

27. $N \times 9 = 72$

28.
$$\begin{array}{r} 5 \\ 7 \\ N \\ 4 \\ 7 \\ 8 \\ 7 \\ 6 \\ +\ 3 \\ \hline 52 \end{array}$$

29. $7 \times B = 28$

30.
$$\begin{array}{r} 4 \\ 14 \\ 144 \\ +\ 444 \\ \hline \end{array}$$

LESSON 69 Estimating Arithmetic Answers • More about Rate

Speed Test: 90 Division Facts, Form B (Test J in Test Booklet)

Estimating We can **estimate** arithmetic answers by using round numbers instead of exact answers to do the arithmetic. Estimating does not give us the exact answer, but it gives us an answer that is close to the exact answer. Estimating is a way to find if our exact answer is **reasonable**.

Example 1 Estimate the sum of 396 and 512.

Solution To estimate, we first change the exact numbers to round numbers. We round 396 to 400. We round 512 to 500. Then we do the arithmetic with the round numbers. We find the sum by adding, so we add 400 and 500.

$$\begin{array}{r} 400 \\ +\ 500 \\ \hline 900 \end{array}$$

The estimated sum of 396 and 512 is **900**. The exact sum of 396 and 512 is 908. The estimated answer is not exact, but it is close to the exact number.

Example 2 Estimate the product of 72 and 5.

Solution We round the two-digit number but not the one-digit number. The estimated product of 72 and 5 is **350**.

$$\begin{array}{r} 70 \\ \times\ 5 \\ \hline 350 \end{array}$$

More about rate

In Lesson 66 we learned that rate problems have the same pattern as equal groups problems. The rate tells how many are in each time group. If we know two numbers in the problem, we can find the third number. We have practiced problems in which we were given the rate and time. We multiplied to find the total. In this lesson we will practice problems in which we are given the total. We will divide to find the number we are not given.

Example 3 Stanley could read 2 pages in 1 minute. How long will it take him to read 18 pages?

Solution This is a rate problem. A rate problem is an equal groups problem. Let's draw the pattern.

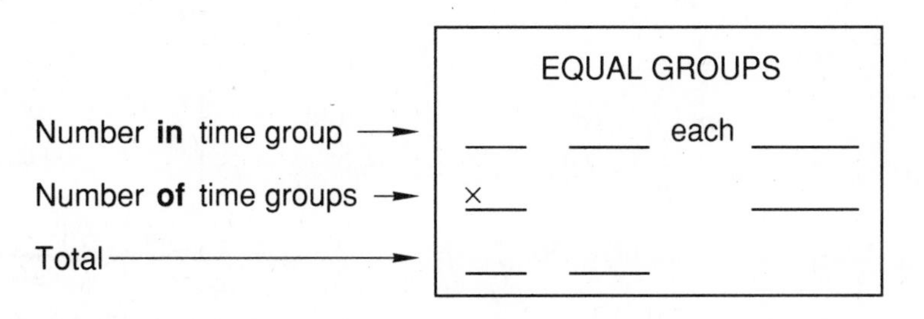

We are told that the rate is 2 pages in 1 minute. This means 2 pages each minute. We are told that the total number of pages is 18.

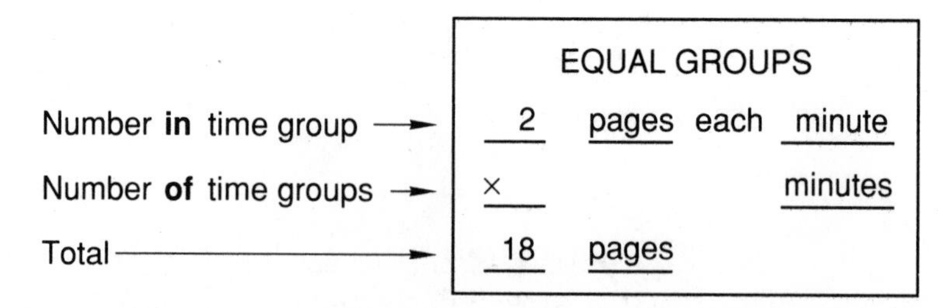

Now we find the missing number. **To find the first or second number in an equal groups pattern, we divide.**

$$2\overline{)18}\quad 9$$

We complete the pattern box and answer the question.

	EQUAL GROUPS		
Number **in** time group →	2	pages each	minute
Number **of** time groups →	× 9		minutes
Total →	18	pages	

It would take Stanley **9 minutes** to read 18 pages.

Example 4 Janice could wrap 21 packages in 3 hours. How many packages could she wrap in 1 hour?

Solution This is a rate problem. We read it carefully. The words "in each" mean "in one." We are not told how many packages she could wrap in 1 hour. This is the number we need to find. The two numbers we are told are the total number of packages and the number of hours.

	EQUAL GROUPS		
Number **in** time group →	___	packages each	hour
Number **of** time groups →	× 3		hours
Total →	21	packages	

We divide to find the missing number.

$$3\overline{)21}\quad 7$$

We complete the pattern and answer the question.

	EQUAL GROUPS	
Number **in** time group →	7 packages each hour	
Number **of** time groups →	× 3 hours	
Total →	21 packages	

Janice could wrap **7 packages in 1 hour.**

Practice Estimate the answer to each arithmetic problem. Then find the exact answer.

a. 59 + 68 + 81

b. 607 + 891

c. 585 − 294

d. 82 − 39

e. 59 × 6

f. 397 × 4

g. Miguel could sharpen 40 pencils in 8 minutes. What was his rate of sharpening pencils?

Problem set 69

1. Forty-two thousand, three hundred five birds sang in the forest. Then thirteen thousand, seven hundred sixty-five birds flew away. How many birds were left in the forest?

2. There were two hundred fourteen parrots, seven hundred fifty-two crows, and two thousand, forty-two blue jays. How many birds were there in all?

3. Maria used 3 pounds of beans to make 36 burritos. How many burritos could she make from just 1 pound of beans?

4. Harry could make 36 tacos in 3 hours. How many tacos could he make in 1 hour?

5. Round to the nearest hundred to estimate the sum of 286 and 415.

6. What is the sum of the fourth multiple of 100, the seventh multiple of 10, and the sixth multiple of 14?

7. Eight hundred thousand, nine hundred seventy-seven is how much greater than five hundred thousand, three hundred forty-two?

8. Hans left on his trip in the evening. He arrived the next morning. How long did he travel?

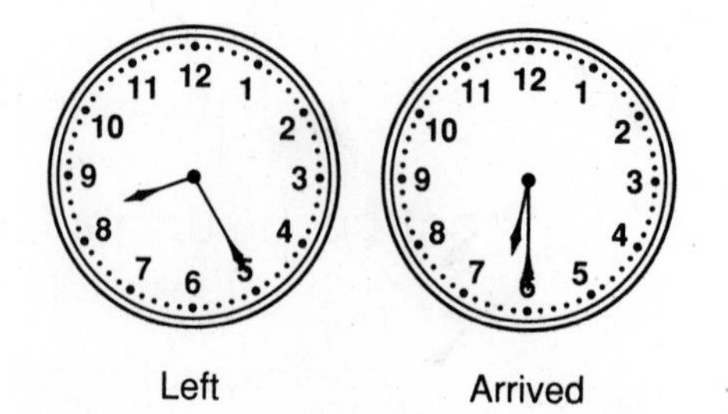

9. $140 + N = 28 \times 5$

10. The product of 6 and 7 is how much greater than the sum of 6 and 7?

11. What is the length of segment BC?

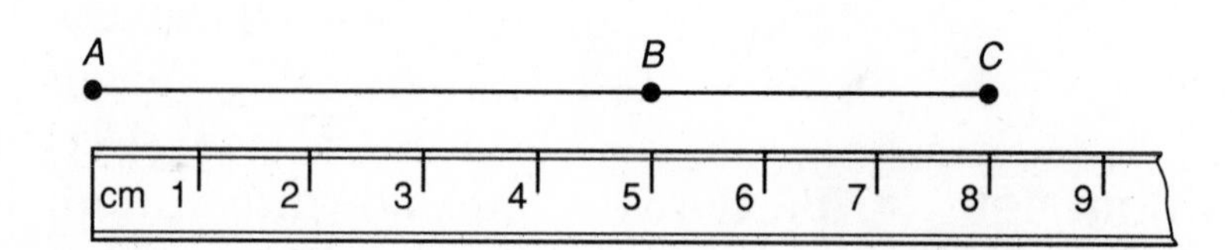

12. Compare: $(32 \div 8) \div 2 \bigcirc 32 \div (8 \div 2)$

13. $6.49 + $12 + $7.59 + 8¢

14. $3742 + 5329 + 8415$

15. $30.00 − $29.17

16. $54,682 - 37,758$

17. $\begin{array}{r} 350 \\ \times \quad 5 \\ \hline \end{array}$

18. $\begin{array}{r} 204 \\ \times \quad 7 \\ \hline \end{array}$

19. $\begin{array}{r} 463 \\ \times \quad 6 \\ \hline \end{array}$

20. $4\overline{)37}$

21. $6\overline{)39}$

22. $3\overline{)28}$

23. $7 \div 2$

24. $48 \div 5$

25. $48 \div 7$

26.
$$\begin{array}{r} 5 \\ 1 \\ 2 \\ 3 \\ 8 \\ 4 \\ +\ 7 \\ \hline \end{array}$$

27.
$$\begin{array}{r} 8 \\ 2 \\ 22 \\ 222 \\ 22 \\ 2 \\ +\quad 2 \\ \hline \end{array}$$

28.
$$\begin{array}{r} 3 \\ 303 \\ 333 \\ 5 \\ 7 \\ 28 \\ +\quad 35 \\ \hline \end{array}$$

29.
$$\begin{array}{r} 4 \\ 2 \\ N \\ 15 \\ 12 \\ 3 \\ +\ 5 \\ \hline 44 \end{array}$$

30.
$$\begin{array}{r} 5 \\ 2 \\ 42 \\ 3 \\ N \\ 16 \\ +\ 4 \\ \hline 75 \end{array}$$

LESSON 70 Remaining Fraction

Speed Test: 90 Division Facts, Form B (Test J in Test Booklet)

If a whole has been divided into parts and we know the size of one part, then we can figure out the size of the other part.

Example 1 (a) What fraction of the circle is shaded?
(b) What fraction of the circle is **not** shaded?

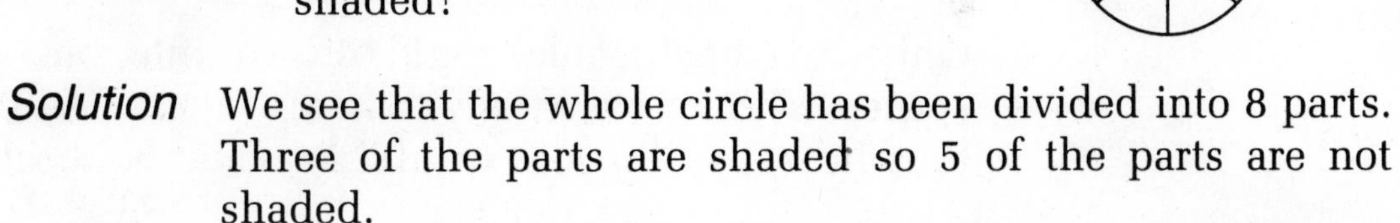

Solution We see that the whole circle has been divided into 8 parts. Three of the parts are shaded so 5 of the parts are not shaded.

(a) The fraction that is shaded is $\frac{3}{8}$.

(b) The fraction that is **not** shaded is $\frac{5}{8}$.

Example 2 The pizza was cut into 8 equal slices. After Bill, Tony, and Jenny each took a slice, what fraction of the pizza was left?

Solution The whole pizza was cut into 8 equal parts. Since 3 of the 8 parts were taken, 5 of the 8 parts remain. The fraction that is left is $\frac{5}{8}$.

Example 3 Two fifths of the crowd cheered. What fraction of the crowd did not cheer?

Solution We think of the crowd as though it were divided into 5 parts. We are told that 2 of the 5 parts cheered. So there were 3 parts that did not cheer. The fraction that did not cheer was $\frac{3}{5}$.

Practice

a. What fraction of this rectangle is not shaded?

b. Three fifths of the race was over. What fraction of the race was left?

Problem set 70

1. The first number was one thousand, three hundred forty-two. The second number was five hundred fourteen. The second number was how much less than the first number?

2. Five little people could crowd into one of the spaces. If there were thirty-five spaces, how many little people could crowd in?

3. Only two big people could crowd into one of the spaces. If there were sixty-four big people standing in line, how many spaces would it take to hold them?

4. The Gilbreth family drank 39 quarts of milk in 3 days. That amounts to how many quarts of milk each day?

5. Carl weighed 88 pounds. He put on his clothes, which weighed 2 pounds, and his shoes, which weighed 1 pound each. Finally he put on a jacket that weighed 3 pounds and stepped on the scale again. Then how much did the scale show that he weighed?

6. What fraction of this rectangle is not shaded?

7. The pumpkin pie was sliced into 6 equal parts. After 1 slice was taken, what fraction of the pie was left?

8. Compare these fractions. Shade parts of circles to show each fraction.

$$\frac{2}{3} \bigcirc \frac{3}{4}$$

9. Use rounded numbers to estimate the sum of 507 and 384.

10. List the months of the year that have exactly 30 days.

11. What is the perimeter of this rectangle?

4 cm

8 cm

12. $\begin{array}{r} \$62.59 \\ +\ \$17.47 \\ \hline \end{array}$

13. $\begin{array}{r} XYZ \\ -\ 417 \\ \hline 268 \end{array}$

14. $\begin{array}{r} 976 \\ 434 \\ 129 \\ 85 \\ +\ 1140 \\ \hline \end{array}$

15. $1000 - (11 \times 9)$

16. $\begin{array}{r} 413{,}216 \\ -\ 127{,}159 \\ \hline \end{array}$

17. $\begin{array}{r} 670 \\ \times \quad 4 \\ \hline \end{array}$

18. $\begin{array}{r} 703 \\ \times \quad 6 \\ \hline \end{array}$

19. $\begin{array}{r} 346 \\ \times \quad 9 \\ \hline \end{array}$

20. $5\overline{)39}$

21. $7\overline{)39}$

22. $4\overline{)39}$

23. $16 \div 3$

24. $26 \div 6$

25. $36 \div 8$

26. $8 \times R = 48$

27. $M \times 6 = 48$

28. $\begin{array}{r} 4 \\ 7 \\ 6 \\ 5 \\ 4 \\ + N \\ \hline 31 \end{array}$

29. $\begin{array}{r} 47 \\ 25 \\ 2 \\ 6 \\ + 14 \\ \hline \end{array}$

30. $\begin{array}{r} 19 \\ 18 \\ 17 \\ 16 \\ + 15 \\ \hline \end{array}$

LESSON 71 Multiplying Three Factors

Speed Test: 90 Division Facts, Form B (Test J in Test Booklet)

To multiply three numbers, we first multiply two of the numbers. Then we multiply the answer we get by the third number.

Example $3 \times 4 \times 5$

Solution First we multiply two numbers and get an answer. Then we multiply that answer by the third number. If we multiply 3×4 first, we get 12. Then we multiply 12×5 and get 60.

First	Then
$3 \times 4 = 12$	$12 \times 5 = \mathbf{60}$

It does not matter which numbers we multiply first. If we multiply 5×4 first, we get 20. Then we multiply 20×3 and get 60 again.

First	Then	
$5 \times 4 = 20$	$20 \times 3 = \mathbf{60}$	← same answer

Practice

a. $2 \times 3 \times 4$

b. $2 \times 4 \times 6$

c. $4 \times 5 \times 6$

d. $3 \times 3 \times 3$

Problem set 71

1. There were twice as many peacocks as peahens. If there were 12 peacocks, how many peahens were there?

2. Beth's dance class begins at 6 p.m. It takes 20 minutes to drive to dance class. What time should she leave home to get to dance class on time?

3. Samantha bought a package of paper for $1.98 and 2 pens for $0.49 each. The tax was 18¢. What was the total price?

4. Nalcomb earns $3 a week for washing the car. How much money does he earn in a year? (There are 52 weeks in a year.)

5. Two thirds of the race was over. What fraction of the race was left?

6. Estimate the difference: $887 - 291$

7. In the equation $9 \times 11 = 100 - y$, the letter y stands for what number?

8. Compare:

$$\frac{2}{4} \bigcirc \frac{4}{8}$$

Draw and shade two rectangles to show the comparison.

9. What is the sum of the eighth multiple of 100, the fourth multiple of 10, and the sixth multiple of 7?

10. According to this calendar, the fourth of July, 2014, is what day of the week?

JULY						2014
S	M	T	W	T	F	S
		1	2	3	4	5
6	7	8	9	10	11	12
13	14	15	16	17	18	19
20	21	22	23	24	25	26
27	28	29	30	31		

11. Four hundred forty-three thousand, two hundred eleven is how much less than six hundred fifty-seven thousand, thirteen?

12. $5 \times 6 \times 7$

13. $4 \times 4 \times 4$

14. $\begin{array}{r} 476{,}385 \\ +\ 259{,}518 \\ \hline \end{array}$

15. $\begin{array}{r} \$20.00 \\ -\ \$17.84 \\ \hline \end{array}$

16. $\begin{array}{r} XZ{,}YMC \\ -\ 19{,}434 \\ \hline 45{,}579 \end{array}$

17. $\begin{array}{r} 417 \\ \times\ \ 8 \\ \hline \end{array}$

18. $\begin{array}{r} 470 \\ \times\ \ 7 \\ \hline \end{array}$

19. $\begin{array}{r} 608 \\ \times\ \ 4 \\ \hline \end{array}$

20. $4\overline{)29}$

21. $8\overline{)65}$

22. $5\overline{)29}$

23. $65 \div 7$

24. $29 \div 5$

25. $65 \div 9$

26. $N \times 7 = 70 - 7$

27.

$$\begin{array}{r} 41{,}635 \\ 478 \\ +\quad N \\ \hline 60{,}318 \end{array}$$

28.

$$\begin{array}{r} 4 \\ 8 \\ 11 \\ 15 \\ 14 \\ 23 \\ +\ 19 \\ \hline \end{array}$$

29.

$$\begin{array}{r} 56 \\ 25 \\ 4 \\ 3 \\ 11 \\ 15 \\ 27 \\ +\ 48 \\ \hline \end{array}$$

30.

$$\begin{array}{r} ABC \\ -\ 793 \\ \hline 909 \end{array}$$

LESSON 72 Polygons

Speed Test: 90 Division Facts, Form B (Test J in Test Booklet)

Polygons are closed, flat shapes formed by straight lines.

Example 1 Which of these shapes is a polygon?

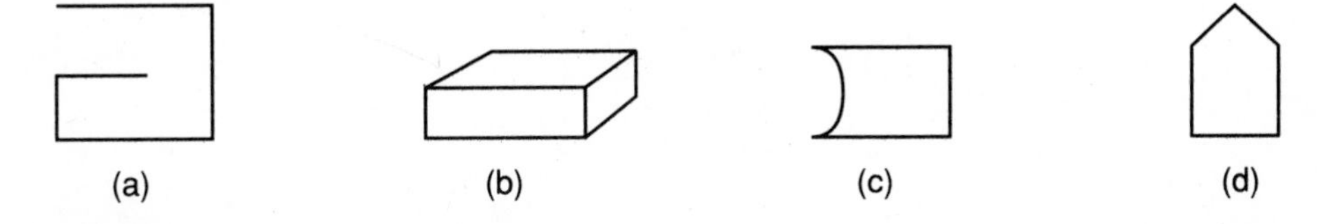

Solution Figure (**a**) is **not a polygon** because it is not closed. Figure (**b**) is **not a polygon** because it is not flat. Figure (**c**) is **not a polygon** because not all of its sides are straight. Figure (**d**) is a **polygon**.

The name of a polygon tells how many sides the polygon has. The sides do not have to have the same lengths.

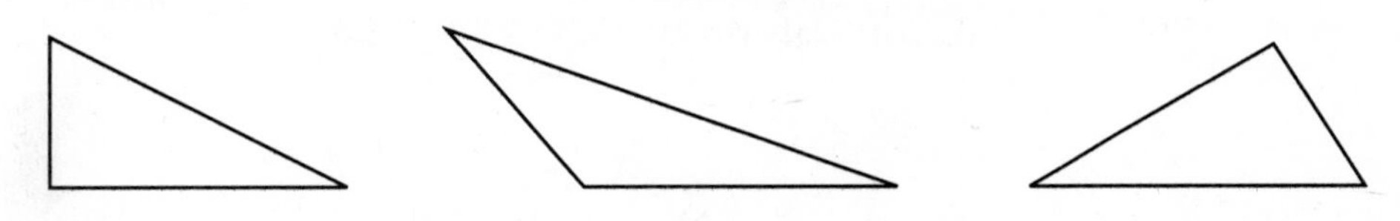

3-sided polygons are **triangles**

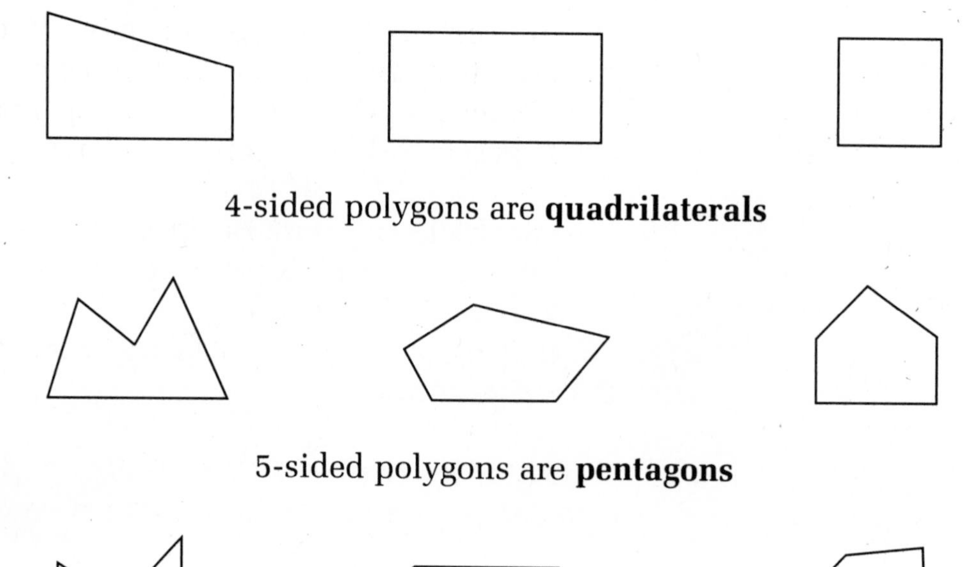

4-sided polygons are **quadrilaterals**

5-sided polygons are **pentagons**

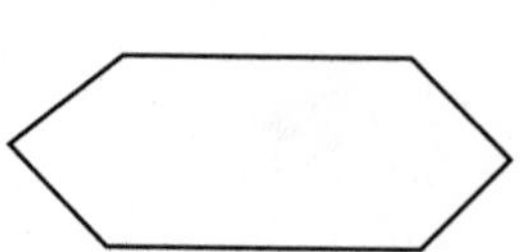

6-sided polygons are **hexagons**

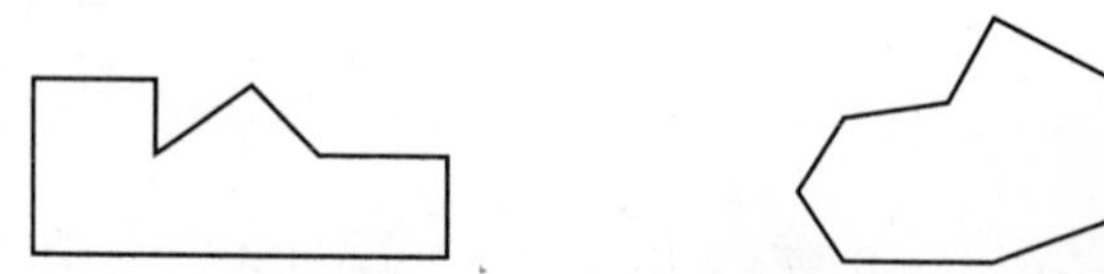

8-sided polygons are **octagons**

Example 2 What kind of a polygon is a square?

Solution A square has 4 sides, so a square is a **quadrilateral**.

Practice Draw an example of each kind of polygon.

a. triangle **b.** quadrilateral **c.** pentagon

d. hexagon **e.** octagon

Problem set 72

1. Three feet equals one yard. A car that is 15 feet long is how many yards long?

2. On the first try, James got forty-two thousand, seven hundred forty-two points. On the second try, James got five thousand, four hundred five points. How many fewer points did he get on the second try?

3. Roberta had six quarters, three dimes, and fourteen pennies. How much money did she have in all?

4. What is the sum of the even numbers that are greater than 10 but less than 20?

5. Estimate the sum of 715 and 594 by rounding these numbers to the nearest hundred before adding.

6. Seven bugs could fit into one small space. There were 56 bugs in all. How many small spaces were needed to hold them all?

7. The arrow points to what number?

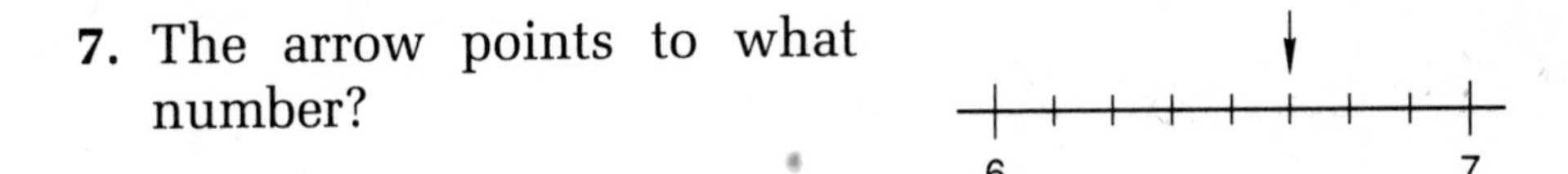

8. The cake was cut into 12 pieces. Seven of the pieces were eaten. What fraction of the cake was left?

9. The product of 4 and 3 is how much greater than the sum of 4 and 3?

10. What is the sum of the fourth multiple of 7, the sixth multiple of 100, and the sixth multiple of 10?

11. (a) What is the name of this polygon?
(b) Each side is the same length. What is the perimeter of this polygon?

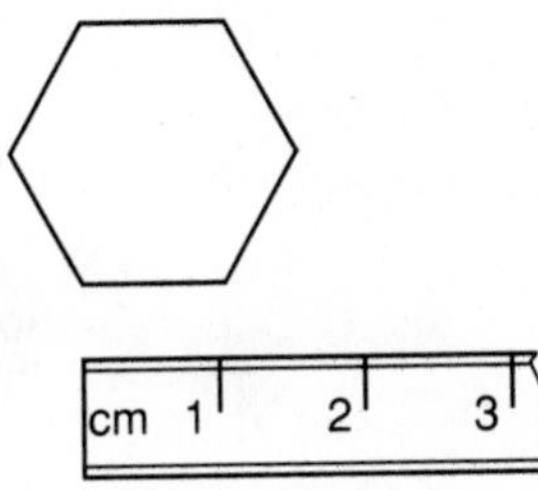

12. Roger could pick 56 flowers in 8 minutes. How many flowers could he pick in 1 minute?

13. Sarah could pick 11 flowers in 1 minute. How many flowers could she pick in 5 minutes?

14. $3 \times 5 \times 7$

15. $5 \times 5 \times 5$

16.
$$\begin{array}{r} 2415 \\ 3621 \\ 157 \\ 642 \\ +\ \ 93 \\ \hline \end{array}$$

17.
$$\begin{array}{r} \$40.00 \\ -\ \$AB.CD \\ \hline \$\ 2.43 \end{array}$$

18.
$$\begin{array}{r} 64{,}150 \\ -\ 48{,}398 \\ \hline \end{array}$$

19. 349×8

20. 760×7

21. $6\overline{)34}$

22. $8\overline{)62}$

23. $5\overline{)24}$

24. $39 \div 7$

25. $33 \div 4$

26. $47 \div 9$

27.
$$\begin{array}{r} 7 \\ 8 \\ 4 \\ 3 \\ 6 \\ 5 \\ 1 \\ 3 \\ +\ N \\ \hline 44 \end{array}$$

28. $5 \times N = 15 + 5$

29.
$$\begin{array}{r} N \\ +\ 419 \\ \hline 658 \end{array}$$

30.
$$\begin{array}{r} 410 \\ -\ \ N \\ \hline 163 \end{array}$$

LESSON 73 Division with Two-Digit Answers, Part 1

Speed Test: 90 Division Facts, Form A (Test I in Test Booklet)

In this lesson we will learn a method, for dividing a two-digit number by a one-digit number. To show the method, we will divide 78 by 3.

$$3\overline{)78}$$

For the first step we ignore the 8 and divide 7 by 3. We write the 2 above the 7. Then we multiply 2×3 and write 6 below the 7. We subtract and write 1.

$$\begin{array}{r} 2 \\ 3\overline{)78} \\ 6 \\ \hline 1 \end{array}$$

Then we "bring down" the 8 as we show here. Together the 1 and 8 form 18.

$$\begin{array}{r} 2 \\ 3\overline{)78} \\ 6 \\ \hline 18 \end{array}$$

Now we divide 18 by 3 and get 6. We write the 6 above the 8 in 78. Then we multiply 6 by 3 and write 18 below 18.

$$\begin{array}{r} 26 \\ 3\overline{)78} \\ 6 \\ \hline 18 \\ 18 \\ \hline 0 \end{array}$$

When we subtract, the remainder is zero.

Example Divide: $3\overline{)87}$

Solution For the first step we ignore the 7. We divide 8 by 3, multiply, and then subtract. Now we bring down the 7. This forms 27. Now we divide 27 by 3, multiply, and subtract again.

$$\begin{array}{r} 29 \\ 3\overline{)87} \\ 6 \\ \hline 27 \\ 27 \\ \hline 0 \end{array}$$

The quotient is **29**, and the remainder is zero.

Practice

a. $3\overline{)51}$ b. $4\overline{)52}$ c. $5\overline{)75}$

d. $3\overline{)72}$ e. $4\overline{)96}$ f. $2\overline{)74}$

Problem set 73

1. Franco liked to run. He ran the course two thousand, three hundred times the first year. He ran the course one thousand, nine hundred forty-two times the second year. How many times did he run the course in all?

2. William Tell's target was one hundred thirteen paces away. If each pace was 3 feet, how many feet away was the target?

3. Tracy's baseball card album will hold five hundred cards. Tracy has three hundred eighty-four cards. How many more cards will fit in the album?

4. The trip lasted 21 days. How many weeks is that?

5. A stop sign has the shape of an octagon. How many sides are on seven stop signs?

6. Find the length of this hairpin to the nearest quarter inch.

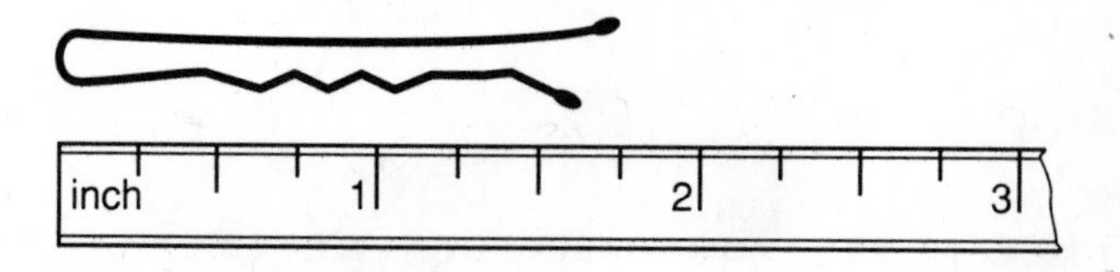

7. Write 406,912 in expanded form. Then use words to write this number.

8. One foot equals 12 inches. If each side of a square is one foot long, what is the perimeter of the square in inches?

9. Estimate the sum of 586 and 797 by rounding the numbers to the nearest hundred and then adding the rounded numbers.

10. Compare:

$$\frac{3}{6} \bigcirc \frac{1}{2}$$

Draw and shade two circles to show the comparison.

11. Four thousand is how much greater than two hundred eighty-two?

12. Some birds sat on the wire at sunup. Then 47 more birds came. Now there are 112 birds sitting on the wire. How many birds sat on the wire at sunup?

13. The girls could do 75 jobs in 25 minutes. How many jobs could they do in 1 minute?

14. $\begin{array}{r} \$32.47 \\ +\ \$67.54 \\ \hline \end{array}$

15. $\begin{array}{r} 33{,}675 \\ +\ 48{,}988 \\ \hline \end{array}$

16. $\begin{array}{r} 536 \\ 29 \\ 974 \\ 88 \\ +\ 4361 \\ \hline \end{array}$

17. $\begin{array}{r} 51{,}036 \\ -\ \ 7{,}648 \\ \hline \end{array}$

18. $\begin{array}{r} N \\ -\ 19{,}755 \\ \hline 14{,}323 \end{array}$

19. $\begin{array}{r} 257 \\ \times\ \ 5 \\ \hline \end{array}$

20. $\begin{array}{r} 709 \\ \times\ \ 3 \\ \hline \end{array}$

21. $\begin{array}{r} 334 \\ \times\ \ 9 \\ \hline \end{array}$

22. $4\overline{)92}$

23. $6\overline{)58}$

24. $2\overline{)36}$

25. $5\overline{)75}$

26. $3\overline{)84}$

27. $\begin{array}{r} 8 \\ 9 \\ 7 \\ 14 \\ 28 \\ +\ N \\ \hline 68 \end{array}$

28. $N \times 4 = 36$

29. $\begin{array}{r} N \\ -\ 312 \\ \hline 109 \end{array}$

30. $\begin{array}{r} 555 \\ +\ \ N \\ \hline 711 \end{array}$

LESSON 74 Division with Two-Digit Answers, Part 2

> **Speed Test:** 90 Division Facts, Form A (Test I in Test Booklet)

If we try to divide 234 by 3, we have a problem because 3 will not divide into 2.

$$3\overline{)234}$$

But 3 will divide into 23 at least 7 times. **We put the 7 above the 3.** Then we multiply. Then we subtract.

$$\begin{array}{r} 7 \\ 3\overline{)234} \\ \underline{21} \\ 2 \end{array}$$

Next we bring down the 4.

$$\begin{array}{r} 7 \\ 3\overline{)234} \\ \underline{21} \\ 24 \end{array}$$

Now we divide 24 by 3. We put the 8 above the 4. Then we multiply and finish by subtracting.

$$\begin{array}{r} 78 \\ 3\overline{)234} \\ \underline{21} \\ 24 \\ \underline{24} \\ 0 \end{array}$$

We see that 3 divides 234 seventy-eight times with no remainder. This means that $3 \times 78 = 234$.

Example Divide: $174 \div 3$

Solution We know that 3 will not divide into 1, but 3 will divide into 17 at least 5 times. We write the 5 above the 7 in 17. Then we multiply. Then we subtract.

$$\begin{array}{r} 5 \\ 3\overline{)174} \\ \underline{15} \\ 2 \end{array}$$

Then we bring down the 4, divide again, and subtract again.

$$\begin{array}{r} 58 \\ 3\overline{)174} \\ \underline{15} \\ 24 \\ \underline{24} \\ 0 \end{array}$$

The quotient is **58** and the remainder is zero, so we know that $3 \times 58 = 174$.

Practice

a. $3\overline{)144}$ **b.** $4\overline{)144}$ **c.** $6\overline{)144}$

d. $225 \div 5$ **e.** $455 \div 7$ **f.** $200 \div 8$

Problem set 74

1. The chef used 3 eggs for each omelet. How many omelets would be made from 24 eggs?

2. Seventy-two young knights met peril in the forest. Twenty-seven fought bravely, but the others fled. How many young knights fled?

3. Brian wore braces for 3 years. How many months is that?

4. Fango could walk 3 miles in 1 hour. How many miles could he walk in 10 hours?

5. Fanga could walk 28 miles in 7 hours. How many miles could she walk in 1 hour?

6. Andrea bought dance shoes for $37.95 and tap shoes for $6.85. How much did she pay for both pairs of shoes?

7. What fraction of this hexagon has not been shaded?

8. Each side of the hexagon in Problem 7 is 1 cm long. What is its perimeter?

9. Jim began walking in the morning. He stopped walking that evening. How long did he walk?

10. Use rounded numbers to estimate the difference between 903 and 395.

11. How long is segment BC?

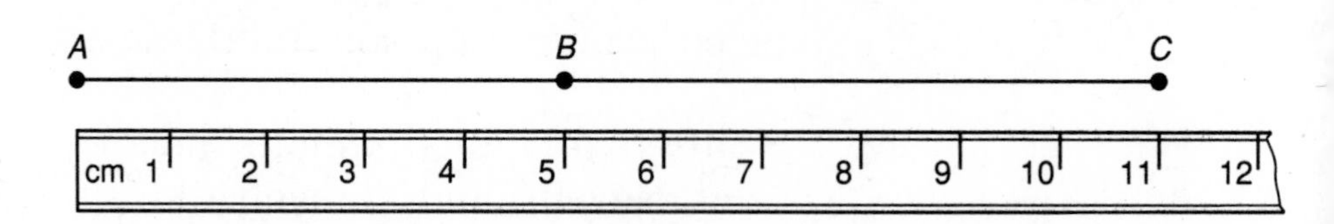

12. How much greater than two hundred ninety-seven thousand, one hundred fifteen is three hundred eighty-four thousand, two hundred?

13. Compare: $(27 \div 9) \div 3 \bigcirc 27 \div (9 \div 3)$

14. $$\begin{array}{r} \$97.56 \\ +\ \$\ 8.49 \\ \hline \end{array}$$

15. $$\begin{array}{r} XYZ,ABC \\ +\ 95{,}784 \\ \hline 970{,}100 \end{array}$$

16. $$\begin{array}{r} 3764 \\ 2945 \\ 301 \\ +\ 7538 \\ \hline \end{array}$$

17. $\begin{array}{r} \$60.00 \\ -\ \$54.78 \\ \hline \end{array}$

18. $\begin{array}{r} 41{,}103 \\ -\ 26{,}347 \\ \hline \end{array}$

19. 840×3

20. 4×564

21. 304×6

22. $4\overline{)136}$

23. $2\overline{)132}$

24. $6\overline{)192}$

25. $168 \div 3$

26. $378 \div 7$

27. $\begin{array}{r} 7 \\ 3 \\ 14 \\ 6 \\ 18 \\ +\ N \\ \hline 54 \end{array}$

28. $N \times 7 = 56$

29. $4 \times N = 56$

30. $\begin{array}{r} ABC \\ -\ 129 \\ \hline 446 \end{array}$

LESSON 75 Division with Two-Digit Answers and a Remainder

> **Speed Test:** 90 Division Facts, Form A (Test I in Test Booklet)

The method we use for dividing has four steps. These steps are divide, multiply, subtract, and bring down. Then we repeat the steps as necessary.

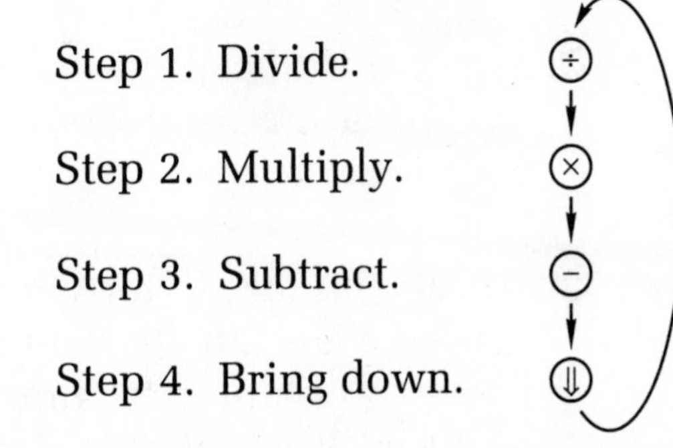

For each step we write a number. When we finish step 4, we go back to step 1 and repeat the steps until there are no digits left to bring down. Then whatever is left after the last subtraction is the remainder. We will continue to write the remainder after the division answer with a small r in front.

Example $5\overline{)137}$

Solution Step 1. We divide $5\overline{)13}$ and write 2.
Step 2. We multiply 2 × 5 and write 10.
Step 3. We subtract 13 − 10 and write 3.
Step 4. We bring down 7 to make 37.

$$\begin{array}{r} 27 \\ 5\overline{)137} \\ -\ 10 \\ \hline 37 \end{array}$$

Now we repeat the same four steps.
Step 1. We divide 37 ÷ 5 and write 7.
Step 2. We multiply 7 × 5 and write 35.
Step 3. We subtract 37 − 35 and write 2.
Step 4. There are no more digits to bring down, so we are finished. We write the 2 as a remainder. Our answer is **27 r 2**.

$$\begin{array}{r} 27 \\ 5\overline{)137} \\ -\ 10 \\ \hline 37 \\ -\ 35 \\ \hline 2 \end{array}$$

Practice **a.** $3\overline{)134}$ **b.** $7\overline{)240}$ **c.** $5\overline{)88}$

d. 259 ÷ 8 **e.** 95 ÷ 4 **f.** 325 ÷ 6

Problem set 75

1. With his bow and arrow, William Tell split 10 apples in half. How many apple halves were there?

2. Every third bead on the necklace was red. There were one hundred forty-one beads in all. How many beads were red?

3. Three hundred forty-two thousand, five hundred eleven was the first number. One hundred seventy-three thousand, two hundred nineteen was the second number. What was the sum of the first number and the second number?

4. Big Fox chased Little Rabbit 20 kilometers north, then 15 kilometers south. How far was Big Fox from where he started? (Draw a diagram.)

5. At 11:45 a.m. Jason glanced at the clock. His doctor's appointment was in $2\frac{1}{2}$ hours. At what time was his appointment?

6. Estimate the product of 89 and 5.

7. The car could go 30 miles on 1 gallon of gas. How far could the car go on 8 gallons of gas?

8. Two sevenths of the crowd cheered wildly. The rest of the crowd stood silently. What fraction of the crowd stood silently?

9. $N + 2 = 3 \times 12$

10. Forty-two glops were required to make a good gloo. Jean needed nine good gloos. How many glops did she need?

11. Compare:

$$\frac{1}{2} \bigcirc \frac{2}{5}$$

Draw and shade two rectangles to show the comparison.

12. 46,278 + 148,095

13. 48 + 163 + 9 + 83 + 3425

14. \$10 − 10¢ **15.** 43,016 − 5987 **16.** 5 × 6 × 8

17. $\begin{array}{r} 486 \\ \times \quad 7 \\ \hline \end{array}$ **18.** $\begin{array}{r} 307 \\ \times \quad 8 \\ \hline \end{array}$ **19.** $\begin{array}{r} 460 \\ \times \quad 9 \\ \hline \end{array}$

20. $2\overline{)153}$ 21. $6\overline{)265}$ 22. $4\overline{)59}$

23. $232 \div 5$ 24. $90 \div 7$ 25. $145 \div 3$

26. Write each amount of money by using a dollar sign and decimal point.
(a) 17¢ (b) 8¢ (c) 345¢

27. $3 \times N = 60$

28. $\begin{array}{r} 465 \\ 968 \\ 742 \\ + 864 \\ \hline \end{array}$ 29. $\begin{array}{r} 416 \\ 213 \\ + \ N \\ \hline 635 \end{array}$ 30. $\begin{array}{r} 467 \\ - XYZ \\ \hline 109 \end{array}$

LESSON 76 Multiplying Money

Speed Test: 90 Division Facts, Form A (Test I in Test Booklet)

To multiply dollars and cents by a whole number, we multiply the digits just like we multiply whole numbers. Then we put a decimal point in the answer so that there are two places after the decimal point. We also write a dollar sign in front.

Example 1 $\begin{array}{r} \$2.35 \\ \times \quad 3 \\ \hline \end{array}$

Solution First we multiply as though the dollar sign and decimal point were not there. We get 705. Then we write a decimal point so that there are two digits after the decimal point. We then write a dollar sign in front. The answer is **$7.05**.

$\begin{array}{r} 1\ 1\ \ \\ \$2.35 \\ \times \quad 3 \\ \hline 7\ 05 \end{array}$

Example 2 25¢ × 9

Solution This problem is in cent form. We can multiply in this form. We multiply 25 × 9 and write a cent sign at the end of the answer. Since 225¢ is more than a dollar, we change the answer to dollar form.

$$\begin{array}{r} 25¢ \\ \times\ \ 9 \\ \hline 225¢ \end{array} = \mathbf{\$2.25}$$

Practice

a. $\begin{array}{r} \$0.75 \\ \times\ \ \ \ 5 \\ \hline \end{array}$ **b.** $\begin{array}{r} \$2.15 \\ \times\ \ \ \ 4 \\ \hline \end{array}$ **c.** $\begin{array}{r} 35¢ \\ \times\ \ 6 \\ \hline \end{array}$

d. 75¢ × 8 **e.** $4.95 × 3 **f.** 39¢ × 5

Problem set 76

1. The first big number was two hundred seventeen thousand, five hundred forty-seven. The second big number was four hundred thousand, nine hundred seventy-five. By how much was the second big number greater than the first big number?

2. One hundred seventy-five beans were equally divided into seven pots. How many beans were in each pot?

3. Five thousand, seven hundred people listened to the concert. Then some went home. Three thousand, forty-two remained. How many people went home?

4. James could go 7 miles on 1 sandwich. He wanted to go 84 miles. How many sandwiches did he need?

5. The starting time was before dawn. The stopping time was in the afternoon. What was the difference in the two times?

Starting

Stopping

6. One hundred forty thousand is how much less than two hundred thousand?

7. What fraction of this pentagon is not shaded?

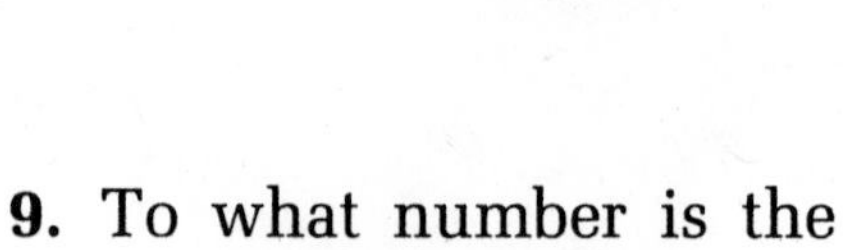

8. According to this calendar, what is the date of the last Saturday in July, 2019?

JULY						2019
S	M	T	W	T	F	S
	1	2	3	4	5	6
7	8	9	10	11	12	13
14	15	16	17	18	19	20
21	22	23	24	25	26	27
28	29	30	31			

9. To what number is the arrow pointing?

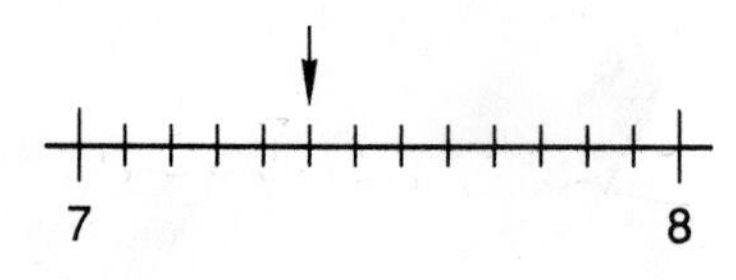

10. Estimate the product of 78 and 4.

11. Compare: $2 \times 2 \times 2 \bigcirc 3 \times 2$

12. $6.25 + $4 + $12.78

13. 142 + 386 + 570 + 864

14. $\begin{array}{r} \$30.25 \\ -\ \$XY.AB \\ \hline \$13.06 \end{array}$

15. $\begin{array}{r} 149{,}384 \\ -\ \ 98{,}765 \\ \hline \end{array}$

16. $\begin{array}{r} 409 \\ \times\ \ 7 \\ \hline \end{array}$

17. 5 × $3.46

18. $0.79 × 6

19. 8 × 39¢

20. $6\overline{)90}$

21. $4\overline{)95}$

22. $8\overline{)356}$

23. 94 ÷ 5

24. 234 ÷ 3

25. 435 ÷ 7

26. $4 \times N = 40$

27. $5 \times N = 80$

28.
$$\begin{array}{r} 87 \\ 46 \\ 29 \\ 55 \\ 47 \\ +\ 28 \\ \hline \end{array}$$

29.
$$\begin{array}{r} 943 \\ 756 \\ 897 \\ +\ 232 \\ \hline \end{array}$$

30.
$$\begin{array}{r} 472 \\ 876 \\ +\quad N \\ \hline 1743 \end{array}$$

LESSON 77 Multiplying by Multiples of 10

Speed Test: 90 Division Facts, Form A (Test I in Test Booklet)

We remember that the multiples of 10 are the numbers we use when we count by tens. The last digit in every multiple of 10 is a zero. The first five multiples of 10 are:

10, 20, 30, 40, 50

To multiply a whole number by a multiple of 10, we may write the multiple of 10 so that the zero "hangs out" to the right as we show here.

$$\begin{array}{r} 34 \\ \times\ 20 \\ \hline \end{array} \quad \leftarrow \quad \text{zero "hangs out" to the right}$$

Next we write a zero in the answer directly below the zero that we let hang out.

$$\begin{array}{r} 34 \\ \times\ 20 \\ \hline 0 \end{array}$$

Then we multiply by the first digit.

$$\begin{array}{r} 34 \\ \times\ 20 \\ \hline 680 \end{array}$$

Example 1 Multiply: 30×34

Solution We set up the problem so that the multiple of 10 is the bottom number. We let the zero "hang out."

$$\begin{array}{r} 34 \\ \times\ 30 \\ \hline \end{array}$$

Next we write a zero in the answer directly below the zero in 30. Then we multiply by the first digit. Our answer is **1020**.

$$\begin{array}{r} \overset{1}{3}4 \\ \times\ 30 \\ \hline 1020 \end{array}$$

Example 2 Multiply $\$1.43 \times 20$.

Solution We set up the problem so that the zero "hangs out." We write a zero below the line, then multiply by the 2. We place the decimal point so that there are two digits after it. Finally we write a dollar sign in front to get **$28.60**.

$$\begin{array}{r} \$\ 1.43 \\ \times\quad 20 \\ \hline \$28.60 \end{array}$$

Practice

a. 75×10 **b.** 10×32

c. $\begin{array}{r} 26 \\ \times\ 20 \\ \hline \end{array}$ **d.** $\begin{array}{r} 164 \\ \times\ 30 \\ \hline \end{array}$ **e.** 50×45

f. 53×40 **g.** $\$1.53 \times 10$ **h.** $20 \times \$2.48$

Problem set 77

1. In the beginning there were only seventeen thousand, five hundred twenty. Then some more came. Now there are one hundred three thousand, two hundred twenty-two. How many more came?

2. It took four spoonfuls to make one complete batch. How many spoonfuls were required to make 40 batches?

3. The first number was forty-two thousand, five hundred. The second number was eighteen thousand, forty-two. How much greater was the first number?

4. Olive ran $\frac{3}{5}$ of the course but walked the rest of the way. What fraction of the course did she walk?

5. Jimmy had an octagon and a pentagon. What was the total of the number of sides in these two polygons?

6. The arrow points to what mixed number on this number line?

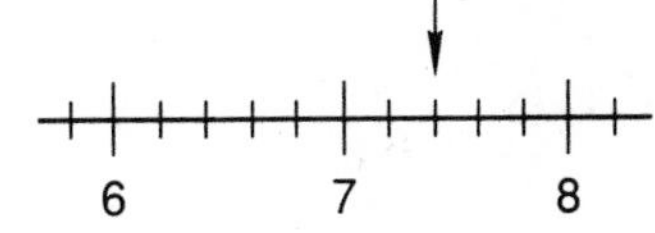

7. Mount Rainier stands four thousand, three hundred ninety-two feet above sea level. Use digits to write that number.

8. Ready got ready to go in the afternoon. Ready got ready to stop the next morning. How much time went by?

9. Mickey could make 35 prizes in 7 minutes. How many prizes could he make in 1 minute?

10. Estimate the sum of 681 and 903.

11. $12 + $8.95 + 75¢

12. 3627 + 5314 + 729

13. $30.00 − $21.49

14. 36,157 − 29,448

15. 43¢ × 8

16. $3.05 × 5

17. $2.63 × 7

18. 47 × 30

19. 60 × 39

20. 85 × 40

21. $5\overline{)96}$

22. $7\overline{)156}$

23. $3\overline{)246}$

24. 215 ÷ 6

25. 156 ÷ 4

26. 195 ÷ 8

27. $7 \times N = 70 + 7$

28.
$$\begin{array}{r} 33 \\ 2 \\ 22 \\ 7 \\ 44 \\ +\ N \\ \hline 110 \end{array}$$

29. $N \times 2 = 80$

30.
$$\begin{array}{r} ABC \\ -\ 347 \\ \hline 656 \end{array}$$

LESSON 78 Millimeters

Speed Test: 100 Multiplication Facts (Test H in Test Booklet)

A centimeter is this long.

———

If we divide a centimeter into 10 smaller lengths, each small length is 1 millimeter long. A dime is about 1 millimeter thick.

10¢ ← 1 millimeter thick

The words "centimeter" and "millimeter" are based on Latin words. *Centum* is the Latin word for hundred. This is why we say there are 100 cents in a dollar. The Latin word for feet is *pede*. A centipede looks like it has 100 feet. **There are 100 centimeters in 1 meter**.

Mille is the Latin word for thousand. A millipede looks like it has 1000 feet. **There are 1000 millimeters in 1 meter. It takes 10 millimeters to equal 1 centimeter**.

Here we show a millimeter scale and a centimeter scale.

Millimeter: mm 10 20 30 40 50 60

Centimeter: cm 1 2 3 4 5 6

We abbreviate the word millimeter with the letters mm.

Example 1 This segment is how many millimeters long?

mm10 20 30 40 50 60

Solution The length of the segment is **35 mm**.

Example 2 This paper clip is 3 cm long. How many millimeters long is it?

Solution Each centimeter is 10 mm. We multiply 3×10 mm and find that the length of the paper clip is **30 mm**.

Practice **a.** The thickness of a dime is about 1 mm. How many dimes would it take to form a stack that is 1 cm high?

b. How long is this segment?

mm10 20 30 40

c. Each side of this square is 1 cm long. What is the perimeter of this square in millimeters?

Problem set 78

1. In the morning one hundred forty-two thousand, seven hundred fifty-three came. In the afternoon ninety-six thousand, five hundred twenty-two came. How many came in all?

2. Tracy has three hundred eighty-four baseball cards. Nathan has two hundred sixty baseball cards. Tracy has how many more cards than Nathan?

3. Forty-two students could get on 1 bus. There were 30 buses. How many students could get on all the buses?

4. To what number is the arrow pointing?

200 400

5. Copy this hexagon and shade one sixth of it.

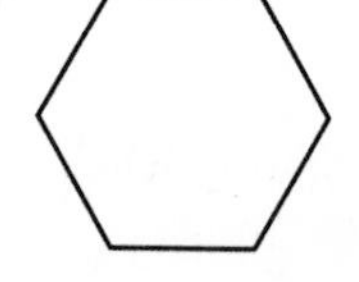

6. (a) This match is how many centimeters long?

(b) This match is how many millimeters long?

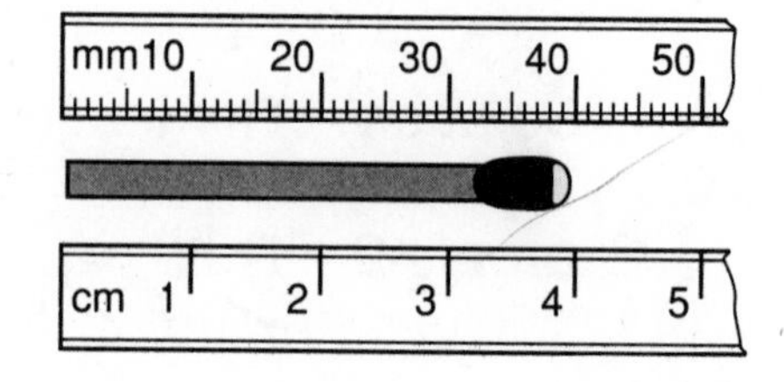

7. Write 751,318 in expanded form. Then use words to write the number.

8. One yard equals 3 feet. If each side of a square is 1 yard long, then what is the perimeter of the square in feet?

9. Estimate the sum of 412, 695, and 379.

10. Segment *AB* is 35 mm long. Segment *AC* is 80 mm long. How long is segment *BC*?

A B C

11. Hugo could go 125 miles in 5 hours. How many miles could Hugo go in 1 hour?

12. Urgo could go 21 miles in one hour. How many miles could Urgo go in 7 hours?

13.
$$\begin{array}{r} \$96.89 \\ +\ \$26.56 \\ \hline \end{array}$$

14.
$$\begin{array}{r} AB,CDE \\ -\ 28,165 \\ \hline 15,246 \end{array}$$

15.
$$\begin{array}{r} 362 \\ 47 \\ 159 \\ 1484 \\ 305 \\ +\ \ 60 \\ \hline \end{array}$$

16.
$$\begin{array}{r} \$10.00 \\ -\ \$\ 1.73 \\ \hline \end{array}$$

17.
$$\begin{array}{r} 36,428 \\ -\ 27,338 \\ \hline \end{array}$$

18. $\begin{array}{r} 78 \\ \times\ 60 \\ \hline \end{array}$

19. $9 \times \$4.63$

20. $80 \times 29¢$

21. $4\overline{)328}$

22. $7\overline{)375}$

23. $5\overline{)320}$

24. $256 \div 8$

25. $250 \div 6$

26. $100 \div 9$

27. $A \times 5 = 25 + 25$

28. $4 \times W = 60$

29. $\begin{array}{r} 6 \\ 2 \\ 8 \\ 4 \\ 7 \\ +\ N \\ \hline 51 \end{array}$

30. $\begin{array}{r} VW,XYZ \\ +\ 30,149 \\ \hline 63,578 \end{array}$

LESSON 79 Fraction-of-a-Group Problems, Part 1

Speed Test: 100 Multiplication Facts (Test H in Test Booklet)

We know that the fraction $\frac{1}{2}$ means that the whole thing has been divided into 2 parts. To find the number in $\frac{1}{2}$ of a group, we divide the total number by 2. To find the number in $\frac{1}{3}$ of a group, we divide the total number by 3. To find the number in $\frac{1}{4}$ of a group, we divide the total number by 4.

Example 1 One half of the carrot seeds sprouted. If 84 seeds were planted, how many sprouted?

Solution To find $\frac{1}{2}$ of a group, we divide by 2. We divide 84 by 2 and find that **42 seeds sprouted.**

$$\begin{array}{r} 42 \\ 2\overline{)84} \\ \underline{8} \\ 04 \\ \underline{4} \\ 0 \end{array}$$

Example 2 One third of the 27 students earned an A on the test. How many students earned an A on the test?

Solution To find $\frac{1}{3}$ of a group, we divide by 3. We divide 27 by 3 and find that **9 students** earned an A on the test.

$$\begin{array}{r} 9 \\ 3\overline{)27} \\ \underline{27} \\ 0 \end{array}$$

Example 3 One fourth of the team's 32 points were scored by Cliff. Cliff scored how many points?

Solution To find $\frac{1}{4}$ of a group, we divide by 4. We divide 32 by 4 and find that Cliff scored **8 points**.

$$\begin{array}{r} 8 \\ 4\overline{)32} \\ \underline{32} \\ 0 \end{array}$$

Practice

a. What is $\frac{1}{3}$ of 60?

b. What is $\frac{1}{2}$ of 60?

c. What is $\frac{1}{4}$ of 60?

d. What is $\frac{1}{5}$ of 60?

Problem set 79

1. Nine hundred forty-two thousand is how much greater than two hundred forty-two thousand, nine hundred sixty-two?

2. There were 150 seats at the Round Table. If 128 seats were taken, how many seats were empty?

3. There were 3 surf riders for each child. How many surf riders were there for 17 children?

4. Stuart bought his lunch Monday through Friday. If each lunch cost $1.25, how much did he spend on lunch for the week?

5. Bruce read four hundred eighty-two pages the first week. He read one thousand, four hundred forty-two pages the second week and only nine hundred eleven pages the third week. How many pages did he read in all?

6. Rhonda read 30 pages a day on Monday, Tuesday, and Wednesday. She read 45 pages on Thursday and 26 pages on Friday. How many pages did she read in all?

7. One half of the cabbage seeds sprouted. If 74 seeds were planted, how many sprouted?

8. What is $\frac{1}{6}$ of 60?

9. The machine made 39 buttons in 1 minute. How many buttons would it make in 1 hour (60 minutes)?

10. The older machine made 160 buttons in 4 minutes. How many buttons could it make in 1 minute?

11. (a) This shirt button is how many centimeters across?
(b) This shirt button is how many millimeters across?

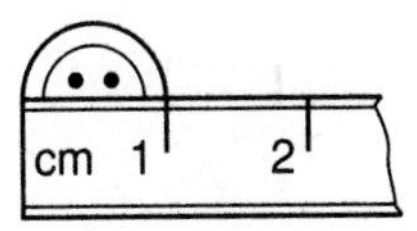

12. Estimate the difference when 506 is subtracted from 893.

13. Segment AB is 27 mm long. Segment BC is 48 mm long. How long is segment AC?

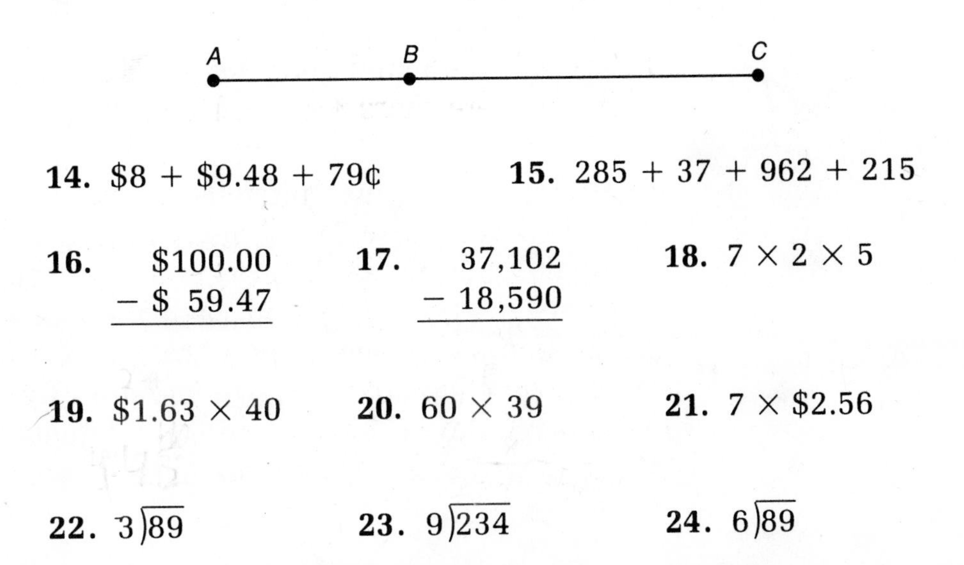

14. \$8 + \$9.48 + 79¢

15. 285 + 37 + 962 + 215

16. $\begin{array}{r} \$100.00 \\ -\ \$\ 59.47 \\ \hline \end{array}$

17. $\begin{array}{r} 37{,}102 \\ -\ 18{,}590 \\ \hline \end{array}$

18. $7 \times 2 \times 5$

19. $\$1.63 \times 40$

20. 60×39

21. $7 \times \$2.56$

22. $3\overline{)89}$

23. $9\overline{)234}$

24. $6\overline{)89}$

25. $243 \div 7$

26. $355 \div 5$

27.
$$\begin{array}{r} 7 \\ 6 \\ 3 \\ 4 \\ 2 \\ 8 \\ + N \\ \hline 54 \end{array}$$

28. $196 \div 8$

29. $5 \times B = 60$

30. $2 \times N = 60$

LESSON 80 Division Answers Ending with Zero

> **Speed Test:** 100 Multiplication Facts (Test H in Test Booklet)

Sometimes two-digit division answers have a zero in them before the division is complete. It is important to continue the division until it is completed. Look at this problem.

> Two hundred pennies are separated into 4 equal piles. How many pennies are in each pile?

This problem can be answered by dividing 200 by 4. We begin by dividing 4 into 20. We put a 5 on top. Then we multiply and then we subtract.

$$\begin{array}{r} 5 \\ 4\overline{)200} \\ \underline{20} \\ 0 \end{array}$$

It may look like the division is complete, but it is not. The answer is not "5 pennies in each pile." That would total only 20 pennies. There is another zero inside the division box that we have not used. **We have not finished dividing.** So we bring down the zero and divide again. Of course, 4 goes into zero 0 times, and zero times zero equals zero.

$$\begin{array}{r} 50 \\ 4\overline{)200} \\ \underline{20} \\ 00 \\ \underline{0} \\ 0 \end{array}$$

The answer is **50 pennies in each pile.**

Sometimes a two-digit division answer will end with a zero, yet there will be a remainder. We show this in the following example.

Example $3\overline{)121}$

Solution We begin by dividing $3\overline{)12}$. After we subtract, we bring down the next digit, which is 1. Now we divide $3\overline{)1}$ (or $3\overline{)01}$, which means the same thing). Since 1 is less than 3, the answer is 0. We write this on top above the 1. Since 0 × 3 is 0, we write 0 below the 1 and subtract. The remainder is 1.

$$\begin{array}{r} 4 \\ 3\overline{)121} \\ \underline{12} \\ 0 \end{array}$$

$$\begin{array}{r} \mathbf{40\ r\ 1} \\ 3\overline{)121} \\ \underline{12} \\ 01 \\ \underline{0} \\ 1 \end{array}$$

Practice **a.** $3\overline{)120}$ **b.** $4\overline{)240}$ **c.** $5\overline{)152}$

d. $4\overline{)121}$ **e.** $3\overline{)91}$ **f.** $2\overline{)41}$

Problem set 80

1. The first number was one hundred forty-two thousand, seven hundred fifty-six. The second number was one hundred eight thousand, five hundred forty-nine. How much larger was the first number?

2. Four hundred eleven people waited to buy tickets. When the rain began, some went home. Then there were only two hundred forty-two in line. How many went home when the rain began?

3. Each cookie contained 5 chocolate chips. How many chocolate chips would be in 115 cookies?

4. Harry could hop 420 spaces in 7 minutes. How many spaces could he hop in 1 minute?

5. What is the value of 5 pennies, 3 dimes, 2 quarters, and 3 nickels?

6. One fourth of the students earned an A. There were 280 students in all. How many students earned an A?

7. What is $\frac{1}{2}$ of 560?

8. Estimate the product of 7 and 82.

9. (a) The line segment shown is how many centimeters long?
   (b) The segment is how many millimeters long?

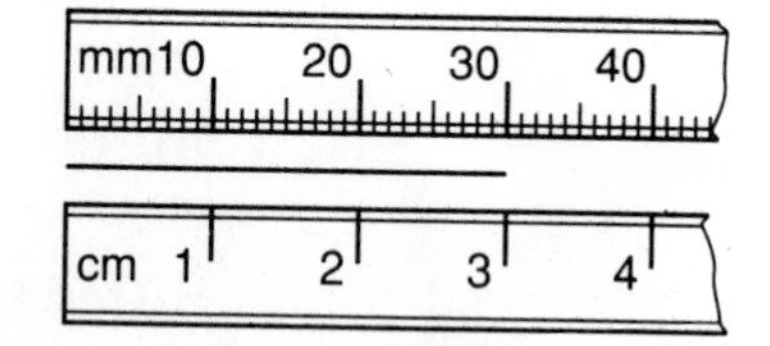

10. It is evening. What time was it 6 hours and 20 minutes ago?

11. Compare:

$$\frac{2}{3} \bigcirc \frac{2}{5}$$

Draw and shade two rectangles to show the comparison.

12. Jenny could hop 72 spaces in 1 minute. How many spaces could she hop in 9 minutes?

13. $\begin{array}{r} \$375.48 \\ +\ \$536.70 \\ \hline \end{array}$

14. $\begin{array}{r} 367{,}419 \\ +\ 90{,}852 \\ \hline \end{array}$

15. $\begin{array}{r} 423 \\ 571 \\ 289 \\ 964 \\ +\ 380 \\ \hline \end{array}$

16. $\begin{array}{r} \$20.00 \\ -\ \$19.39 \\ \hline \end{array}$

17. $\begin{array}{r} 310{,}419 \\ -\ 250{,}527 \\ \hline \end{array}$

18. $\begin{array}{r} 608 \\ \times \quad 7 \\ \hline \end{array}$

19. $\begin{array}{r} 86 \\ \times \ 40 \\ \hline \end{array}$

20. $\begin{array}{r} 59¢ \\ \times \ 8 \\ \hline \end{array}$

21. $3\overline{)180}$

22. $8\overline{)241}$

23. $5\overline{)323}$

24. $184 \div 6$

25. $\begin{array}{r} 7 \\ 6 \\ 4 \\ 5 \\ 9 \\ 4 \\ 3 \\ 7 \\ + N \\ \hline 56 \end{array}$

26. $\begin{array}{r} 8 \\ 5 \\ 3 \\ 4 \\ 7 \\ 8 \\ 2 \\ 3 \\ + N \\ \hline 56 \end{array}$

27. $279 \div 4$

28. $423 \div 7$

29. $9 \times M = 27 + 72$

30. $N \times 6 = 90$

LESSON 81 Finding Information to Solve Problems

> **Speed Test 1:** 100 Multiplication Facts (Test H in Test Booklet)

Part of the problem-solving process is finding the information needed to solve the problem. We may find information in graphs, tables, books, or other places. In this lesson we will practice solving problems in which we need to choose the information needed to solve the problem.

Example Read this information. Then answer the questions.

> The school elections were held on Friday, February 2. Kim, Dan, and Miguel ran for president. Dan got 146 votes, and Kim got 117 votes. Miguel got 35 more votes than Kim.

(a) How many votes did Miguel get?

(b) Who won the election?

(c) Speeches were given on the Tuesday before the election. What was the date on which the speeches were given?

Solution (a) We read that Miguel got 35 more votes than Kim and that Kim got 117 votes. We add 35 and 117 and find that Miguel got **152 votes**.

(b) Miguel got the most votes, so the winner was **Miguel**.

(c) We read that the election was on Friday, February 2. The Tuesday before that is 3 days before that. We count back 3 days: February 1, January 31, January 30. The speeches were given on Tuesday, **January 30**.

Practice Read this information. Then answer the questions.

Tom did yard work on Saturday. He worked for 3 hours in the morning and 4 hours in the afternoon. He was paid $4 for every hour he worked.

a. How many hours did Tom work in all?

b. How much money did Tom earn in the morning?

c. How much money did Tom earn in all?

Problem set 81

1. Of the one thousand, six hundred forty-two fair maidens that Sir Launcelot rescued, only one hundred twenty-one remembered to thank him. How many did not thank Sir Launcelot?

2. Christie's car goes 18 miles on each gallon of gas. How far can it go on 10 gallons of gas?

3. Humpty Dumpty weighed 160 pounds. If Humpty Dumpty broke into 8 equal pieces, how much did each piece weigh?

4. Soccer practice lasts for an hour and a half. If practice starts at 3:15 p.m., at what time does it end?

5. One third of the team's 36 points were scored by Cliff. How many points were scored by Cliff?

6. Seven hundred forty-two thousand, five hundred seventy-three is how much more than forty-six thousand, six hundred ten?

7. A dime is about 1 millimeter thick. A roll of 50 dimes is about how many centimeters long?

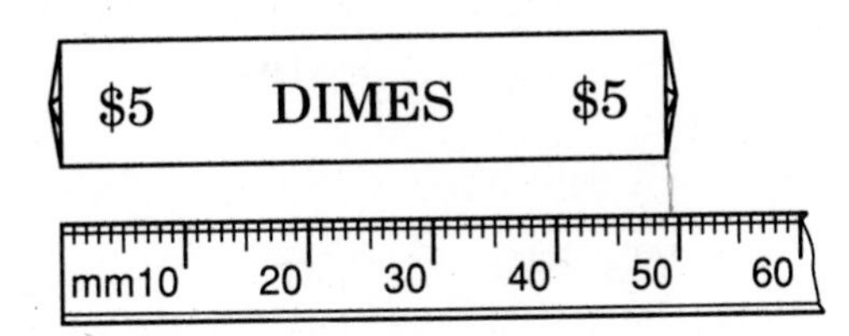

8. According to this calendar, the year 1902 began on what day of the week?

DECEMBER						1901
S	M	T	W	T	F	S
1	2	3	4	5	6	7
8	9	10	11	12	13	14
15	16	17	18	19	20	21
22	23	24	25	26	27	28
29	30					

9. Write 746,357 in expanded form. Then use words to write the number.

10. A meter equals 100 centimeters. If each side of a square is 1 meter, then what is the perimeter of the square in centimeters?

11. $1.68 + 32¢ + $6.37 + $5

12. 43 + 24 + 8 + 67 + 4327

13. $10 − $8.63

14. 361,420 − 23,169

15. 5 × 4 × 5

16. 359 × 70

17. 50 × 74

18. $2\overline{)161}$

19. $5\overline{)400}$

20. $9\overline{)462}$

21. 216 ÷ 3

22. 159 ÷ 4

23. $478 \div 7$

24. $126 \div 3$

25. $324 \div 6$

26. $5 \times N = 120$

27.
$$\begin{array}{r} 473 \\ -\ N \\ \hline 274 \end{array}$$

28.
$$\begin{array}{r} 9 \\ 5 \\ 6 \\ 4 \\ 7 \\ 8 \\ 5 \\ 7 \\ +\ N \\ \hline 64 \end{array}$$

29.
$$\begin{array}{r} 4 \\ 3 \\ 7 \\ 6 \\ 8 \\ 2 \\ 9 \\ 1 \\ +\ N \\ \hline 64 \end{array}$$

30.
$$\begin{array}{r} N \\ -\ 437 \\ \hline 943 \end{array}$$

LESSON 82 Measuring Liquids

Speed Test: 100 Multiplication Facts (Test H in Test Booklet)

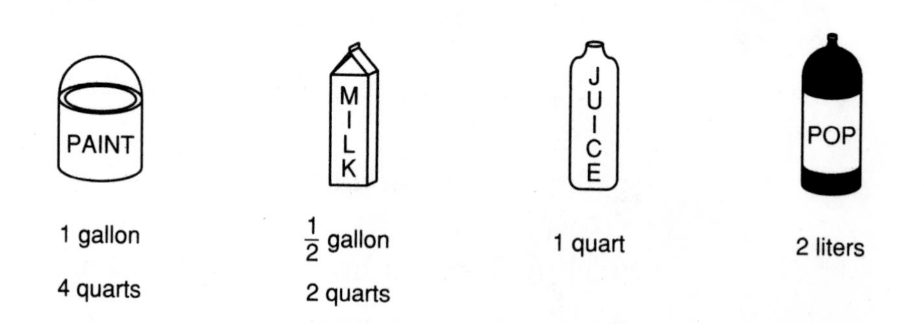

Liquids like paint, milk, juice, and soda pop are measured in **liquid units. Ounces, pints, quarts,** and **gallons** are liquid units in the U.S. system. **Liters** and **milliliters** are liquid units in the metric system. The chart at the top of the next page shows the number of units needed to equal the next larger unit.

Units of Liquid Measure

U.S. UNITS	METRIC UNITS
16 ounces = 1 pint 2 pints = 1 quart 4 quarts = 1 gallon	1000 milliliters = 1 liter

A liter is about the same amount as a quart.

The abbreviations for units of liquid measure are as follows:

oz	ounce	L	liter
pt	pint	mL	milliliter
qt	quart		
gal	gallon		

Example 1 Two liters of water is how many milliliters?

Solution From the chart we see that 1 liter is 1000 milliliters. Two liters would be twice as much: 2 liters = **2000 milliliters.**

Example 2 A half gallon of milk is how many quarts?

Solution One gallon of milk is 4 quarts. So a half gallon is half of 4 quarts. Since half of 4 is 2, a half gallon is **2 quarts**.

Practice

a. A quart is a "quarter" of a gallon. How many "quarters" make a whole?

b. Two pints equals a quart. How many ounces are in 2 pints?

c. Three liters is how many milliliters?

Choose the most reasonable measure.

d. Would a glass of milk be about 8 oz or 8 qt?

e. Would a can of pop be about 350 mL or 350 L?

Problem set 82

Use this information to answer questions 1–3.

Thirty students are going on a field trip. Each car can hold five students. The field trip will cost each student $5.

1. How many cars are needed for the field trip?

2. Altogether, how much money will be collected?

3. Don has saved $3.25. How much more does he need to go on the field trip?

4. During the summer the swim team practiced $3\frac{1}{2}$ hours a day. If practice started at 6:30 a.m., at what time did it end?

5. Half of the 48 pencils were sharpened. How many were not sharpened?

6. What number is $\frac{1}{4}$ of 60?

7. One gallon of water will be poured into 1-quart bottles. How many 1-quart bottles can be filled?

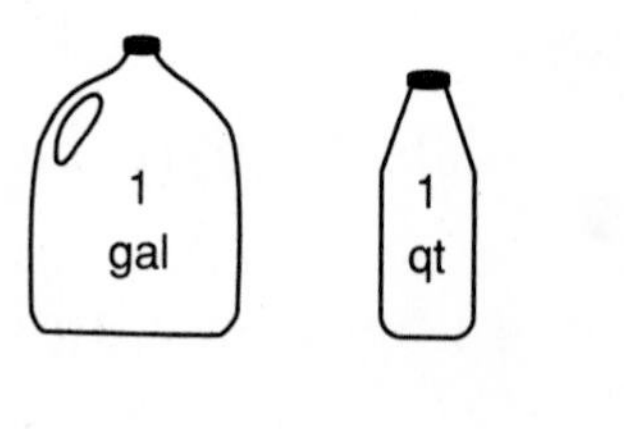

8. Each side of a regular polygon has the same length. This is a regular hexagon. How many **millimeters** is the perimeter of this hexagon?

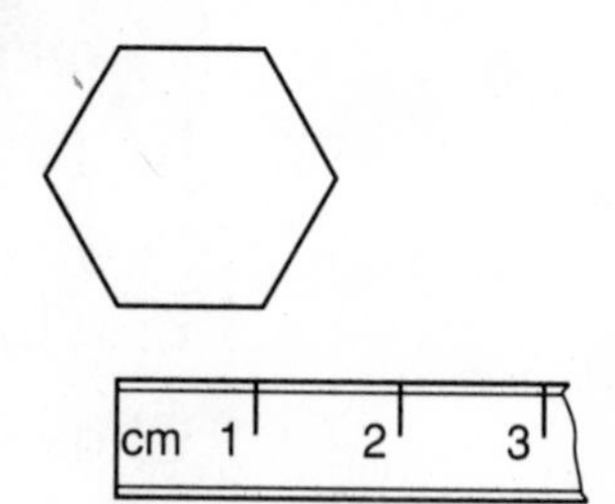

9. A mile is five thousand, two hundred eighty feet. The Golden Gate Bridge is four thousand, two hundred feet long. The Golden Gate Bridge is how many feet less than 1 mile long?

10. Estimate the product of 394 and 5.

11. What number is halfway between 300 and 400?

12. $\begin{array}{r} XYZ \\ -\ 476 \\ \hline 325 \end{array}$

13. $\begin{array}{r} 37{,}156 \\ +\ 214{,}390 \\ \hline \end{array}$

14. $\begin{array}{r} 1534 \\ 367 \\ 29 \\ 6 \\ +\ 142 \\ \hline \end{array}$

15. $\begin{array}{r} \$100.00 \\ -\ \$\ 31.53 \\ \hline \end{array}$

16. $\begin{array}{r} 251{,}546 \\ -\ 37{,}156 \\ \hline \end{array}$

17. $\begin{array}{r} 346 \\ \times\ 7 \\ \hline \end{array}$

18. $\begin{array}{r} 96 \\ \times\ 30 \\ \hline \end{array}$

19. $\begin{array}{r} \$0.59 \\ \times\ 8 \\ \hline \end{array}$

20. $7\overline{)633}$

21. $5\overline{)98}$

22. $3\overline{)150}$

23. $329 \div 6$

24. $274 \div 4$

25. $247 \div 8$

26. $A \times 6 = 12 + 6$

27. $\begin{array}{r} 7 \\ 5 \\ 6 \\ 4 \\ 9 \\ 3 \\ 2 \\ +\ N \\ \hline 54 \end{array}$

28. $\begin{array}{r} 5 \\ 2 \\ 7 \\ 9 \\ 6 \\ 4 \\ 8 \\ +\ N \\ \hline 61 \end{array}$

29. $5 \times M = 135$

30. $\begin{array}{r} N \\ +\ 423 \\ \hline 618 \end{array}$

LESSON 83 Fraction of a Set

Speed Test: 64 Multiplication Facts (Test G in Test Booklet)

There are seven circles in the set below. Three of the circles are shaded. The fraction of the set that is shaded is $\frac{3}{7}$.

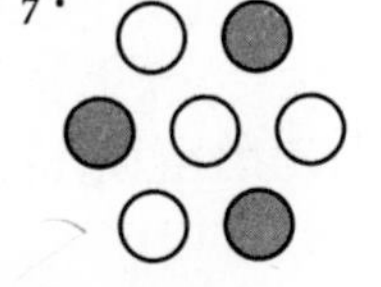

$\frac{3}{7}$ Three circles are shaded.
There are seven circles in all.

The number of members in the set is the bottom number (denominator) of the fraction. The number of members named is the top number (numerator) of the fraction.

Example 1 What fraction of the triangles is not shaded?

Solution There are 9 triangles in all (bottom number). There are 5 that are not shaded (top number). The fraction not shaded is $\frac{5}{9}$.

Example 2 In a class of 25 students, there are 12 girls and 13 boys. What fraction of the class is girls?

Solution The total number of members in the class is 25 (bottom number). The number of girls is 12 (top number). The fraction of the class that is girls is $\frac{12}{25}$.

Practice

a. What fraction of the set is shaded?

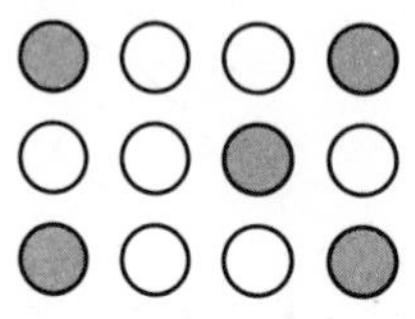

b. What fraction of the set is not shaded?

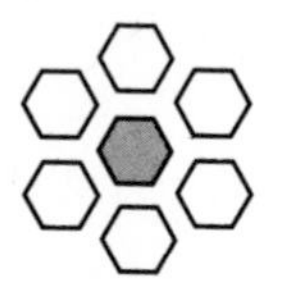

c. In a class of 27 students, there are 14 girls and 13 boys. What fraction of the class is boys?

d. What fraction of the letters in the word ALABAMA is made up of A's?

Problem set 83

1. Michael caught sixty-two crawfish in the creek. James caught seven crawfish, and Julie and John each caught twelve crawfish. Altogether, how many crawfish did these young people catch?

2. The Matterhorn is fourteen thousand, six hundred ninety-two feet high. Mont Blanc is fifteen thousand, seven hundred eighty-one feet high. How much taller is Mont Blanc than the Matterhorn?

3. There are 25 squares on a bingo card. How many squares are there on 4 bingo cards?

4. Ninety-six books were placed on 4 shelves so that there were the same number of books on each shelf. How many books were on each shelf?

Books

5. One half of the 780 fans stood and cheered. How many fans stood and cheered?

6. What are the next three numbers in this sequence?

2000, 3000, 4000, ________, ________, ________, . . .

7. What fraction of this set is not shaded?

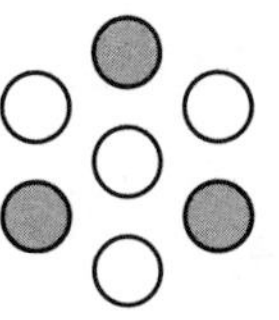

8. The 2-liter bottle contained how many milliliters of beverage?

9. Estimate the sum of 493 and 387.

10. What is the perimeter of this rectangle?

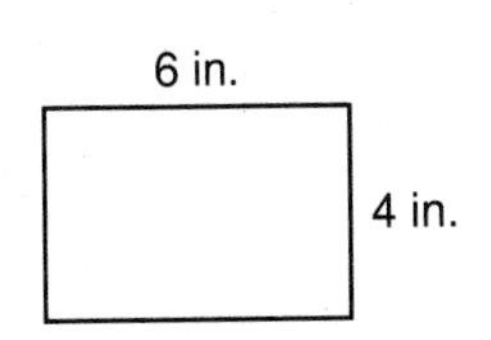

11. Write 47,310 in expanded form. Then use words to write this number.

12. $6.15 − ($0.57 + $1.20)

13. 3746 + 2357 + 489

14. 462 − Y = 205

15. 43,160 − 8,459

16. 8 × 8 × 8

17. $3.54 × 6

18. 80×57

19. 704×9

20. $9\overline{)354}$

21. $7\overline{)285}$

22. $5\overline{)439}$

23. $515 \div 6$

24. $361 \div 4$

25. $784 \div 8$

26. $50 = 5 \times R$

27.
$$\begin{array}{r} 4 \\ 7 \\ 8 \\ 5 \\ 7 \\ 6 \\ 4 \\ 3 \\ +\ N \\ \hline 56 \end{array}$$

28. $7 \times N = 140$

29.
$$\begin{array}{r} 476 \\ +\ N \\ \hline 732 \end{array}$$

30.
$$\begin{array}{r} N \\ -\ 214 \\ \hline 638 \end{array}$$

LESSON 84 Pictographs and Bar Graphs

Speed Test: 64 Multiplication Facts (Test G in Test Booklet)

A **graph** is a picture that shows number information about the topic of the graph. There are many types of graphs. In this lesson we will look at **pictographs** and **bar graphs**.

Example 1 James made a pictograph to show the number of gummy bears of a certain color that he found in one package. According to this graph, what was the total number of yellow and green gummy bears in the package?

Solution From the pictograph we can count the number of gummy bears of each color. There were 4 yellow gummy bears and 9 green gummy bears. To find the total of yellow and green gummy bears, we add 4 and 9 to get **13**.

Example 2 This bar graph shows the same information as the pictograph, but in a different form. Use the bar graph to answer each question.

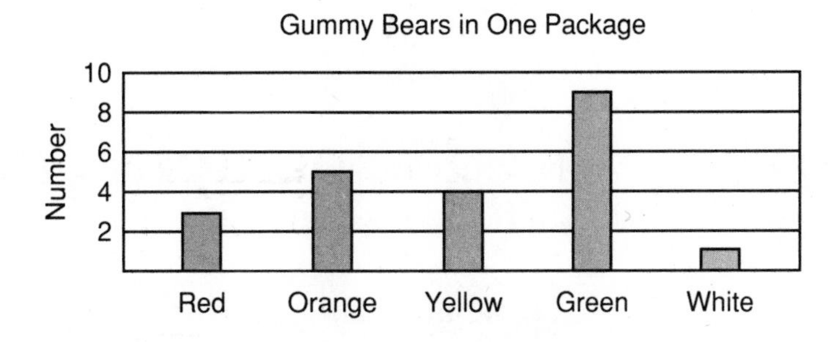

(a) How many red gummy bears were there?

(b) How many more gummy bears were green than were orange?

Solution (a) The top of the bar showing the number of red gummy bears is halfway between 2 and 4 on the scale. So there were **3 red gummy bears.**

(b) There were 9 green gummy bears and 5 orange gummy bears. By subtracting, we find there were **4 more green gummy bears** than orange gummy bears.

Practice Use the information in the graphs from the lesson to answer each question.

a. How many gummy bears were in the package?

b. The total of which two colors of gummy bears equals the number of green gummy bears?

c. The total of which three colors of gummy bears equals the number of green gummy bears?

Problem set 84

1. Pears cost 59¢ per pound. How much would 4 pounds of pears cost?

2. Forty-two thousand, three hundred seventy-two people came to the first game. Seventy-five thousand, two people came to the second game. How many more people came to the second game?

3. There were three hundred twelve books on the floor. Frankie put one fourth of the books on a table. How many books did Frankie put on the table?

4. Jana began work at exactly 3:15 p.m. She worked until 2:30 the next morning. How long did she work?

5. To what number is the arrow pointing?

201 202

6. Estimate the sum of 272 and 483. Begin by rounding both numbers to the nearest hundred.

7. What fraction of this set is shaded?

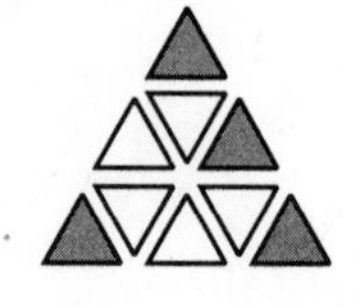

8. One quart of milk is how many ounces?

9. Write 914,718 in expanded form. Then use words to write the number.

Use the information in this bar graph to answer questions 10 and 11.

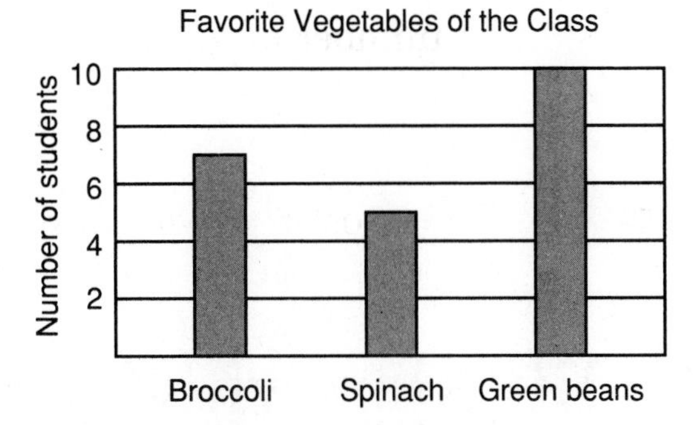

10. Spinach was the favorite vegetable of how many students?

11. Altogether, how many students said broccoli or spinach was their favorite vegetable?

12. $\begin{array}{r} \$86.47 \\ +\ \$47.98 \\ \hline \end{array}$

13. $\begin{array}{r} 362{,}458 \\ +\ 179{,}682 \\ \hline \end{array}$

14. $\begin{array}{r} 2358 \\ 4715 \\ 317 \\ 2103 \\ +\quad 62 \\ \hline \end{array}$

15. $\begin{array}{r} N \\ -\ \$17.53 \\ \hline \$21.49 \end{array}$

16. $\begin{array}{r} 17{,}362 \\ -\ \ 9{,}854 \\ \hline \end{array}$

17. $\begin{array}{r} \$6.95 \\ \times \quad 8 \\ \hline \end{array}$

18. $\begin{array}{r} 46 \\ \times\ 70 \\ \hline \end{array}$

19. $\begin{array}{r} 460 \\ \times \quad 9 \\ \hline \end{array}$

20. $8\overline{)716}$

21. $2\overline{)161}$

22. $7\overline{)434}$

23. $275 \div 9$

24. $513 \div 6$

25. $267 \div 3$

26. $3 \times A = 30 + 30$

27. $N \times 4 = 120$

28.
$$\begin{array}{r} 4 \\ 7 \\ 3 \\ 5 \\ 6 \\ 5 \\ 7 \\ 2 \\ +\ N \\ \hline 43 \end{array}$$

29.
$$\begin{array}{r} 9 \\ 5 \\ 6 \\ 4 \\ 8 \\ N \\ 2 \\ 8 \\ +\ 3 \\ \hline 56 \end{array}$$

30.
$$\begin{array}{r} 5 \\ 4 \\ 7 \\ 3 \\ N \\ 1 \\ 6 \\ 4 \\ +\ 6 \\ \hline 45 \end{array}$$

LESSON 85 Division with Three-Digit Answers

Speed Test: 64 Multiplication Facts (Test G in Test Booklet)

We have practiced division problems that have two-digit answers. In this lesson we will practice division problems that have three-digit answers.

We remember that the method we use for dividing has four steps.

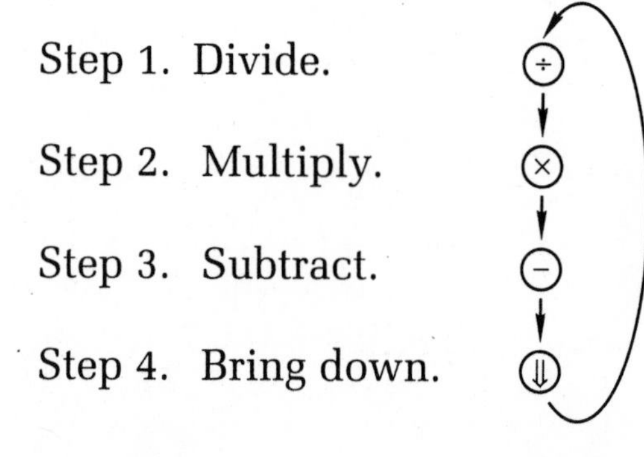

For each step we write a number. When we finish step 4, we go back to step 1 and repeat the steps until there are no digits left to bring down.

Example $3\overline{)794}$

Solution Step 1. Divide $3\overline{)7}$ and write 2.
Step 2. Multiply 2 × 3 and write 6.
Step 3. Subtract 6 from 7 and write 1.
Step 4. Bring down the 9, making 19.

Repeat Step 1. Divide $3\overline{)19}$ and write 6.
Step 2. Multiply 6 × 3 and write 18.
Step 3. Subtract 18 from 19 and write 1.
Step 4. Bring down the 4, making 14.

$$\begin{array}{r} 264 \text{ r } 2 \\ 3\overline{)794} \\ \underline{6} \\ 19 \\ \underline{18} \\ 14 \\ \underline{12} \\ 2 \end{array}$$

Repeat Step 1. Divide $3\overline{)14}$ and write 4.
Step 2. Multiply 4 × 3 and write 12.
Step 3. Subtract 12 from 14 and write 2.
Step 4. There are no digits to bring down.
We are finished dividing. We write 2 as a remainder for a final answer of **264 r 2**.

Practice **a.** Copy this diagram and name the four steps of division.

÷ → × → − → ⇓ (arrow back to ÷)

b. $4\overline{)974}$ **c.** $5\overline{)794}$

d. $6\overline{)1512}$ **e.** $7\overline{)3745}$ **f.** $8\overline{)5000}$

Problem set 85

1. What number is seven thousand, three hundred ninety-six less than eleven thousand, eight hundred seventy-three?

2. Shannon has five days to read a 200-page book. If she wants to read the same number of pages each day, how many pages should she read each day?

3. Julie ordered a book for $2.99, a dictionary for $3.99, and a set of maps for $4.99. What was the price for all three items?

4. The prince searched 7 weeks for the princess. For how many days did he search?

5. One third of the books were placed on the first shelf. What fraction of the books were not placed on the first shelf?

6. What number is halfway between 2000 and 3000?

7. What fraction of the letters in the word HIPPOPOTAMI is made up of P's?

8. Mary ran a 5-kilometer race. Five kilometers is how many meters?

9. $12 + 13 + 5 + N = 9 \times 8$

10. What is the perimeter of this triangle?

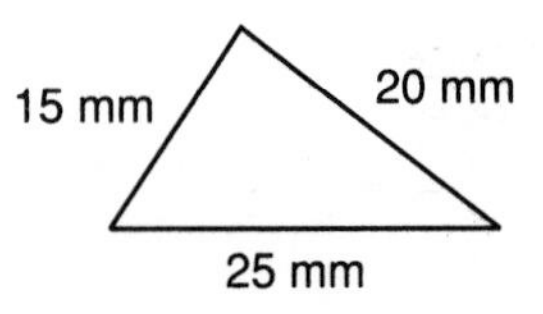

11. The length of segment AB is 36 mm. The length of segment AC is 82 mm. What is the length of segment BC?

A B C

12. $25 − ($19.71 + 98¢)

13. 365 + 470 + 903 + 18

14. $5.00 − $2.92

15. 36,214 − 579

16. $5 \times 6 \times 9$

17. 50×63

18. 478×6

19. $3\overline{)435}$

20. $7\overline{)867}$

21. $5\overline{)1365}$

22. $453 \div 6$ **23.** $543 \div 4$ **24.** $482 \div 8$

25. $N \times 6 = 120$ **26.** $4 \times W = 132$

27.
$$\begin{array}{r} 4 \\ 8 \\ 7 \\ 6 \\ 4 \\ N \\ 3 \\ 6 \\ +\ 5 \\ \hline 55 \end{array}$$

28.
$$\begin{array}{r} 4 \\ 6 \\ 5 \\ 7 \\ +\ N \\ \hline 35 \end{array}$$

29.
$$\begin{array}{r} 2 \\ 14 \\ 3 \\ 7 \\ +\ N \\ \hline 35 \end{array}$$

30.
$$\begin{array}{r} 473 \\ 285 \\ +\ N \\ \hline 963 \end{array}$$

LESSON 86 Ounces, Pounds, and Tons

Speed Test: 64 Multiplication Facts (Test G in Test Booklet)

In Lesson 82 we used the word "ounce" to describe an amount of liquid. The word ounce can also describe an amount of weight. A liquid ounce tells how much volume. A weight ounce tells how much weight. A liquid ounce of water weighs about one weight ounce. The units of weight of the U.S. system are **ounces**, **pounds**, and **tons**. A ton equals 2000 pounds.

16 ounces = 1 pound

2000 pounds = 1 ton

Ounce is abbreviated **oz**. Pound is abbreviated **lb**. Some boxes of cereal weight 24 ounces. Some boys weigh 98 pounds. Some cars weigh 1 ton.

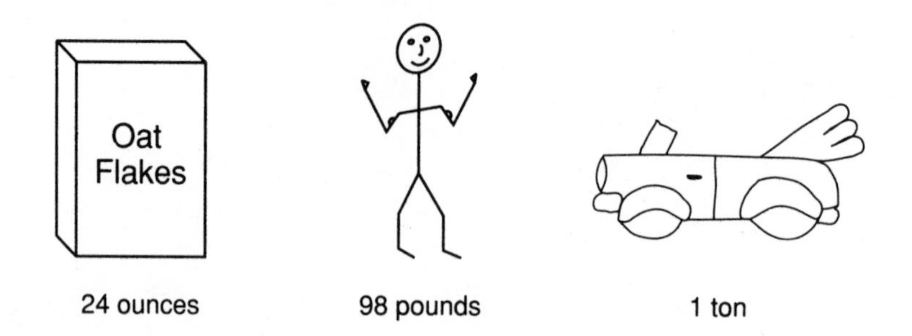

Example 1 This book weighs about 2 pounds. Two pounds is how many ounces?

Solution Each pound is 16 ounces, so 2 pounds is 2×16 ounces, which is **32 ounces**.

Example 2 The rhinoceros weighed 3 tons. Three tons is how many pounds?

Solution Each ton is 2000 pounds, so 3 tons is 3×2000 pounds, which is **6000 pounds**.

Problem set 86 Use the information in this pictograph to answer questions 1–3.

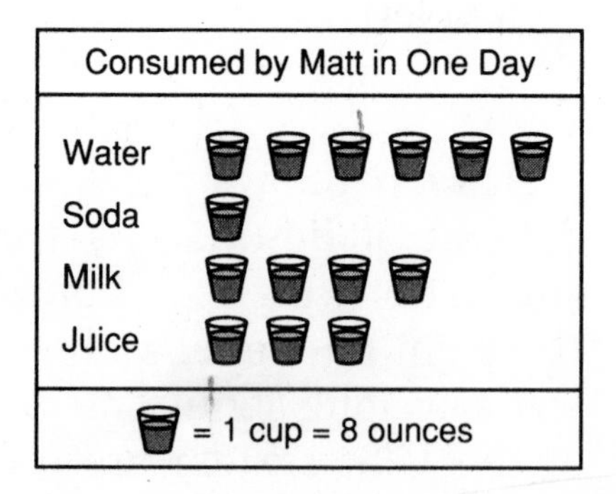

1. How many pints of liquid did Matt drink in 1 day?

2. Matt drank twice as much water as he did what other beverage?

3. Of which beverage did he drink exactly 1 quart?

4. There were 4 rooms. One fourth of the 56 guests gathered in each room. How many guests were in each room?

5. Which of these arrows could be pointing to 2500?

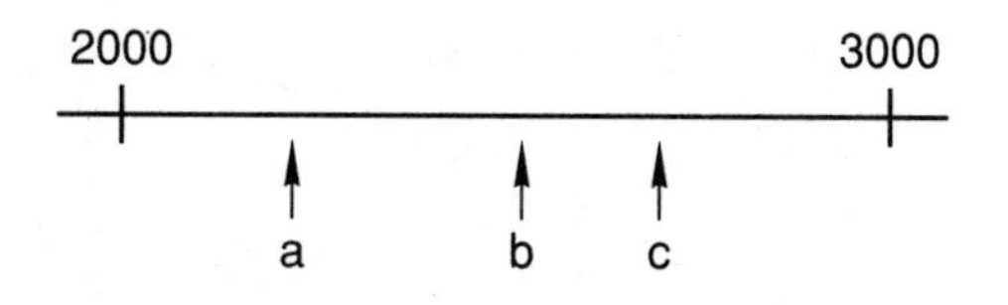

6. Estimate the sum of 682, 437, and 396.

7. What fraction of this set is shaded?

8. McGillicuddy weighed 9 pounds when he was born. How many ounces is that?

9. (a) The segment is how many centimeters long?
   (b) The segment is how many millimeters long?

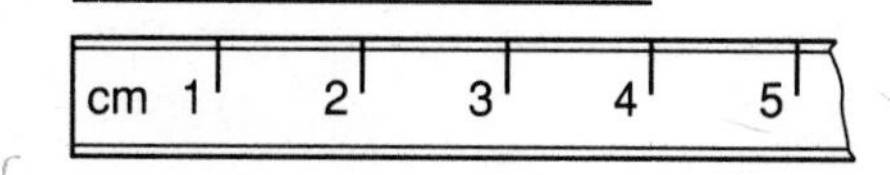

10. Four hundred forty-two thousand, nine hundred seventy-five is how much greater than three hundred thousand, twenty-two?

11. If each side of a hexagon is 1 foot long, how many inches is its perimeter?

12. $\begin{array}{r} 42{,}718 \\ -\ \mathit{TX{,}YZK} \\ \hline 26{,}054 \end{array}$

13. $\begin{array}{r} 93{,}417 \\ +\ \ 8{,}915 \\ \hline \end{array}$

14. $\begin{array}{r} 1307 \\ 638 \\ 5219 \\ 138 \\ +\ \ \ 16 \\ \hline \end{array}$

15. $\begin{array}{r} \$100.00 \\ -\ \$\ 86.32 \\ \hline \end{array}$

16. $\begin{array}{r} 405{,}158 \\ -\ 396{,}370 \\ \hline \end{array}$

17. 567×8

18. 30×84

19. $\$2.08 \times 4$

20. $4\overline{)1500}$

21. $6\overline{)933}$

22. $8\overline{)4537}$

23. $452 \div 5$

24. $378 \div 9$

25. $960 \div 7$

26. $4 \times N = 100$

27. $5 \times B = 100$

28.
$$\begin{array}{r} 4 \\ 7 \\ 3 \\ 5 \\ 4 \\ 8 \\ + N \\ \hline 45 \end{array}$$

29.
$$\begin{array}{r} N \\ - 463 \\ \hline 214 \end{array}$$

30.
$$\begin{array}{r} 756 \\ - N \\ \hline 462 \end{array}$$

LESSON 87 Grams and Kilograms

Speed Test: 64 Multiplication Facts (Test G in Test Booklet)

Grams and kilograms are metric units of weight.* Recall that the prefix "kilo-" means 1000. Thus a kilogram is 1000 grams.

1000 grams = 1 kilogram

This book weighs about 1 kilogram, so one of these pages weighs more than a gram. A dollar bill weighs about 1 gram. Gram is abbreviated **g**. Kilogram is abbreviated **kg**.

---

*There is a technical difference between the terms "weight" and "mass" that will be clarified in other coursework. In this book we will use the word weight to include the meaning of both terms.

Example 1 Choose the more sensible measure.

(a) A pair of shoes
1 g   1 kg

(b) A chicken
3 g   3 kg

(c) A quarter
5 g   5 kg

Solution (a) **1 kg**   (b) **3 kg**   (c) **5 g**

Example 2 Jason's rabbit weighs 4 kilograms. Four kilograms is how many grams?

Solution Each kilogram is 1000 grams. Four kilograms is 4×1000 grams, which is **4000 grams**.

Practice Choose the more sensible measure in Problems (a)–(c).

a. Tennis ball
57 g   57 kg

b. Cat
6 g   6 kg

c. Bowling ball
7 g   7 kg

d. Seven kilograms is how many grams?

Problem set 87

1. Katie hit the ball seven hundred forty-two times on Monday, two hundred sixty-seven times on Tuesday, and eight hundred forty-seven times on Wednesday. How many times did she hit the ball in the 3 days?

2. Mickey bought apples at 5 cents per pound at the sale. She spent 95 cents. How many pounds of apples did she buy?

3. Bill placed 243 paint cans on the shelf. Ninety-five cans fell off during the earthquake. How many cans stayed on the shelf?

4. Pamela listened to half of a 90-minute tape. How many minutes of the tape did she hear?

5. One fourth of the guests gathered in the living room. What fraction of the guests did not gather in the living room?

6. Which of these arrows could be pointing to 2750?

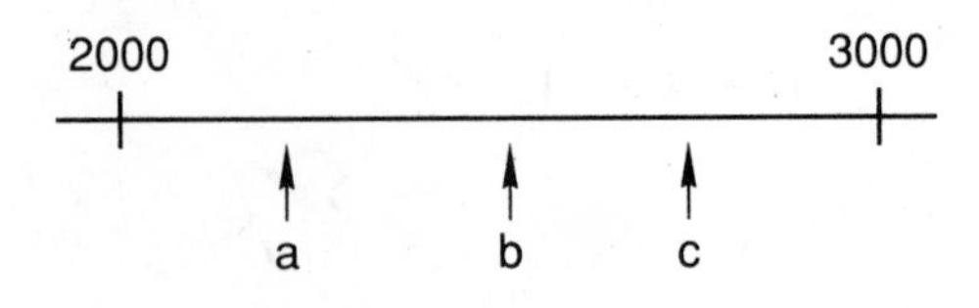

7. The five letters that are always vowels are a, e, i, o, and u. What fraction of the first 10 letters of the alphabet is vowels?

8. Debbie weighed 3 kilograms when she was born. How many grams is that?

9. Estimate the product of 396 and 7.

10. It is afternoon. What time was it 14 hours ago?

11. Compare:

$$\frac{3}{4} \bigcirc \frac{4}{5}$$

Draw and shade two rectangles to show the comparison.

12. $4329 + 157 + (6385 - 9)$

13. $17.54 + 49¢ + $15

14. $36,147 - 16,280$

15. $50.00 − $42.60

16. 398×6

17. 47×60

18. $8 \times \$6.25$

19. $2\overline{)567}$

20. $6\overline{)3456}$

21. $4\overline{)978}$

22. $970 \div 5$

23. $372 \div 3$

24. $491 \div 7$

25. $8 \times N = 120$

26. $F \times 9 = 108$

27.
$$\begin{array}{r} 6 \\ 5 \\ 3 \\ 8 \\ 7 \\ 1 \\ 6 \\ +\ 4 \\ \hline \end{array}$$

28.
$$\begin{array}{r} 7 \\ 8 \\ 5 \\ 4 \\ N \\ 2 \\ 7 \\ +\ 3 \\ \hline 54 \end{array}$$

29.
$$\begin{array}{r} 9 \\ 7 \\ 1 \\ 3 \\ N \\ 6 \\ 5 \\ +\ 2 \\ \hline 54 \end{array}$$

30.
$$\begin{array}{r} 8 \\ 2 \\ 5 \\ 6 \\ 4 \\ N \\ 2 \\ +\ 8 \\ \hline 54 \end{array}$$

LESSON **88** Tables

Speed Test: 90 Division Facts, Form A (Test I in Test Booklet)

We have studied graphs that present number information in picture form. Another way of presenting number information is in a **table**.

Example Use the information in the table to answer questions (a) and (b).

Height of Major Mountains

MOUNTAIN	FEET	METERS
Everest	29,028	8842
McKinley	20,320	6194
Kilimanjaro	19,340	5895
Matterhorn	14,691	4478
Pike's Peak	14,110	4301
Fuji	12,389	3776

(a) The Matterhorn is how many meters taller than Pike's Peak?

(b) Mount McKinley is how many feet taller than Mount Kilimanjaro?

Solution We compare the heights by subtracting.

(a) We use the numbers from the meters column.

Matterhorn	4478 m
Pike's Peak	− 4301 m
	177 m

(b) We use the numbers from the feet column.

McKinley	20,320 ft
Kilimanjaro	− 19,340 ft
	980 ft

Practice Use the information from the table given in the example to answer questions (a) and (b).

a. Mount Kilimanjaro is how many meters taller than Mount Fuji?

b. Mount Everest is how many feet taller than the Matterhorn?

Problem set 88 Use the information in this table to answer questions 1–3.

Yearly Average Rainfall

CITY	RAINFALL IN INCHES
Boston	43 in.
Chicago	35 in.
Denver	16 in.
Houston	49 in.
San Francisco	20 in.

1. Which cities listed in the table average less than 2 feet of rain per year?

2. One year Houston received 62 inches of rain. This was how much more than its yearly average?

3. About how many inches of rain would Denver receive in 3 average years?

4. One fifth of the 60 eggs were placed in each box. How many eggs were placed in each box?

60 eggs

5. Which of these arrows could be pointing to 2250?

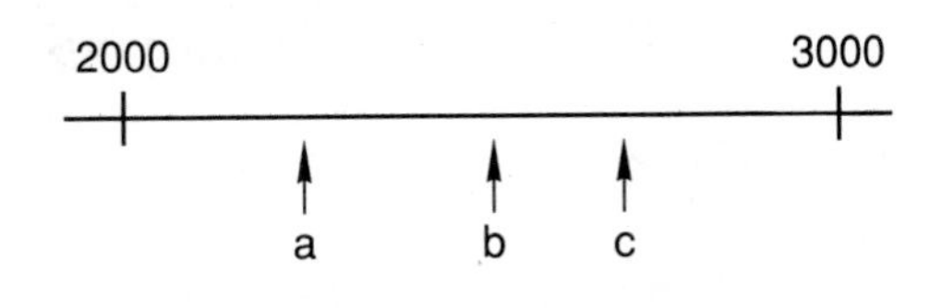

6. Estimate the sum of 427, 533, and 764.

7. What fraction of this set is not shaded?

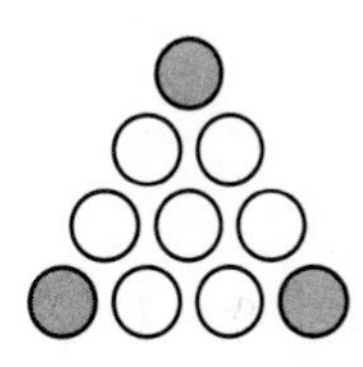

8. Four hundred forty-seven thousand, seven hundred sixty-three wore red. Two hundred forty-six thousand, seven hundred two did not wear red. How many were there in all?

9. Forty-two oranges could be shipped in 1 box. Luther had 30 boxes. How many oranges could he ship?

10. Only 5 apples would fit in one small box. If Roger had 145 apples, how many boxes did he need?

11. Gwenn could go through 140 in 20 minutes. How many could she go through in 1 minute?

12. Peanut found 42 every hour. How many did Peanut find in 7 hours?

13. $5 \times 75 = 75 + N$

14. $\begin{array}{r} 37{,}246 \\ +\ 159{,}087 \\ \hline \end{array}$

15. $\begin{array}{r} \$98.54 \\ +\ \$\ \ 9.85 \\ \hline \end{array}$

16. $\begin{array}{r} 135 \\ 2416 \\ 350 \\ 4139 \\ +\ \ \ 67 \\ \hline \end{array}$

17. $\begin{array}{r} 431{,}567 \\ -\ \ 17{,}295 \\ \hline \end{array}$

18. $\begin{array}{r} \$20.10 \\ -\ \$16.45 \\ \hline \end{array}$

19. 380×4

20. 97×80

21. $\$8.63 \times 7$

22. $5\overline{)3840}$

23. $8\overline{)7000}$

24. $6\overline{)3795}$

25. $376 \div 2$

26. $287 \div 7$

27. $416 \div 9$

28. $4 \times P = 160$

29. $\begin{array}{r} N \\ +\ 3219 \\ \hline 6507 \end{array}$

30. $\begin{array}{r} N \\ -\ 4129 \\ \hline 1009 \end{array}$

LESSON 89 Division with Zeros in Three-Digit Answers

> **Speed Test:** 90 Division Facts, Form A (Test I in Test Booklet)

Recall that the method we have used for dividing numbers has four steps.

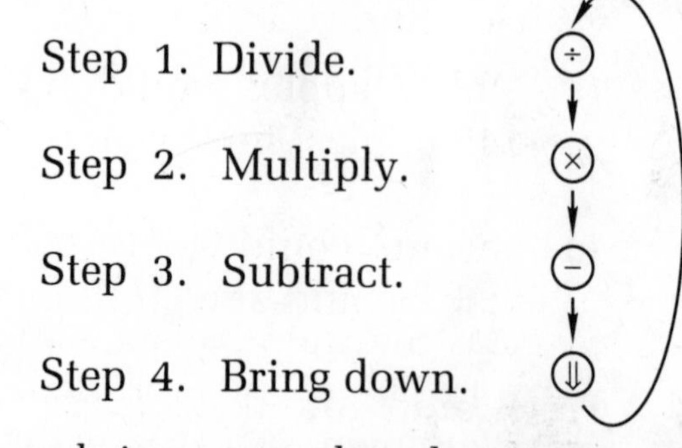

Every time we bring a number down, we return to step 1 and divide again. Sometimes the answer to the next division is zero, and we end up with a zero in the answer.

Example 1 $3\overline{)618}$

Solution Step 1. Divide $3\overline{)6}$ and write 2.
Step 2. Multiply 2×3 and write 6.
Step 3. Subtract 6 from 6 and write 0.
Step 4. Bring down the 1, making 01 (which is 1).

Repeat Step 1. Divide $3\overline{)1}$ and write 0.
Step 2. Multiply 0×3 and write 0.
Step 3. Subtract 0 from 1 and write 1.
Step 4. Bring down the 8, making 18.

Repeat Step 1. Divide $3\overline{)18}$ and write 6.
Step 2. Multiply 6×3 and write 18.
Step 3. Subtract 18 from 18 and write 0.
Step 4. There are no digits to bring down. The division is complete. The remainder is zero.

$$\begin{array}{r} \mathbf{206} \\ 3\overline{)618} \\ \underline{6} \\ 01 \\ \underline{0} \\ 18 \\ \underline{18} \\ 0 \end{array}$$

Example 2 $4\overline{)1483}$

Solution Step 1. Divide $4\overline{)14}$ and write 3.
Step 2. Multiply 3×4 and write 12.
Step 3. Subtract 12 from 14 and write 2.
Step 4. Bring down 8, making 28.

Repeat Step 1. Divide $4\overline{)28}$ and write 7.
Step 2. Multiply 7×4 and write 28.
Step 3. Subtract 28 from 28 and write 0.
Step 4. Bring down 3, making 03 (which is 3).

Repeat Step 1. Divide $4\overline{)03}$ and write 0.
Step 2. Multiply 0×4 and write 0.
Step 3. Subtract 0 from 3 and write 3.
Step 4. There are no digits to bring down. We are finished dividing. We write 3 as a remainder.

$$\begin{array}{r} \mathbf{370\ r\ 3} \\ 4\overline{)1483} \\ \underline{12} \\ 28 \\ \underline{28} \\ 03 \\ \underline{0} \\ 3 \end{array}$$

Practice

a. List the four steps of division and draw the division diagram.

b. $4\overline{)815}$

c. $5\overline{)4152}$

d. $6\overline{)5432}$

e. $7\overline{)845}$

Problem set 89

1. Thirty-six thousand, one hundred twelve is how much more than twenty-nine thousand, three hundred twenty?

2. Monty's time for running the race was 12 seconds less than Ivan's time. Monty's time was 58 seconds. What was Ivan's time?

3. The llama weighed 4 times as much as its 95-pound load. How much did the llama weigh?

4. How many 6-inch-long sticks can be cut from a 72-inch-long stick of sugar cane?

5. One fifth of the leaves had fallen. What fraction of the leaves had not fallen?

6. Which of these arrows could be pointing to 5263?

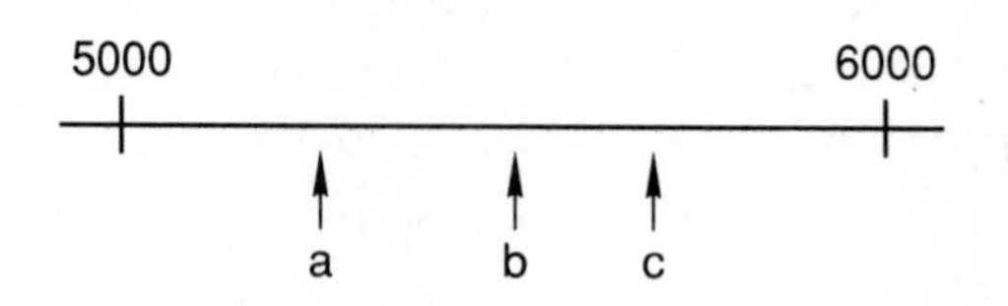

7. What fraction of the year is made up of months that have 31 days?

8. The prefix "kilo-" means what number?

9. Estimate the sum of 393, 589, and 241.

10. The sides of this triangle are equal in length. How many millimeters is the perimeter of the triangle?

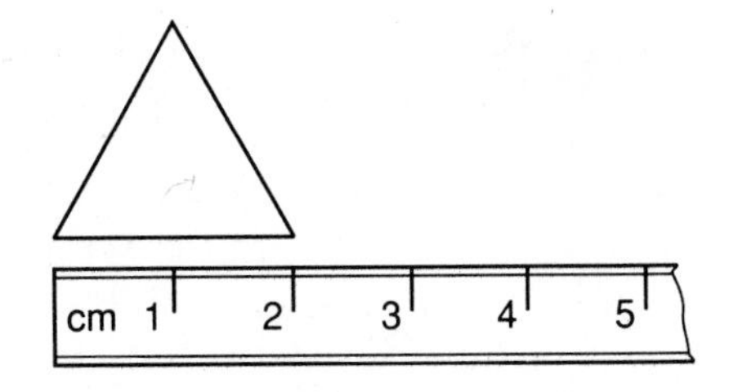

11. 3 liters equals how many milliliters?

12. Wilma could run 5 miles in 1 hour. How long would it take her to run 40 miles?

13. $\begin{array}{r} 103{,}279 \\ +\ 97{,}814 \\ \hline \end{array}$

14. $\begin{array}{r} \$36.14 \\ +\ \$27.95 \\ \hline \end{array}$

15. $\begin{array}{r} 3615 \\ 294 \\ 18 \\ 378 \\ +\ 4290 \\ \hline \end{array}$

16. $\begin{array}{r} 39{,}420 \\ -\ 29{,}516 \\ \hline \end{array}$

17. $\begin{array}{r} \$60.50 \\ -\ \$\quad N \\ \hline \$43.20 \end{array}$

18. $\begin{array}{r} 604 \\ \times\quad 9 \\ \hline \end{array}$

19. $\begin{array}{r} 87 \\ \times\ 60 \\ \hline \end{array}$

20. $\begin{array}{r} \$6.75 \\ \times\quad 4 \\ \hline \end{array}$

21. $3\overline{)618}$

22. $5\overline{)2026}$

23. $\begin{array}{r} N \\ +\ 1467 \\ \hline 2459 \end{array}$

24. $600 \div 4$

25. $543 \div 6$

26. $472 \div 8$

27. $\begin{array}{r} 4 \\ 2 \\ 3 \\ N \\ 7 \\ +\ 2 \\ \hline 27 \end{array}$

28. $\begin{array}{r} 5 \\ 2 \\ 3 \\ 1 \\ 4 \\ +\ N \\ \hline 27 \end{array}$

29. $9 \times W = 81 + 18$

30. $N \times 2 = 150$

LESSON 90 Rounding to the Nearest Thousand

Speed Test: 90 Division Facts, Form A (Test I in Test Booklet)

When we round a number to the nearest ten, we get a multiple of 10. The multiples of 10 are the numbers we get when we count by tens. **The multiples of 10 have a zero as the last digit.**

10, 20, 30, 40, 50, 60, ...

When we round a number to the nearest hundred, we get a multiple of 100. The multiples of 100 are the numbers we get when we count by hundreds. **The multiples of 100 have zeros as the last two digits.**

100, 200, 300, 400, 500, 600, ...

When we round a number to the nearest thousand, we get a multiple of 1000. The multiples of 1000 are the numbers we get when we count by thousands. **The multiples of 1000 have zeros as the last three digits.**

1000, 2000, 3000, 4000, 5000, 6000, ...

We will use a number line to help us understand rounding to the nearest thousand.

Example Round 2781 to the nearest thousand.

Solution The number 2781 is between the thousand numbers 2000 and 3000. Halfway between 2000 and 3000 is 2500. As we see on the number line below, 2781 is nearer 3000 than 2000.

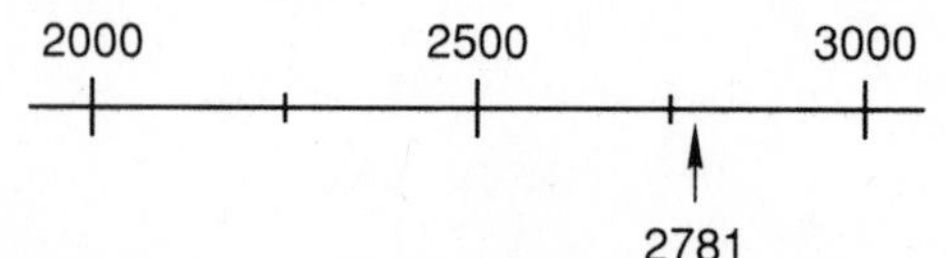

Rounding 2781 to the nearest thousand gives us **3000**.

Practice Round each number to the nearest thousand.

a. 5263 **b.** 4986 **c.** 7814 **d.** 8176

Problem set 90

1. Cecilia skated 27 times around the rink forward and 33 times around the rink backward. In all, how many times did she skate around the rink?

2. Nectarines cost 68¢ per pound. What is the price for 3 pounds of nectarines?

3. In bowling, the sum of Amber's score and Beth's score was equal to Sarah's score. If Sarah's score was 113 and Beth's score was 55, what was Amber's score?

4. One third of the 84 students were assigned to each room. How many students were assigned to each room?

Students

5. Round 2250 to the nearest thousand.

6. What fraction of the letters in ARIZONA are not A's?

7. The African elephant weighed 7 tons. How many pounds is that?

8. The tip of this shoelace is how many millimeters long?

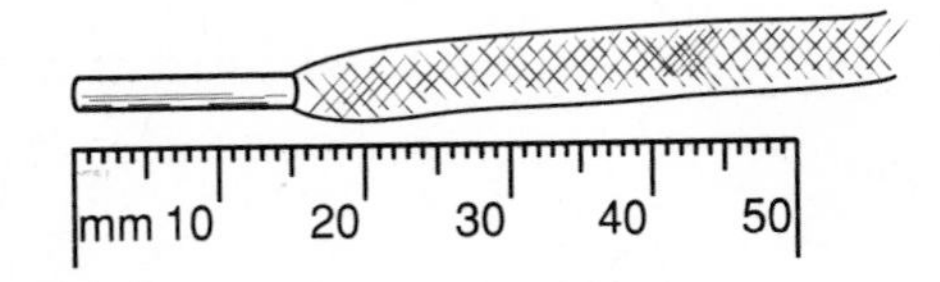

9. Pick the more reasonable measure:
(a) Box of cereal: 400 g or 400 kg
(b) Pail of water: 10 mL or 10 liters

10. According to this calendar, what is the date of the last Tuesday in February, 2019?

FEBRUARY 2019

S	M	T	W	T	F	S
					1	2
3	4	5	6	7	8	9
10	11	12	13	14	15	16
17	18	19	20	21	22	23
24	25	26	27	28		

11. Forty-two thousand, seven hundred twenty-four is how much greater than thirty-four thousand, nine hundred eighty-seven?

12. How much time is it from 7:05 a.m. to 4:57 p.m.?

13. Michelle could pack 75 packages in 3 hours. How many packages could she pack in 1 hour?

14. $6743 - (507 \times 6)$

15. $70.00 − $63.17

16. $3 \times 7 \times 0$

17. $8.15 × 6

18. 67×40

19. 43,162 + 5,917

20. $2\overline{)1216}$

21. $6\overline{)4321}$

22. $8\overline{)4537}$

23. $963 \div 3$

24. $254 \div 5$

25. $657 \div 9$

26. $200 = 4 \times B$

27. $D \times 7 = 105$

28.
$$\begin{array}{r} 473 \\ 286 \\ + \ N \\ \hline 943 \end{array}$$

29.
$$\begin{array}{r} 1 \\ 12 \\ 3 \\ 14 \\ 5 \\ +\ 26 \\ \hline \end{array}$$

30.
$$\begin{array}{r} 2 \\ 33 \\ 4 \\ 25 \\ 6 \\ +\ 27 \\ \hline \end{array}$$

LESSON 91 Line Graphs

Speed Test: 90 Division Facts, Form A (Test I in Test Booklet)

If we connect points on a graph with lines, we can make a

line graph. Line graphs are often used to show how something has changed over a period of time.

Example Sean kept a record of his test scores on a line graph. According to his line graph, how many correct answers did he have on Test 3?

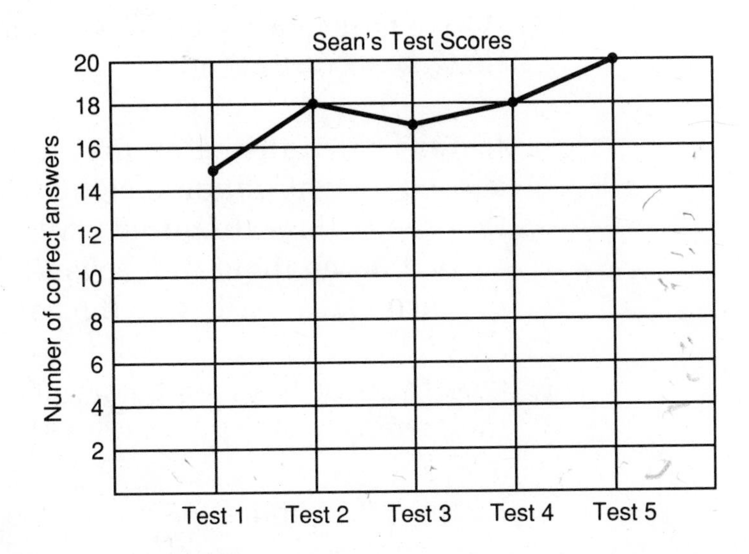

Solution This line graph shows at a glance whether Sean's scores are going up or going down. To read this graph, we compare the level of a point with the scale on the left. The dot above Test 3 is halfway between 16 and 18 on the scale. This means that Sean's score on Test 3 was **17 correct answers**.

Practice Use the graph of Sean's test scores from the example to answer each question.

a. From Test 1 to Test 2, Sean's score improved by how many answers?

b. There are 20 questions on a test. How many questions did Sean miss on Test 4?

c. On which test did Sean have a perfect score?

d. On Sherry's first five tests her scores were 18, 17, 20, 18, and 17. Draw a line graph that shows this information.

Problem set 91

1. There were 35 students in the class but only 28 math books. How many more math books did they need?

2. Each of the 7 children slid down the water slide 11 times. How many times did they slide in all?

3. A bowling lane is 60 feet long. How many yards is that?

4. Justin carried the baton four hundred forty yards. Eric carried it eight hundred eighty yards. Joe carried it one thousand, three hundred twenty yards, and Braulio carried it one thousand, seven hundred sixty yards. In all, how many yards was the baton carried?

5. One third of the members voted no. What fraction of the members did not vote no?

6. Round 6821 and 4963 to the nearest thousand. Then add the rounded numbers.

7. What fraction of the days of the week start with the letter S?

8. Together Bob's shoes weigh about 1 kilogram. Each shoe weighs about how many grams?

Use the information in this line graph to answer questions 9 and 10.

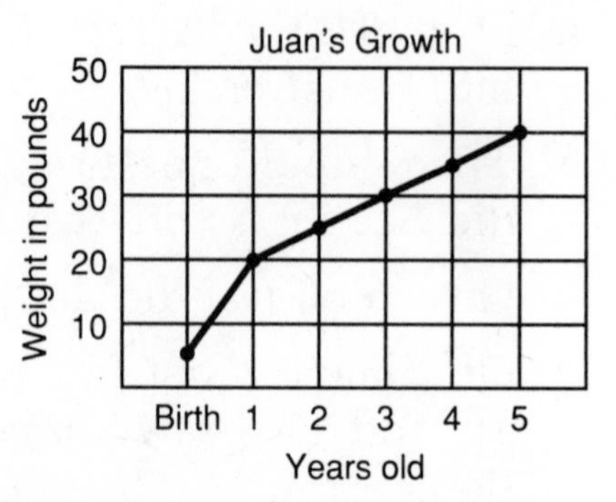

9. About how many pounds did Juan weigh on his second birthday?

10. About how many pounds did Juan gain between his third and fifth birthdays?

11. Julia could put 7 balls in 1 bucket. She had 2800 balls. How many buckets did she need to hold them all?

12. Mary could knit 2 sweaters in 1 month. How many months would it take her to knit 42 sweaters?

13. $\begin{array}{r} \$60.75 \\ +\ \$95.75 \\ \hline \end{array}$

14. $\begin{array}{r} \$16.00 \\ -\ \$15.43 \\ \hline \end{array}$

15. $\begin{array}{r} \$63.14 \\ 2.75 \\ 39.81 \\ 4.62 \\ +\ 0.87 \\ \hline \end{array}$

16. $\begin{array}{r} 96{,}376 \\ -\ 89{,}382 \\ \hline \end{array}$

17. $\begin{array}{r} 320 \\ \times\ 30 \\ \hline \end{array}$

18. $\begin{array}{r} 465 \\ \times\ 7 \\ \hline \end{array}$

19. $\begin{array}{r} \$0.98 \\ \times\ 6 \\ \hline \end{array}$

20. $4\overline{)563}$

21. $9\overline{)2714}$

22. $7\overline{)2800}$

23. $425 \div 6$

24. $600 \div 8$

25. $625 \div 5$

26. $3 \times R = 150$

27. $T \times 6 = 150$

28. $\begin{array}{r} 1 \\ 7 \\ 2 \\ 6 \\ 9 \\ 4 \\ +\ N \\ \hline 37 \end{array}$

29. $\begin{array}{r} 5 \\ 2 \\ N \\ 7 \\ 8 \\ 5 \\ +\ 4 \\ \hline 37 \end{array}$

30. $\begin{array}{r} 9 \\ 8 \\ 7 \\ 6 \\ 8 \\ 5 \\ +\ N \\ \hline 48 \end{array}$

LESSON 92

Dividing Money

Speed Test: 90 Division Facts, Form A (Test I in Test Booklet)

To divide dollars and cents by a whole number, we divide the digits just like we divide whole numbers. **The decimal point in the answer is placed directly above the decimal point inside the division box.** We write a dollar sign in front of the answer.

Example \$8.40 ÷ 3

Solution We do not need to write the dollar sign in the division box. We put a decimal point in the answer directly above the decimal point in the problem. Then we divide. Then we put a dollar sign in front of the answer.

$$3\overline{)8.40}$$ ← decimal point (the decimal point is placed above the division box)

$$\begin{array}{r} 2.80 \\ 3\overline{)8.40} \\ \underline{6} \\ 2\,4 \\ \underline{2\,4} \\ 00 \\ \underline{0} \end{array}$$

\$2.80
↑
dollar sign

Practice

a. \$4.65 ÷ 3 **b.** \$9.40 ÷ 4 **c.** \$10.08 ÷ 9

d. \$4.80 ÷ 5 **e.** \$15.40 ÷ 7 **f.** \$23.58 ÷ 6

Problem set 92

1. Ali Baba brought home 30 bags. Each bag contained 320 gold coins. How many coins were there in all?

2. The movie was 3 hours long. If it started at 11:10 a.m., at what time did it end?

3. Tony is reading a 212-page book. If he has finished page 135, how many pages does he still have to read?

4. Brad, Jan, and Jordan each scored one third of the team's 42 points. They each scored how many points?

Points

5. Round 4286 to the nearest thousand.

6. Four hundred forty-nine thousand, seven hundred thirty-three is how much greater than two hundred seventeen thousand, nine hundred sixty-five?

7. What fraction of the letters in the following word is made up of I's?

ANTIDISESTABLISHMENTARIANISM

Use the information given to answer questions 8–10.

In the first 8 games of this season, the Rio Hondo football team won 6 games and lost 2 games. They won their next game by a score of 24 to 20. The team plays 12 games in all.

8. In the first 9 games of the season, how many games did Rio Hondo win?

9. Rio Hondo won its ninth game by how many points?

10. What is the greatest number of games Rio Hondo could win this season?

11. Compare: $3 \times 4 \times 5 \bigcirc 5 \times 4 \times 3$

12.
$$\begin{array}{r} XY{,}ZPM \\ -\ 13{,}728 \\ \hline 25{,}723 \end{array}$$

13.
$$\begin{array}{r} N \\ +\ 95{,}486 \\ \hline 568{,}372 \end{array}$$

14.
$$\begin{array}{r} \$\ 8.53 \\ 12.47 \\ 5.25 \\ 0.67 \\ +\ 14.37 \\ \hline \end{array}$$

15.
$$\begin{array}{r} 101{,}482 \\ -\quad N \\ \hline 3{,}134 \end{array}$$

16.
$$\begin{array}{r} \$57.84 \\ -\ \$19.92 \\ \hline \end{array}$$

17. $\begin{array}{r} 638 \\ \times\ 50 \\ \hline \end{array}$

18. $\begin{array}{r} 472 \\ \times\ 9 \\ \hline \end{array}$

19. $\begin{array}{r} \$6.09 \\ \times\ 6 \\ \hline \end{array}$

20. $3\overline{)921}$

21. $5\overline{)678}$

22. $4\overline{)2401}$

23. $\$12.60 \div 5$

24. $\$14.34 \div 6$

25. $\$46.00 \div 8$

26. $234 = 9 \times N$

27. $5 \times W = 500$

28. $\begin{array}{r} 5 \\ 4 \\ 3 \\ A \\ 4 \\ 7 \\ 6 \\ 5 \\ +\ 9 \\ \hline 62 \end{array}$

29. $\begin{array}{r} 3 \\ 5 \\ 6 \\ 7 \\ 8 \\ 6 \\ +\ N \\ \hline 43 \end{array}$

30. $\begin{array}{r} 7 \\ 9 \\ 3 \\ 2 \\ 4 \\ N \\ +\ 6 \\ \hline 43 \end{array}$

LESSON 93 Illustrating Fraction-of-a-Group Problems

Special Test: 64 Multiplication Facts (Test 6 in Test Booklet)

Drawing a picture of fraction-of-a-group problems can help us understand these problems. We will illustrate the three examples from Lesson 79.

Example 1 Illustrate this fraction-of-a-group statement:

One half of the carrot seeds sprouted. There were 84 carrot seeds in all.

Solution We draw a rectangle for the whole group. One half of the seeds sprouted, so we divide the rectangle in half and divide 84 in half: 84 ÷ 2 = 42. We write 42 in each half. One half sprouted and one half did not sprout. We write that information in front of each half of the box. We have finished the illustration.

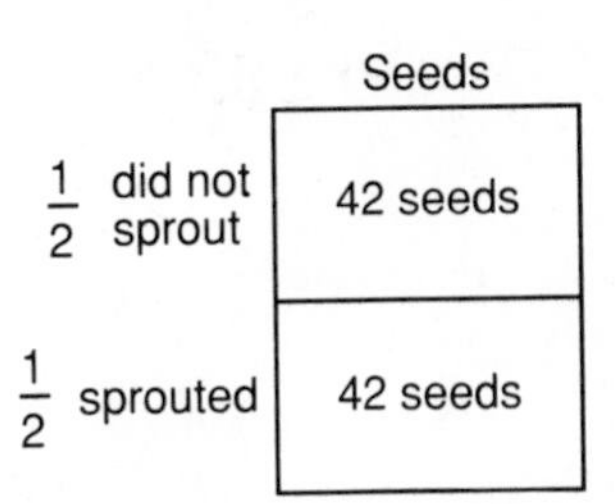

Example 2 Illustrate this fraction-of-a-group statement:

One third of the 27 students earned an A on the test.

Solution We draw a rectangle for the whole group. One third earned an A so we divide the rectangle into thirds and divide 27 into thirds: 27 ÷ 3 = 9. One third earned an A so the rest did not. The rest is $\frac{2}{3}$. We write "$\frac{1}{3}$ earned an A" and "$\frac{2}{3}$ did not earn an A."

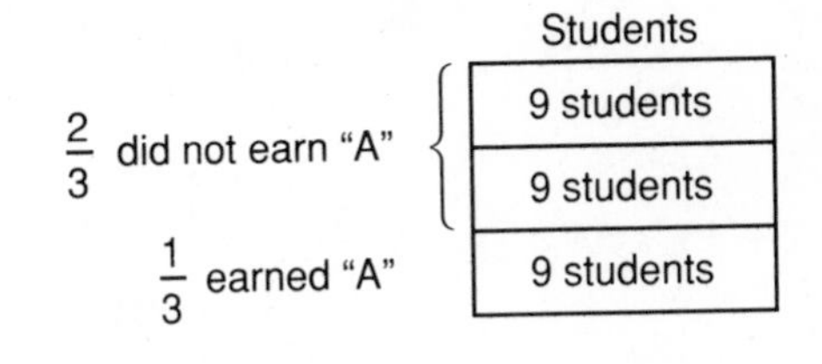

Example 3 Illustrate this fraction-of-a-group statement:

One fourth of the team's 32 points were scored by Cliff.

Solution We draw a rectangle for all the points. One fourth of the points were scored by Cliff, so we divide the rectangle into fourths and divide 32 into fourths. Since 32 ÷ 4 = 8, we write "8 points" in each fourth. Since $\frac{1}{4}$ of the points were scored by Cliff, $\frac{3}{4}$ of the points were not scored by Cliff. We write "$\frac{1}{4}$ scored by Cliff" in front of one box. We write "$\frac{3}{4}$ not scored by Cliff" in front of the remaining three boxes.

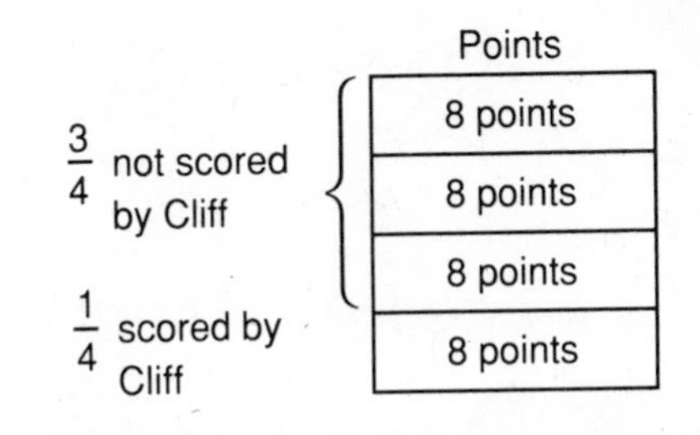

Practice Illustrate each fraction-of-a-group statement.

a. One half of the 28 students earned an A.

b. One third of the 12 lights were on.

c. One fourth of the 60 gummy bears were green.

d. One fifth of the 30 bicycles were red.

Problem set 93

1. If it is not a leap year, what is the total number of days in January, February, and March?

2. The shoemaker's wife made each of the 12 elves a pair of pants and 2 shirts. How many pieces of clothing did she make?

3. John did seven more chin-ups than Larry did. If John did eighteen chin-ups, how many chin-ups did Larry do?

4. Jan drove 200 miles on 8 gallons of gas. Her car averaged how many miles on each gallon of gas?

5. How much is one third of one hundred seventy-four?

6. The tally for 8 is ~~////~~ ///. What is the tally for 9?

7. If each side of an octagon is 1 centimeter, what is its perimeter in millimeters?

8. $3 + 5 + N = 7 \times 8$

9. Illustrate this fraction-of-a-group statement: One third of the 18 marbles were cat's-eyes.

10. Robert picked 476 peaches in 1 day. How many peaches could he pick in 6 days?

11. Mary picked 3444 peaches in 7 days. How many peaches could she pick in 1 day?

12. $\begin{array}{r} ABC,DEF \\ +\ 27,415 \\ \hline 271,052 \end{array}$

13. $\begin{array}{r} \$96.47 \\ +\ \$\ 5.85 \\ \hline \end{array}$

14. $\begin{array}{r} 35 \\ 45 \\ 25 \\ 65 \\ 115 \\ 20 \\ +\ 25 \\ \hline \end{array}$

15. $\begin{array}{r} 46,216 \\ -\ 7,515 \\ \hline \end{array}$

16. $\begin{array}{r} \$39.24 \\ -\ \$17.58 \\ \hline \end{array}$

17. $3 \times 6 \times 9$

18. 462×3

19. $\begin{array}{r} 36 \\ \times\ 50 \\ \hline \end{array}$

20. $\begin{array}{r} 476 \\ \times\ 7 \\ \hline \end{array}$

21. $4\overline{)524}$

22. $6\overline{)4216}$

23. $5\overline{)\$26.30}$

24. $\$3.70 \div 2$

25. $786 \div 3$

26. $4902 \div 7$

27. $\begin{array}{r} 4 \\ 3 \\ 2 \\ 7 \\ 6 \\ 8 \\ +\ N \\ \hline 47 \end{array}$

28. $Y \times 8 = 400$

29. $5 \times G = 400$

30. $\begin{array}{r} 4163 \\ -\ N \\ \hline 2876 \end{array}$

LESSON 94 Multiplying by Tens, Hundreds, Thousands

Speed Test: 64 Multiplication Facts (Test G in Test Booklet)

When we multiply a whole number by 10, the answer has the same digits as the number that was multiplied and the answer has an additional zero at the end.

$$\begin{array}{r} 123 \\ \times\ 10 \\ \hline 1230 \end{array}$$

When we multiply a whole number by 100, there are two additional zeros at the end.

$$\begin{array}{r} 123 \\ \times \quad 100 \\ \hline 12{,}300 \end{array}$$

When we multiply a whole number by 1000, there are three additional zeros at the end.

$$\begin{array}{r} 123 \\ \times \quad 1{,}000 \\ \hline 123{,}000 \end{array}$$

We can use this pattern to multiply mentally by 10 or 100 or 1000.

Example Multiply mentally:
(a) 37×10 (b) 612×100 (c) 45×1000

Solution (a) We write 37 with one zero at the end: **370**

(b) We write 612 with two zeros at the end: **61,200**

(c) We write 45 with three zeros at the end: **45,000**

Practice Multiply mentally.

a. 365×10 **b.** 52×100 **c.** 7×1000

d. 360×10 **e.** 420×100 **f.** 250×1000

Problem set 94 Use the information in the graph to answer questions 1–3.

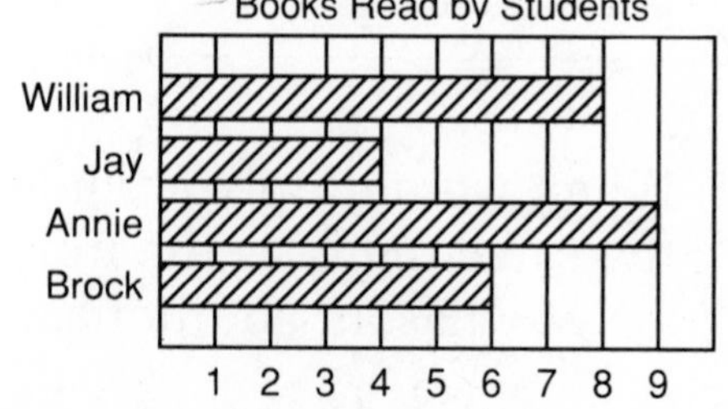

1. Which student has read exactly twice as many books as Jay?

2. Brock's goal is to read 10 books. How many more books does he need to read to reach his goal?

3. If the books Annie has read have an average of 160 pages each, how many pages has she read?

4. Jim saw some pentagons. The pentagons had a total of 100 sides. How many pentagons did Jim see?

5. Four hundred seventy-five thousand, nine hundred forty-two soldiers were in the first army. Nine hundred thirty-one thousand, five hundred sixty-seven soldiers were in the first two armies. How many soldiers were in the second army?

6. What is the sum of the fifth multiple of 10, the third multiple of 8, and the fourth multiple of 7?

7. A pitcher of orange juice is about
(a) 2 ounces (b) 2 liters (c) 2 gallons

8. Draw a triangle so that two sides are perpendicular.

9. Illustrate this fraction-of-a-group statement: One fourth of the 48 gems were rubies.

10. Jan started the trip in the morning. It was dark when she stopped. How long did she travel?

11. $\begin{array}{r} 463{,}271 \\ +\ 175{,}349 \\ \hline \end{array}$

12. $\begin{array}{r} 728 \\ +\ ABC \\ \hline 1205 \end{array}$

13. $\begin{array}{r} 68{,}418 \\ -\ 47{,}615 \\ \hline \end{array}$

14. $\begin{array}{r} \$30.00 \\ -\ \$14.75 \\ \hline \end{array}$

15. 36×10 **16.** 100×42 **17.** 27×1000

18. $\begin{array}{r} 317 \\ \times \quad 4 \\ \hline \end{array}$ **19.** $\begin{array}{r} 206 \\ \times \quad 5 \\ \hline \end{array}$ **20.** $\begin{array}{r} 37 \\ \times \ 40 \\ \hline \end{array}$

21. $3\overline{)492}$ **22.** $5\overline{)860}$ **23.** $6\overline{)\$9.30}$

24. $168 \div 8$ **25.** $\$20.00 \div 8$ **26.** $2315 \div 4$

27. $8 \times Z = 200$ **28.** $N \times 4 = 200$

29. $\begin{array}{r} 4 \\ 7 \\ 8 \\ 6 \\ 5 \\ 9 \\ +\,N \\ \hline 47 \end{array}$ **30.** $\begin{array}{r} 2 \\ 8 \\ 7 \\ 3 \\ 5 \\ 4 \\ +\,N \\ \hline 36 \end{array}$

LESSON 95 Multiplying Round Numbers Mentally

Speed Test: 64 Multiplication Facts (Test G in Test Booklet)

If we have memorized the multiplication facts, we can multiply round numbers "in our head." To do this, we multiply the first digits and count zeros. Study the multiplication below.

$$\begin{array}{r} 40 \\ \times\ 30 \\ \hline 1200 \end{array}$$

40 and 30: two zeros

1200: 4×3 (the 12) and two zeros

We can find the product of 40 and 30 by multiplying 4 and 3 and then attaching two zeros.

Example 1 Multiply mentally: 60×80

Solution We multiply 6 times 8 and get 48. There is 1 zero in 60 and 1 zero in 80. We attach 2 zeros to 48 and get 4800.

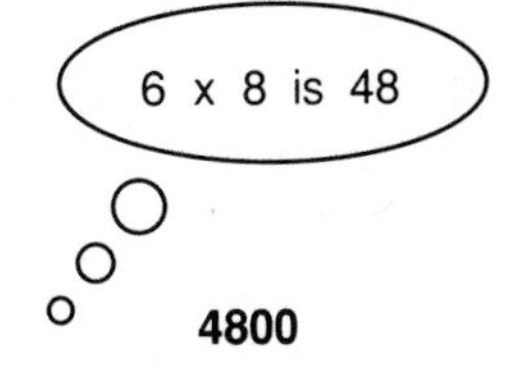

Example 2 Multiply mentally: 30×700

Solution We multiply 3 times 7 and get 21. There are 3 zeros in the problem. We attach 3 zeros to 21 and get 21,000.

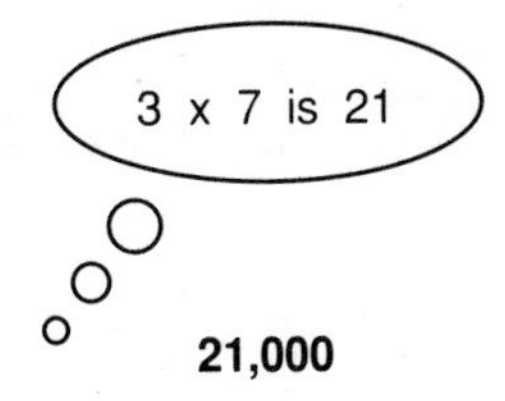

Example 3 Multiply mentally: 400×700.

Solution Four times 7 is 28. We attach 4 zeros and get 280,000.

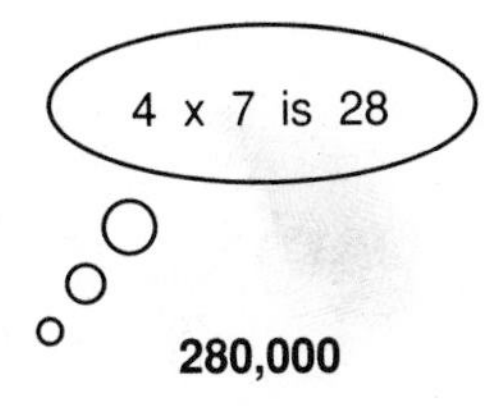

Practice Multiply mentally.

a. 70×80 **b.** 40×50 **c.** 40×600 **d.** 30×800

Problem set 95

1. It takes Jennifer 20 minutes to walk to school. At what time should she start for school if she wants to arrive at 8:10 a.m.?

2. Before her haircut Rapunzel weighed 125 pounds. After her haircut she weighed 118 pounds. What was the weight of her hair?

3. Lucy bought a hamburger for \$2.89, fries for \$0.89, and a drink for 79¢. How much did she pay?

4. Find N if $23 + 12 + N = 80 \times 2$.

5. According to this calendar, October 30, 1902, was what day of the week?

OCTOBER						1902
S	M	T	W	T	F	S
						1
2	3	4	5	6	7	8
9	10	11	12	13	14	15
16	17	18	19	20	21	22
23	24	25	26	27	28	29
30	31					

6. The tally for 16 is ~~||||~~ ~~||||~~ ~~||||~~ /. What is the tally for 17?

7. Round three thousand, seven hundred eighty-two to the nearest thousand.

8. The limousine weighed 2 tons. How many pounds is 2 tons?

9. Illustrate this fraction-of-a-group statement: One fifth of the 45 horses were pintos.

10. A few were present at 9 a.m. Then four thousand, seven hundred eighty-two more came. Now there were six thousand, two present. How many were present at 9 a.m.?

11. Debby could run 21 laps in 3 hours. How many laps could she run in 1 hour?

12. Mick could run 42 laps in 1 hour. How many laps could he run in 5 hours?

13. $\begin{array}{r} 17,414 \\ +\ 89,875 \\ \hline \end{array}$

14. $\begin{array}{r} \$36.47 \\ +\ \$\ 9.68 \\ \hline \end{array}$

15. $\begin{array}{r} 6 \\ 8 \\ 17 \\ 23 \\ 110 \\ 25 \\ +\ 104 \\ \hline \end{array}$

16. $\begin{array}{r} 31,425 \\ -\ 17,633 \\ \hline \end{array}$

17. $\begin{array}{r} \$30.00 \\ -\ \$13.45 \\ \hline \end{array}$

18. $\begin{array}{r} 476 \\ \times\ \ 7 \\ \hline \end{array}$

19. $\begin{array}{r} 804 \\ \times\ \ 5 \\ \hline \end{array}$

20. 100×23 **21.** 60×30 **22.** 70×200

23. $3\overline{)\$6.27}$ **24.** $7\overline{)820}$ **25.** $6\overline{)333}$

26. $625 \div 5$ **27.** $4334 \div 8$ **28.** $1370 \div 2$

29. $7 \times C = 210$ **30.** $M \times 3 = 210$

LESSON 96 Multiplying Two Two-Digit Numbers, Part 1

Speed Test: 64 Multiplication Facts (Test G in Test Booklet)

When we multiply by a two-digit number, we take three steps. First we multiply by the ones' digit. Next we multiply by the tens' digit. Then we add the products together. To multiply 34 by 12, we multiply 34 by 2 and multiply 34 by 10. Then we add.

$$\begin{array}{rl} 34 \times 2 = 68 & \text{product} \\ 34 \times 10 = 340 & \text{product} \\ \hline 34 \times 12 = 408 & \text{total} \end{array}$$

It is easier to write the numbers one above the other when we multiply.

$$\begin{array}{r} 34 \\ \times\ 12 \\ \hline \end{array}$$

First we multiply 34 by 2 and write the answer with the last digit in the ones' column.

$$\begin{array}{r} 34 \\ \times\ 12 \\ \hline 68 \end{array}$$

Next we multiply 34 by 1. **We record the last digit of this multiplication in the tens' column, as we show on the left.** Then we add the results of the two multiplications and get 408.

$$\begin{array}{r} 34 \\ \times\ 12 \\ \hline 68 \\ 34 \\ \hline \mathbf{408} \end{array} \qquad \begin{array}{r} 34 \\ \times\ 12 \\ \hline 68 \\ 340 \\ \hline \mathbf{408} \end{array}$$

On the right, we show another way. We put the zero after the 34 to show that 10 times 34 is 340. Then we add. We get the same answer both ways.

Example Multiply: $\begin{array}{r} 31 \\ \times\ 23 \\ \hline \end{array}$

Solution We multiply 31 × 3 and write the last digit of the product in the ones' column.

$$\begin{array}{r} 31 \\ \times\ 23 \\ \hline 93 \end{array}$$

Now we multiply 31 × 2. Since this 2 is in the tens' column we write the last digit of the product in the tens' column. Then we add.

$$\begin{array}{r} 31 \\ \times\ 23 \\ \hline 93 \\ 62 \\ \hline \mathbf{713} \end{array} \quad \text{or} \quad \begin{array}{r} 31 \\ \times\ 23 \\ \hline 93 \\ 620 \\ \hline \mathbf{713} \end{array}$$

Practice

a. $\begin{array}{r} 32 \\ \times\ 23 \\ \hline \end{array}$ **b.** $\begin{array}{r} 24 \\ \times\ 32 \\ \hline \end{array}$ **c.** $\begin{array}{r} 43 \\ \times\ 12 \\ \hline \end{array}$ **d.** $\begin{array}{r} 34 \\ \times\ 21 \\ \hline \end{array}$

e. $\begin{array}{r} 32 \\ \times\ 32 \\ \hline \end{array}$ **f.** $\begin{array}{r} 23 \\ \times\ 14 \\ \hline \end{array}$ **g.** $\begin{array}{r} 13 \\ \times\ 32 \\ \hline \end{array}$ **h.** $\begin{array}{r} 33 \\ \times\ 33 \\ \hline \end{array}$

Problem set 96

Use the information given to answer questions 1–3.

Freeman rode his bike 2 miles from his house to Daniel's house. Together they rode 4 miles to the lake. Daniel caught 8 fish. At 3:30 p.m., they rode back to Daniel's house. Then Freeman rode home.

1. Altogether, how far did Freeman ride his bike?

2. It took Freeman an hour and a half to get home from the lake. At what time did he get home?

3. Daniel caught twice as many fish as Freeman. How many fish did Freeman catch?

4. Eighty-eight horseshoes are enough to shoe how many horses?

5. Estimate the sum of 4876 and 3149. Round each number to the nearest thousand before adding.

6. This is the tally for what number? 𝍸 𝍸 ////

7. What is the perimeter of a pentagon if each side is 20 centimeters long?

8. Find the length of this segment to the nearest quarter inch.

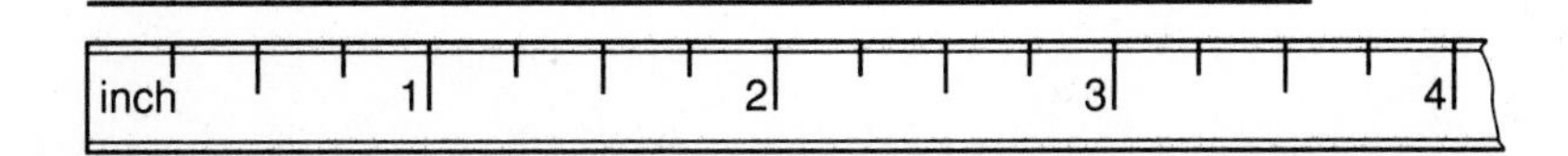

9. Illustrate this fraction-of-a-group statement: One half of the 18 players were on the field.

10. 363 + 429 + (572 − 260) **11.** \$2.58 + \$98.95

12. 21,316 − 14,141 **13.** \$20 − 20¢

14. 31×21 **15.** 32×31 **16.** 14×32

17. $\begin{array}{r} 11 \\ \times\ 11 \\ \hline \end{array}$

18. $\begin{array}{r} 12 \\ \times\ 14 \\ \hline \end{array}$

19. 30×800

20. $7\overline{)1000}$

21. $3\overline{)477}$

22. $5\overline{)2535}$

23. $\$64.80 \div 9$

24. $716 \div 4$

25. $352 \div 8$

26. $M \times 4 = 112$

27. $5 \times N = 240$

28. $\begin{array}{r} 4 \\ 7 \\ 3 \\ 5 \\ 2 \\ 5 \\ 9 \\ 4 \\ +\ 3 \\ \hline \end{array}$

29. $\begin{array}{r} 478 \\ -\ 299 \\ \hline \end{array}$

30. $\begin{array}{r} N \\ -\ 478 \\ \hline 643 \end{array}$

LESSON 97 Division Word Problems with Remainders

Speed Test: 64 Multiplication Facts (Test G in Test Booklet)

We have been practicing division word problems in which there are no remainders. In this lesson we will begin practicing division word problems in which there are remainders. These problems should be read with special care so that we know exactly what the question is.

Example The packer placed 100 bottles into boxes that held 6 bottles each.
(a) How many boxes were **filled?**
(b) How many bottles were **left over?**
(c) How many boxes were needed to hold **all the bottles?**

Solution These three questions have three different answers. To answer all three questions, we begin by dividing 100 by 6.

$$\begin{array}{r} 16 \text{ r } 4 \\ 6\overline{)100} \\ \underline{6} \\ 40 \\ \underline{36} \\ 4 \end{array}$$

This means that the 100 bottles can be separated into 16 groups of 6. Then there will be 4 extra bottles.

(a) The result 16 r 4 means that **16 boxes can be filled**, but there are still 4 more bottles.

(b) The result 16 r 4 means that after filling 16 boxes, there are still **4 bottles left over**.

(c) The result 16 r 4 means that after filling 16 boxes, there are still 4 more bottles. These 4 bottles will not fill another box, but another box is needed to hold them. Thus in order to hold all the bottles, **17 boxes are needed.**

Practice Tomorrow 32 students are going on a field trip. Each car can carry 5 students.

a. How many cars can be filled?

b. How many cars will be needed?

Rafik found 31 quarters in his bank. He made stacks of 4 quarters each.

c. How many stacks of 4 quarters did he make?

d. How many extra quarters did he have?

e. If he made a "short stack" with the extra quarters, how many stacks were there in all?

Problem set 97

1. Peter packed 6 ping-pong balls in each package. There were 100 ping-pong balls to pack.
(a) How many packages could he fill?
(b) How many ping-pong balls were left over?

2. What number is one hundred twenty-three less than three hundred twenty-one?

3. The llama traveled 684 kilometers through the desert in 4 months. How far did the llama travel in 1 month?

4. Twenty-four inches is how many feet?

5. What is the length of segment YZ in millimeters? In centimeters?

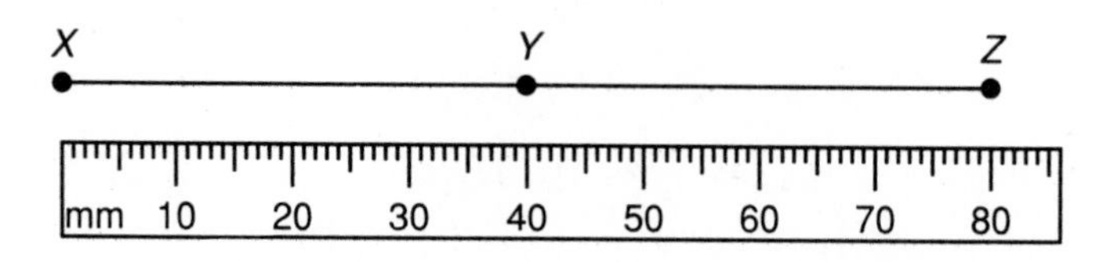

6. Find N if $14 + N + 28 + 4 = 7 \times 7$.

7. It is morning. What time will it be 5 hours and 20 minutes from now?

8. Write the number 217,528 in expanded form. Then use words to write the number.

9. Illustrate this fraction-of-a-group statement: One fifth of the 25 band members missed the note.

10. $\begin{array}{r} 49{,}249 \\ +\ XYZ{,}MPQ \\ \hline 222{,}527 \end{array}$

11. $\$6.35 + \$14.25 + \$0.97 + \$5$

12. $16{,}456 - 7638$

13. $\$10.00 - (46¢ + \$1.30)$

14. 28×1000

15. $\begin{array}{r} 13 \\ \times\ 13 \\ \hline \end{array}$

16. $\begin{array}{r} 12 \\ \times\ 11 \\ \hline \end{array}$

17. $\begin{array}{r} \$8.67 \\ \times\quad 9 \\ \hline \end{array}$

18. $\begin{array}{r} 31 \\ \times 31 \\ \hline \end{array}$

19. $\begin{array}{r} 16 \\ \times\ 31 \\ \hline \end{array}$

20. $7\overline{)3542}$ **21.** $6\overline{)\$33.00}$ **22.** $8\overline{)4965}$

23. $482 \div 5$ **24.** $3714 \div 9$ **25.** $2600 \div 3$

26. $3 \times N = 6 \times 8$

27. $\begin{array}{r} 4 \\ 2 \\ 3 \\ 5 \\ N \\ 8 \\ 6 \\ +\ 3 \\ \hline 45 \end{array}$

28. $\begin{array}{r} 5 \\ 7 \\ 3 \\ N \\ 6 \\ 4 \\ 2 \\ +\ 1 \\ \hline 47 \end{array}$

29. $W \times 4 = 1000$

30. $\begin{array}{r} N \\ -\ 472 \\ \hline 385 \end{array}$

LESSON 98 Mixed Numbers and Improper Fractions

Speed Test: 90 Division Facts, Form A (Test I in Test Booklet)

Here we show a picture of $1\frac{1}{2}$ shaded circles. Each whole circle has been divided into two half circles. We see that $1\frac{1}{2}$ is the same as **three halves**, which is written $\frac{3}{2}$.

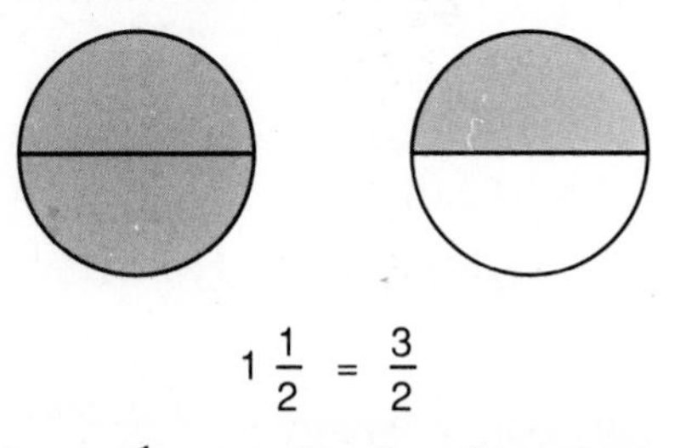

$1\frac{1}{2} = \frac{3}{2}$

The mixed number $1\frac{1}{2}$ equals the fraction $\frac{3}{2}$. The fraction $\frac{3}{2}$ is greater than 1. Fractions that are greater than or equal to 1 are called **improper fractions**. In this lesson we will draw pictures to show which mixed numbers and improper fractions are equal to each other.

Example Draw circles to show that $2\frac{3}{4}$ equals $\frac{11}{4}$.

Solution We begin by drawing 3 circles and shading $2\frac{3}{4}$.

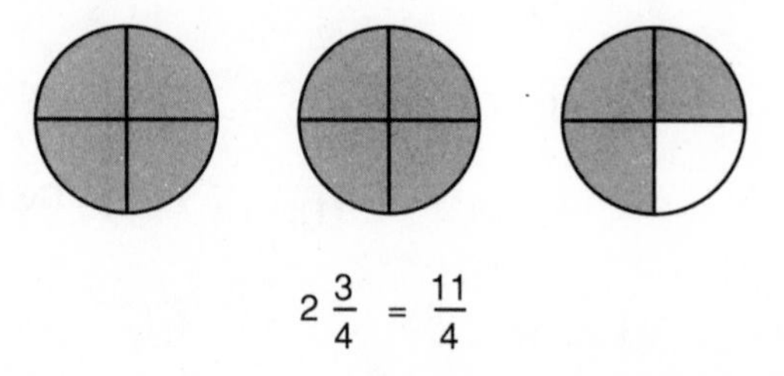

$2\frac{3}{4} = \frac{11}{4}$

We see that the fraction part of $2\frac{3}{4}$ is **fourths. So we divide all the circles into fourths**. We count 11 shaded fourths. This shows that $2\frac{3}{4}$ equals $\frac{11}{4}$.

Practice **a.** Draw circles to show that $1\frac{3}{4} = \frac{7}{4}$.

b. Draw circles to show that $2\frac{1}{2} = \frac{5}{2}$.

c. Draw circles to show that $1\frac{1}{3} = \frac{4}{3}$.

Problem set 98

1. The coach divided 33 players as equally as possible into 4 teams.
   (a) How many teams had exactly 8 players?
   (b) How many teams had 9 players?

2. On the package there were two 25-cent stamps, two 20-cent stamps, and one 15-cent stamp. Altogether, how much did the stamps that were on the package cost?

3. Don read 20 pages each day. How many pages did he read in 2 weeks?

4. The Frog Prince leaped 27 feet to get out of the well. How many yards did he leap?

5. What is the perimeter of this triangle in centimeters?

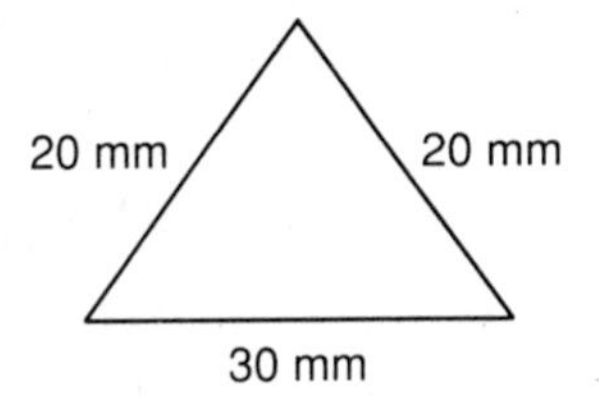

6. This is the tally for what number?

~~////~~ ~~////~~ ~~////~~ ///

7. This dropperful of water is about
   (a) 2 milliliters
   (b) 2 liters
   (c) 2 pints

8. Find N if $87 + 0 = 87 \times N$.

9. Illustrate this fraction-of-a-group statement: One third of the 24 students finished early.

10. $\begin{array}{r} YZ{,}PQR \\ +\ 289{,}517 \\ \hline 334{,}952 \end{array}$

11. $\begin{array}{r} \$478.63 \\ +\ \$\ \ 32.47 \\ \hline \end{array}$

12. $\begin{array}{r} 137{,}140 \\ -\ 129{,}536 \\ \hline \end{array}$

13. $\begin{array}{r} \$60.00 \\ -\ \$24.38 \\ \hline \end{array}$

14. 70×90

15. $\begin{array}{r} 11 \\ \times\ 13 \\ \hline \end{array}$

16. $\begin{array}{r} 12 \\ \times\ 12 \\ \hline \end{array}$

17. $\begin{array}{r} \$4.76 \\ \times\ \ \ \ 8 \\ \hline \end{array}$

18. $\begin{array}{r} 21 \\ \times\ 13 \\ \hline \end{array}$

19. $\begin{array}{r} 17 \\ \times\ 21 \\ \hline \end{array}$

20. $4\overline{)3000}$

21. $5\overline{)635}$

22. $7\overline{)426}$

23. $8\overline{)3614}$

24. $2736 \div 6$

25. $\$10.00 \div 4$

26. Draw and shade circles to show that $1\frac{1}{2}$ equals $\frac{3}{2}$.

27.
$$\begin{array}{r} N \\ -\ 472 \\ \hline 325 \end{array}$$

28.
$$\begin{array}{r} 516 \\ 423 \\ +\ \ N \\ \hline 1100 \end{array}$$

29.
$$\begin{array}{r} 4 \\ 24 \\ 34 \\ 4 \\ 4 \\ +\ N \\ \hline 86 \end{array}$$

30.
$$\begin{array}{r} 519 \\ +\ \ N \\ \hline 840 \end{array}$$

LESSON 99 Multiplying Two Two-Digit Numbers, Part 2

Speed Test: 90 Division Facts, Form A (Test I in Test Booklet)

We remember that three steps are required to multiply by a two-digit number:

1. Multiply by the ones' digit. Record the last digit of this product in the ones' column.
2. Multiply by the tens' digit. **Record the last digit of this product in the tens' column.**
3. Add to find the total.

In this lesson we will do multiplications that require carrying.

Example 1 Multiply:
$$\begin{array}{r} 46 \\ \times\ 27 \\ \hline \end{array}$$

Solution First we multiply 7×46. This product is 322.

$$\begin{array}{r} {\scriptstyle 4} \\ 46 \\ \times\ 27 \\ \hline 322 \end{array}$$

Now we multiply 2 × 46. First we multiply 2 × 6 and get 12. **We record the 2 in the tens' column** and carry the 1.

$$\begin{array}{r} {\scriptstyle 1} \\ {\scriptstyle 4} \\ 46 \\ \times\ 27 \\ \hline 322 \\ 2 \end{array}$$

Then 2 × 4 equals 8, plus 1 equals 9. So the second product equals 92. Then we add the products and get 1242.

$$\begin{array}{r} {\scriptstyle 1} \\ {\scriptstyle 4} \\ 46 \\ \times\ 27 \\ \hline 322 \\ 92 \\ \hline \mathbf{1242} \end{array}$$

Example 2 Multiply: $\begin{array}{r} 46 \\ \times 72 \\ \hline \end{array}$

Solution First we multiply 46 × 2 and get 92.

$$\begin{array}{r} {\scriptstyle 1} \\ 46 \\ \times\ 72 \\ \hline 92 \end{array}$$

Next we multiply 46 × 7 and get 322. Then we add the products.

$$\begin{array}{r} \overset{4}{\overset{1}{46}} \\ \times \ 72 \\ \hline 92 \\ 322\ \ \\ \hline \mathbf{3312} \end{array}$$

Practice

a. $\begin{array}{r} 38 \\ \times\ 26 \\ \hline \end{array}$ b. $\begin{array}{r} 49 \\ \times\ 82 \\ \hline \end{array}$ c. $\begin{array}{r} 84 \\ \times\ 67 \\ \hline \end{array}$ d. $\begin{array}{r} 65 \\ \times\ 48 \\ \hline \end{array}$

e. 48×24 f. 83×61 g. 48×75

Problem set 99

Use the information in the graph to answer questions 1–3.

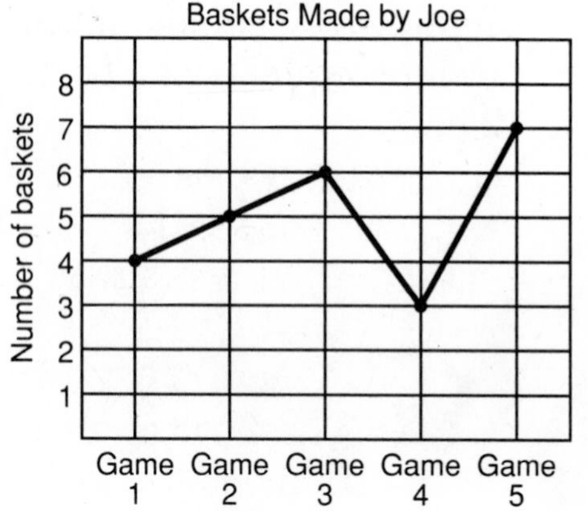

1. Altogether, how many baskets did Joe make in the first game?

2. Each basket is worth 2 points. How many points did Joe score in Game 5?

3. How many more points did Joe score in Game 5 than in Game 3?

4. The 3-pound melon cost \$1.44. What was the cost per pound?

5. What is the product of the fourth multiple of 5 and the sixth multiple of 7?

6. The first number was seven hundred forty-two thousand, nine hundred fifteen. The second number was five hundred forty-two thousand, seventeen. By how much was the first number greater than the second number?

7. If the perimeter of a square is 1 foot, how many inches long is each side?

8. A dollar bill weighs about 1 gram. How many dollar bills would it take to weigh 1 kilogram?

9. Illustrate this fraction-of-a-group statement: One fourth of the 64 clowns had red noses.

10. Kerri knew that her trip would take 7 hours and 42 minutes. If she left at 9:15 p.m., what time would she arrive?

11. Jill's boat would hold 42 containers. Each container would hold 8 big fish. How many big fish could Jill put in her 42 containers?

12. Kyle counted 572 birds in 4 hours. If he counted the same number of birds every hour, how many birds did he count in 1 hour?

13. $\begin{array}{r} 37{,}134 \\ +\ 52{,}891 \\ \hline \end{array}$

14. $\begin{array}{r} \$XYZ.AB \\ +\ \$\ 67.24 \\ \hline \$211.03 \end{array}$

15. $\begin{array}{r} 26 \\ 32 \\ 54 \\ 213 \\ 60 \\ 143 \\ +\ \ 8 \\ \hline \end{array}$

16. $\begin{array}{r} 98{,}428 \\ -\ \ 9{,}824 \\ \hline \end{array}$

17. $\begin{array}{r} \$85.00 \\ -\ \$68.65 \\ \hline \end{array}$

18. 47×100

19. 60×700

20. $\begin{array}{r} 328 \\ \times\ \ 4 \\ \hline \end{array}$

21. $\begin{array}{r} 43 \\ \times\ 32 \\ \hline \end{array}$

22. $\begin{array}{r} 25 \\ \times\ 35 \\ \hline \end{array}$

23. $5\overline{)4317}$ **24.** $8\overline{)\$44.00}$ **25.** $6\overline{)3963}$

26. $426 \div 3$ **27.** $2524 \div 4$

28. $3565 \div 7$

29. Draw and shade circles to show that $2\frac{1}{2}$ equals $\frac{5}{2}$.

30.
$$\begin{array}{r} 4 \\ 3 \\ 27 \\ 35 \\ 8 \\ + N \\ \hline 112 \end{array}$$

LESSON 100 Decimal Place Value · Tenths

Speed Test: 90 Division Facts, Form A (Test I in Test Booklet)

Part of this square is shaded. We show three ways to write how much of the square is shaded.

One tenth is shaded.

$\frac{1}{10}$ is shaded.

0.1 is shaded.

One tenth can be written with words.

One tenth can be written as a fraction: $\frac{1}{10}$

One tenth can be written as a decimal number: 0.1

We use **place value** to write tenths as a decimal number. We remember that the dot is called a **decimal point**. The digit just to the right of the decimal point is in the tenths' place. The digit in the tenths' place tells how many tenths. When we use a decimal point to help us write a number, we call the number a **decimal number** or just a **decimal**.

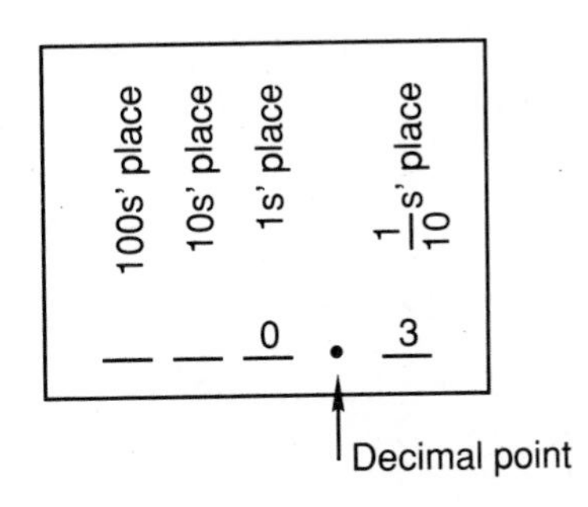

The digit in the tenths' place in 0.3 is 3. We read this number as "three tenths." We say the digit in the tenths' place and then say the word "tenths." It is customary but not necessary to write the zero in front of the decimal point.

.3, 0.3, and $\frac{3}{10}$ are read as "three tenths"

.1, 0.1, and $\frac{1}{10}$ are read as "one tenth"

.5, 0.5, and $\frac{5}{10}$ are read as "five tenths"

Example 1 Write the shaded part of this square as a fraction and as a decimal number.

Solution The square is divided into 10 parts. Three of these parts are shaded. Three tenths of the square is shaded.

$\frac{3}{10}$ is the fraction 0.3 is the decimal number

Example 2 Which of these fractions equals 0.7?

(a) $\frac{1}{7}$ (b) $\frac{7}{8}$ (c) $\frac{7}{10}$

Solution The decimal number 0.7 is understood to have a denominator of 10. The decimal number 0.7 is seven tenths. The correct choice is (**c**) because

$$0.7 = \frac{7}{10}$$

Practice

a. What fraction names the shaded part of this square?

b. What decimal number names the shaded part of the square?

c. What fraction names the part that is not shaded?

d. What decimal number names the part that is not shaded?

e. Write a fraction equal to 0.3.

f. Write a decimal number equal to $\frac{4}{10}$.

g. Write seven tenths as a fraction and as a decimal number.

Problem set 100

1. Three quarters, 4 dimes, 2 nickels, and 7 pennies is how much money?

2. Trudy separated the 37 math books as equally as possible into 4 stacks.
(a) How many stacks had exactly 9 books?
(b) How many stacks had 10 books?

3. Martin paid $1 for a folder and got 52¢ back in change. How much did the folder cost?

4. Frank wrote each of his 12 spelling words 5 times. In all, how many words did he write?

5. Round 5456 to the nearest thousand. Round 2872 to the nearest thousand. Find the sum of the two rounded numbers.

6. What is the tally for 10?

7. Part of this square is shaded.
   (a) Use a fraction to write the shaded part.
   (b) Use a decimal number to write the shaded part.

8. Illustrate this fraction-of-a-group statement: One sixth of the 48 crayons are broken.

9. Segment *AB* is 32 mm long. Segment *BC* is 26 mm long. Segment *AD* is 91 mm long. How many millimeters long is segment *CD*?

A B C D

10. 45,245 + 7,589

11. \$63.75 + \$184.75

12. 39,279 − 9,816

13. \$60.00 − (\$49.38 + 75¢)

14. $\begin{array}{r} \$6.08 \\ \times \quad 8 \\ \hline \end{array}$

15. $\begin{array}{r} 47 \\ \times \; 24 \\ \hline \end{array}$

16. $\begin{array}{r} 36 \\ \times \; 62 \\ \hline \end{array}$

17. 53 × 30

18. 63 × 37

19. 100 × 32

20. $4\overline{)3456}$

21. $8\overline{)6912}$

22. $7\overline{)\$50.40}$

23. 483 ÷ 6

24. 1530 ÷ 2

25. 2939 ÷ 9

26. Draw and shade circles to show that $1\frac{1}{4}$ equals $\frac{5}{4}$.

27. $\begin{array}{r} N \\ +\ 977 \\ \hline 5164 \end{array}$

28. $\begin{array}{r} 4 \\ 6 \\ 7 \\ 3 \\ 5 \\ +\ N \\ \hline 52 \end{array}$

29. $\begin{array}{r} 5 \\ 5 \\ 7 \\ 18 \\ 5 \\ +\ N \\ \hline 55 \end{array}$

30. $\begin{array}{r} 4 \\ 8 \\ 9 \\ 2 \\ 6 \\ +\ N \\ \hline 37 \end{array}$

LESSON 101 Naming Hundredths with Decimal Numbers

Speed Test: 90 Division Facts, Form A (Test I in Test Booklet)

In the preceding lesson we noticed that the first place to the right of the decimal point is the tenths' place. If a 1 is written in the tenths' place, it has a value of one tenth.

$$0.1 \quad \text{means} \quad \frac{1}{10}$$

If a 3 is written in the tenths' place, it has value of three tenths.

$$0.3 \quad \text{means} \quad \frac{3}{10}$$

The second place to the right of the decimal point is the **hundredths' place**. If a 1 is written in the hundredths' place, it has a value of one hundredth.

$$0.01 \quad \text{means} \quad \frac{1}{100}$$

If a 3 is written in the hundredths' place, it has a value of three hundredths.

$$0.03 \qquad \text{means} \qquad \frac{3}{100}$$

If we write 34 to the right of the decimal point, we mean thirty-four hundredths.

$$0.34 \qquad \text{means} \qquad \frac{34}{100}$$

When we write the fraction $\frac{34}{100}$, we **show** that the denominator is 100. When we write the decimal number 0.34, we **understand** that the denominator is 100. It is customary but not necessary to write a zero in front of the decimal point. The values of .34 and 0.34 are equal: .34 = 0.34.

Example 1 Part of this square is shaded.

(a) Use a fraction to write the shaded part.

(b) Use a decimal number to write the shaded part.

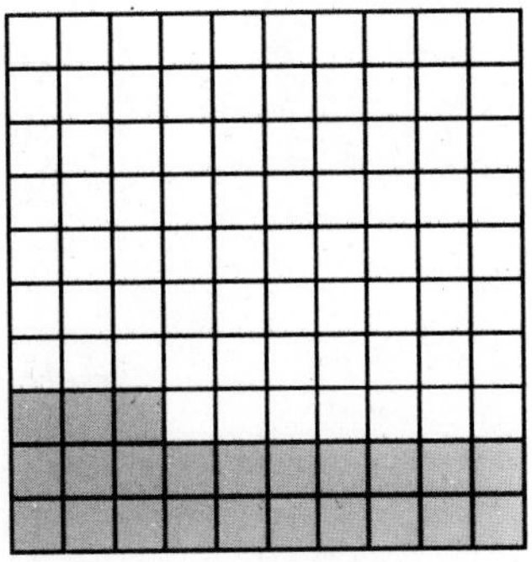

Solution The square is divided into hundredths. Twenty-three hundredths are shaded.

(a) Twenty-three hundredths as a fraction is $\mathbf{\frac{23}{100}}$.

(b) Twenty-three hundredths as a decimal number is **0.23.**

Example 2 (a) Write a fraction that equals 0.37.

(b) Write a decimal number that equals $\frac{29}{100}$.

Solution (a) The decimal number 0.37 is written with **two places** to the right of the decimal point. That means the denominator is 100. Thus, 0.37 equals

$$\frac{37}{100}$$

(b) A fraction with a denominator of 100 can be written as a decimal with two places to the right of the decimal point. So $\frac{29}{100}$ equals **0.29**.

Practice Write a fraction for each decimal number.

a. 0.37 **b.** 0.07 **c.** 0.99 **d.** 0.03

Write a decimal number for each fraction.

e. $\frac{61}{100}$ **f.** $\frac{9}{100}$ **g.** $\frac{25}{100}$ **h.** $\frac{4}{100}$

i. What decimal number names the shaded part of this square?

j. What decimal number names the part that is not shaded?

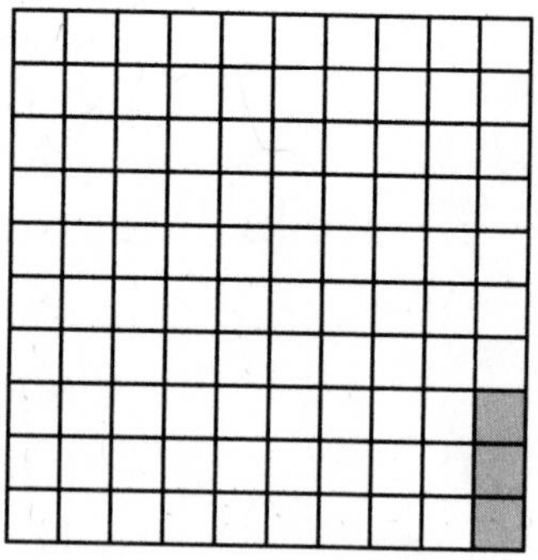

Problem set 101 Use the information given to answer questions 1–3.

Mary invited 14 friends for lunch. She plans to make 12 tuna sandwiches, 10 bologna sandwiches, and 8 roast beef sandwiches.

1. How many sandwiches will she make in all?

2. Including Mary, each person can have how many sandwiches?

3. If she cuts each tuna sandwich in half, how many halves will there be?

4. Five pounds of grapes cost \$2.95. What was the cost per pound?

5. If each side of a hexagon is 4 inches long, what is the perimeter of the hexagon in feet?

6. Nine hundred forty-seven thousand, five hundred eighty-seven is how much greater than two hundred seventy-five thousand, three hundred?

7. Part of this square is shaded.
(a) Use a fraction to write the shaded part.
(b) Use a decimal number to write the shaded part.

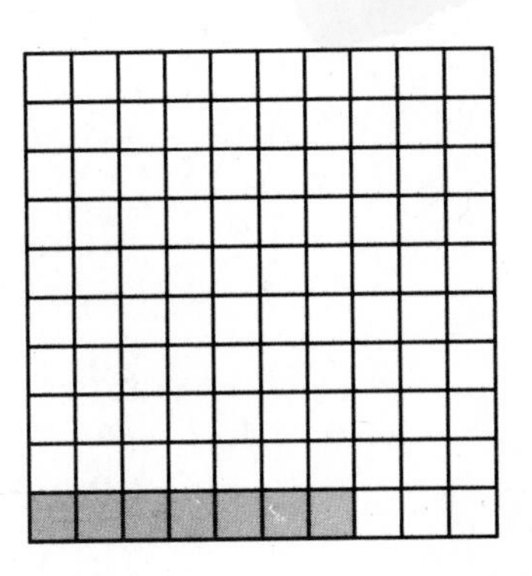

8. Use words to write $7572\frac{1}{8}$.

9. Illustrate this fraction-of-a-group statement: One fifth of the 80 chariots lost wheels in the chase.

10. Franca began the trip when it was still dark. She finished the trip the next afternoon. How long did the trip take?

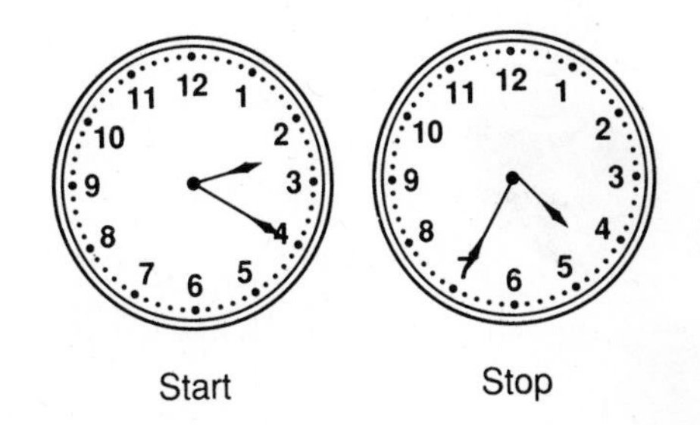

11. James traveled 301 miles in 7 hours. How far did he travel in 1 hour?

12. Mary could travel 47 miles in 1 hour. How far could she travel in 13 hours?

13. 37,625 + 9,485

14. \$25 + \$2.75 + \$15.44 + 27¢

15. 21,631 − 5,716

16. \$100.00 − \$89.85

17. 60×900

18. 42×30

19. 21×17

20. $\begin{array}{r} 36 \\ \times\ 74 \\ \hline \end{array}$

21. $\begin{array}{r} 48 \\ \times\ 25 \\ \hline \end{array}$

22. $\begin{array}{r} \$4.79 \\ \times\quad 6 \\ \hline \end{array}$

23. $9\overline{)2718}$

24. $5\overline{)4815}$

25. $6\overline{)4829}$

26. \$50.00 ÷ 8

27. 2121 ÷ 7

28. $\begin{array}{r} 7 \\ 8 \\ 5 \\ N \\ 6 \\ 2 \\ +\ 3 \\ \hline 47 \end{array}$

29. 2612 ÷ 3

30. Draw and shade rectangles to show that $1\frac{2}{3}$ equals $\frac{5}{3}$.

LESSON 102 Estimating Two-Digit Multiplication Answers

> **Speed Test:** 90 Division Facts, Form A (Test I in Test Booklet)

Sometimes we make mistakes when we multiply. If we estimate the answer before we multiply, we can tell if our answer is reasonable. Estimation can help prevent mistakes.

Example 1 Jim multiplied 43×29 and got 203. Could this answer be correct?

Solution Let's estimate the product by multiplying the rounded numbers 40 and 30.

$$40 \times 30 = 1200$$

Jim's answer of 203 and our estimate of 1200 are very different, so **Jim's answer of 203 is probably not correct.** He should check his multiplication.

Example 2 Estimate the product of 38 and 53. Then find an exact answer.

Solution First we multiply the rounded numbers.

$$40 \times 50 = 2000$$

Now we will get an exact answer.

$$\begin{array}{r} 38 \\ \times\ 53 \\ \hline 114 \\ 190 \\ \hline 2014 \end{array}$$

Our estimate of the product was **2000**, so our answer of **2014** is reasonable.

Practice Estimate the product of each multiplication. Then find the exact answer.

a. 58×23 **b.** 49×51 **c.** 61×38

d. 82×21 **e.** 38×49 **f.** 59×51

Problem set 102

1. Ninety-one students are divided as equally as possible among 3 classrooms.
(a) How many classrooms have exactly 30 students?

(b) How many classrooms have 31 students?

2. In 1970 it cost 6¢ to mail a letter. How much did it cost to mail 20 letters in 1970?

3. What number is seven hundred ninety more than two hundred ten?

4. George Washington was born in 1732 and died in 1799. How many years did he live?

5. A $1 bill weighs about 1 gram. How much would a $5 bill weigh?

6. This is the tally for what number?

卌 卌 卌 ||||

7. Part of this square is shaded.
(a) Use a fraction to write the shaded part.
(b) Use a decimal number to write the shaded part.

8. Estimate the product of 49 and 62.

9. Illustrate this fraction-of-a-group statement: One half of the 32 chess pieces were still on the board.

10. Mary left home at 10:52 a.m. She traveled for 7 hours and 20 minutes. What time was it when she arrived?

11. Mark traveled 42 miles in 1 hour. If he kept going at the same speed, how far did he travel in 15 hours?

12. The Fairy Queen flew 820 miles in 5 hours. How far did she fly in 1 hour?

13. $3714 + 238 + 46 + 7$ **14.** \$15.27 + \$85.75

15. \$18.00 − \$15.63 **16.** $10{,}141 - (363 + 99)$

17. $4 \times 6 \times 8$ **18.** 30×90 **19.** \$7.50 × 8

20. $\begin{array}{r} 34 \\ \times\ 28 \\ \hline \end{array}$ **21.** $\begin{array}{r} 54 \\ \times\ 23 \\ \hline \end{array}$ **22.** $\begin{array}{r} 74 \\ \times\ 40 \\ \hline \end{array}$

23. $4\overline{)\$6.36}$ **24.** $5\overline{)800}$ **25.** $6\overline{)1845}$

26. $4735 \div 8$ **27.** $2615 \div 9$

28. $3 + 5 + 1 + 4 + 7 + N + 8 + 2 + 3 + 5 + 2 = 58$

29. $1800 \div 3$

30. Draw and shade circles to show that $2\frac{1}{4}$ equals $\frac{9}{4}$.

LESSON 103 Two-Step Word Problems

Speed Test: 100 Multiplication Facts (Test H in Test Booklet)

Many math problems require more than one step to solve. Writing down what is given or drawing a picture is often helpful in solving these problems.

Example 1 Jim is 5 years older than Tad. Tad is 2 years younger than Blanca. Blanca is 9 years old. How old is Jim?

Solution We will write down the information we have and see if that helps.

Blanca = 9
Tad = 2 years younger than Blanca
Jim = 5 years older than Tad

We see that Blanca is 9. Tad is 2 years younger than Blanca, so Tad is 7. Jim is 5 years older than Tad, so **Jim is 12 years old**.

Example 2 One gallon of milk costs \$1.96 and one box of cereal costs \$2.12. Bert paid with a \$5 bill. How much change did he get?

Solution We begin by finding the total cost.

$$\begin{array}{rl} \$1.96 & \text{milk} \\ + \$2.12 & \text{cereal} \\ \hline \$4.08 & \text{cost} \end{array}$$

To find the change, we subtract the cost from \$5.00.

$$\begin{array}{r} \$5.00 \\ - \$4.08 \\ \hline \$0.92 \end{array}$$

Bert got **92 cents** in change.

Practice Nancy paid for 4 pounds of peaches with a \$5 bill. She got back \$3. What was the cost of 1 pound of peaches?

Problem set 103

1. Gabriel gave a \$5 bill to pay for a half gallon of milk that cost \$1.06 and a box of cereal that cost \$2.39. How much money should he get back?

2. Eighty-one animals crossed the bridge. One third of them were billy goats. The rest were bears. How many bears crossed the bridge?

3. Johnny planted 8 rows of apple trees. There were 15 trees in each row. How many trees did he plant?

4. Four pounds of bananas cost the Fairy Queen one hundred fifty-six trinkets. Each pound of bananas cost how many trinkets?

5. The scale shows a weight of how many grams?

6. Write the tally for 16.

7. Part of this square is shaded.
(a) Use a fraction to write the shaded part.
(b) Use a decimal number to write the shaded part.

8. Estimate the product of 32 and 48. Then find the exact product.

9. Illustrate this fraction-of-a-group statement: One third of the 24 camels were Bactrian.

10. Quite a few people in the north heard the gong sound. Seven thousand, forty-two people in the south heard the gong sound. If nine thousand, twelve people in all heard the gong sound, how many people in the north heard the gong sound?

11. Five hundred seventy people sat in the bleachers. Then some went home. Only twenty-seven were left. How many went home?

12. Matthew could travel 496 miles in 8 hours. How far could he travel in 1 hour?

13. $\begin{array}{r} N \\ +\quad 731 \\ \hline 398{,}548 \end{array}$

14. $\begin{array}{r} \$46.39 \\ +\ \$54.60 \\ \hline \end{array}$

15. $\begin{array}{r} 37 \\ 81 \\ 45 \\ 139 \\ 7 \\ 15 \\ +\ \ 60 \\ \hline \end{array}$

16. $\begin{array}{r} 96{,}410 \\ -\ 9{,}641 \\ \hline \end{array}$

17. $\begin{array}{r} \$37.81 \\ -\ \$16.79 \\ \hline \end{array}$

18. 63×1000

19. 80×50

20. $\begin{array}{r} 52 \\ \times\ 15 \\ \hline \end{array}$

21. $\begin{array}{r} 36 \\ \times\ 27 \\ \hline \end{array}$

22. $\begin{array}{r} 59 \\ \times\ 32 \\ \hline \end{array}$

23. $2\overline{)714}$ **24.** $6\overline{)789}$ **25.** $4\overline{)2363}$

26. $2835 \div 7$ **27.** $1000 \div 3$

28.
$$\begin{array}{r} 3 \\ 5 \\ 8 \\ 7 \\ 4 \\ N \\ +\ 6 \\ \hline 53 \end{array}$$

29. $\$29.00 \div 5$

30. Draw and shade rectangles to show that $1\frac{2}{5}$ equals $\frac{7}{5}$.

LESSON 104 Fraction-of-a-Group Problems, Part 2

> **Speed Test:** 100 Multiplication Facts (Test H in Test Booklet)

To divide a group into 5 equal parts, we divide the number in the group by 5. Suppose we have 30 marbles. If we divide 30 by 5, the answer is 6. This means that there are 6 marbles in each of the 5 groups.

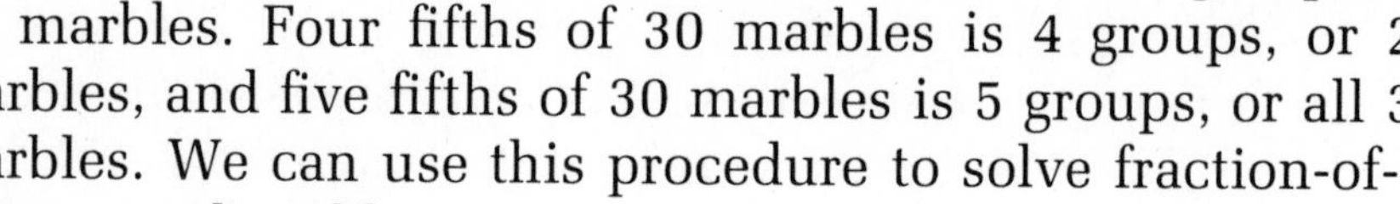

One fifth of 30 marbles is 1 of these groups, or 6 marbles. Two fifths of 30 marbles is 2 of these groups, or 12 marbles. Three fifths of 30 marbles is 3 of these groups, or 18 marbles. Four fifths of 30 marbles is 4 groups, or 24 marbles, and five fifths of 30 marbles is 5 groups, or all 30 marbles. We can use this procedure to solve fraction-of-a-group word problems.

Example 1 There were 30 leprechauns in the forest. Three fifths of them wore green jackets. How many leprechauns wore green jackets?

Solution The word "fifths" tells us there were 5 equal groups. We need to find the number in each group. To do this, we divide 30 by 5.

$$5\overline{)30}\quad = 6$$

So there are 6 leprechauns in each group.

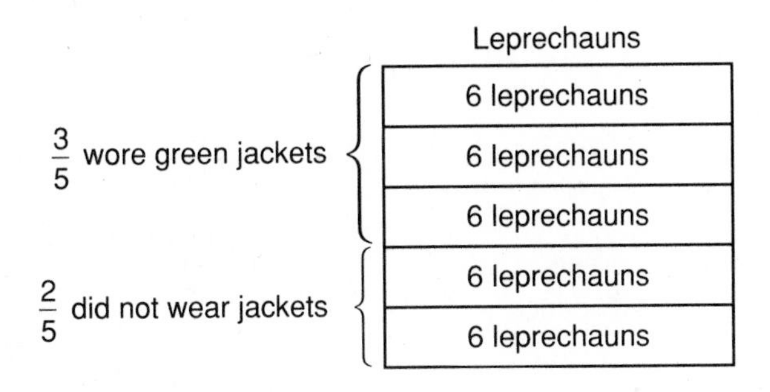

This figure shows that one fifth of 30 is 6. In three fifths, there are 3 times 6 leprechauns. That is, **18 leprechauns wore green jackets.**

Example 2 Two thirds of the 24 elves worked in the toy factory. How many elves worked in the toy factory?

Solution First we divide 24 by 3 and find that the number of elves in **one** third is 8.

Elves

$\frac{2}{3}$ worked in the factory: 8 elves, 8 elves

$\frac{1}{3}$ didn't work in the factory: 8 elves

Then we multiply 8 by 2 and find that the number of elves in **two** thirds is 16. Thus we have found that **16 elves** worked in the toy factory.

Practice Solve and illustrate each problem.

a. Three fourths of the 24 checkers were still on the board. How many checkers were still on the board?

b. Two fifths of the 30 leprechauns guarded the treasure. How many leprechauns guarded the treasure?

c. Three eighths of the 40 students had perfect scores. How many students had perfect scores?

Problem set 104 Use the information in this tally sheet to answer questions 1–3.

Results of Class Election

Irma	卌 II
Brad	卌 I
Thanh	卌 III
Marisol	卌 卌 II

Tally of number of votes received

1. Who was second in the election?

2. Who received twice as many votes as Brad?

3. Altogether, how many votes were cast?

4. Two fifths of the 20 balloons were yellow. How many balloons were yellow? Illustrate the problem.

5. Eight hundred ninety-six thousand, seven hundred ninety-eight is how much greater than four hundred forty-four thousand, eleven?

6. Part of this group is shaded.
(a) Write the shaded part as a fraction
(b) Write the shaded part as a decimal.

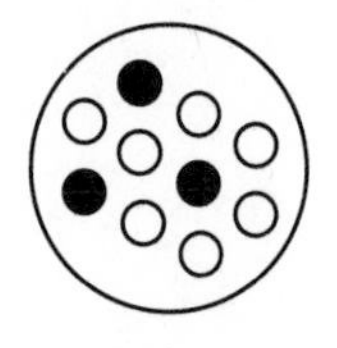

7. Estimate the product of 88 and 59.

8. Sue's birthday is May 2. Her birthday will be on what day of the week in the year 2047?

May						2047
S	**M**	**T**	**W**	**T**	**F**	**S**
	1	2	3	4	5	6
7	8	9	10	11	12	13
14	15	16	17	18	19	20
21	22	23	24	25	26	27
28	29	30				

9. Segment AB is 17 mm long. Segment CD is 36 mm long. Segment AD is 89 mm long. How long is segment BC?

10. \$32.63 + \$42 + \$7.56

11. 37,415 + 62,514

12. \$86.45 − (\$74.50 + \$5)

13. 49,329 − 19,519

14. 83×40

15. 1000×53

16. $9 \times 9 \times 9$

17. $\begin{array}{r} 32 \\ \times\ 16 \\ \hline \end{array}$

18. $\begin{array}{r} 67 \\ \times\ 32 \\ \hline \end{array}$

19. $\begin{array}{r} \$18.95 \\ \times\quad\ 4 \\ \hline \end{array}$

20. $3\overline{)625}$

21. $4\overline{)714}$

22. $6\overline{)1385}$

23. $916 \div 5$

24. $3748 \div 9$

25. $\$28.56 \div 8$

26. This circle shows that $\frac{2}{2}$ equals 1. Draw a circle that shows $\frac{3}{3}$ equals 1.

27. $\begin{array}{r} 1 \\ 2 \\ 3 \\ 4 \\ 5 \\ 4 \\ 3 \\ 2 \\ 1 \\ 6 \\ +\ 7 \\ \hline \end{array}$

28. $\begin{array}{r} 2 \\ 3 \\ 4 \\ 4 \\ 3 \\ 2 \\ 5 \\ 7 \\ 9 \\ 2 \\ +\ 1 \\ \hline \end{array}$

29. $\begin{array}{r} 8 \\ 2 \\ 1 \\ 3 \\ 5 \\ 7 \\ 2 \\ 4 \\ 2 \\ 7 \\ +\ 5 \\ \hline \end{array}$

30. $\begin{array}{r} 9 \\ 2 \\ 1 \\ 3 \\ 7 \\ 2 \\ 6 \\ N \\ 2 \\ 4 \\ +\ 1 \\ \hline 45 \end{array}$

LESSON 105 The Bar as a Division Symbol

Speed Test: 100 Multiplication Facts (Test H in Test Booklet)

We have used a division symbol to show division. We have also used a division box. To show that 12 is to be divided by 3, we can write either

$$12 \div 3 \qquad \text{or} \qquad 3\overline{)12}$$

We can also use a **division bar** to show division. A division bar is a short line segment. The number above the line is to be divided by the number below the line. To show that 12 is to be divided by 3, we can write

$$\frac{12}{3}$$

Example Divide: $\frac{150}{6}$

Solution We read this from top to bottom, "One hundred fifty divided by six." We draw a division box with 150 inside and divide. The answer is **25**.

$$\begin{array}{r} 25 \\ 6\overline{)150} \\ \underline{12} \\ 30 \\ \underline{30} \\ 0 \end{array}$$

Practice a. $\frac{364}{7}$ b. $\frac{18}{3}$ c. $\frac{300}{6}$ d. $\frac{816}{4}$

Problem set 105

1. Freddie is 2 years older than Ivan. Ivan is twice as old as Becky. Becky is 6 years old. How old is Freddie?

2. What is the total number of days in July, August, and September?

3. It cost \$1.39 to mail the package. Marty put three 25-cent stamps on the package. How much more postage does it need?

4. Thirty-two desks were arranged as equally as possible in 6 rows.
   (a) How many rows had exactly 5 desks?
   (b) How many rows had 6 desks?

5. Two thirds of the 21 riders rode bareback. How many riders rode bareback? Illustrate the problem.

6. (a) What decimal number names the shaded part of this square?
   (b) What decimal number names that part that is not shaded?

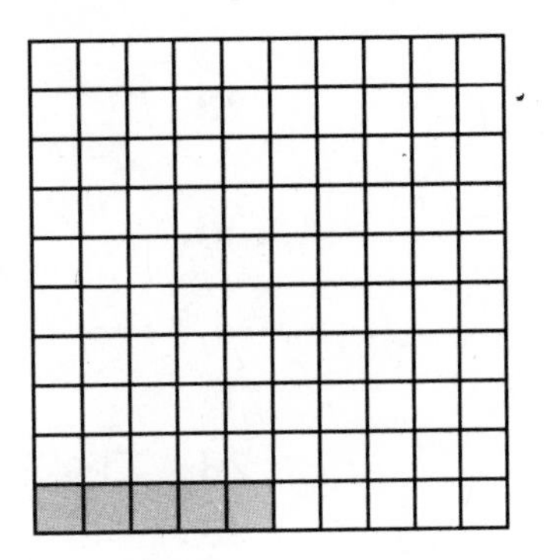

7. Write the number 314,002 in expanded form. Then use words to write the number.

8. Round 3874 to the nearest thousand.

9. Beth opened a liter of milk and poured half of it into a pitcher. How many milliliters of milk did she pour into the pitcher?

10. The sun was up when Mark started. It was dark when he stopped. How much time had gone by?

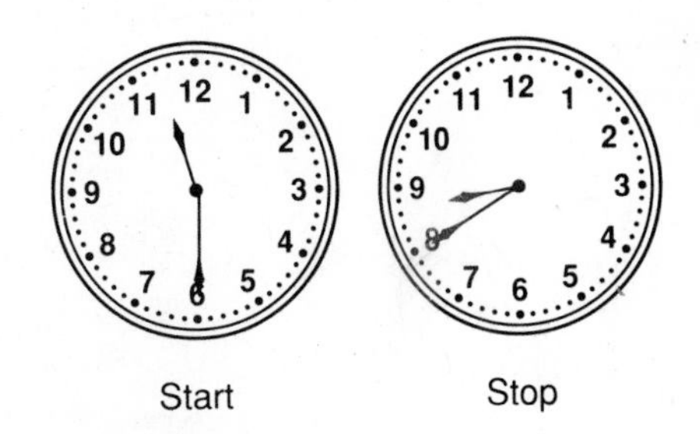

11. The packer could pack 14 boxes in 1 hour. How many boxes could be packed in 13 hours?

12. Mickey drove 368 miles in 8 hours. If she drove the same number of miles each hour, how far could she drive in 1 hour?

13. $\begin{array}{r} 496{,}325 \\ + \quad 3{,}680 \\ \hline \end{array}$

14. $\begin{array}{r} 1{,}345 \\ +\ 27{,}581 \\ \hline \end{array}$

15. $\begin{array}{r} \$12.45 \\ 1.30 \\ 2.00 \\ 0.25 \\ 0.04 \\ 0.32 \\ +\ 1.29 \\ \hline \end{array}$

16. $\begin{array}{r} \$36.00 \\ -\ \$30.78 \\ \hline \end{array}$

17. $\begin{array}{r} 471{,}053 \\ -\ 164{,}363 \\ \hline \end{array}$

18. $8.56 × 7

19. 60 × 300

20. $\begin{array}{r} 47 \\ \times\ 36 \\ \hline \end{array}$

21. $\begin{array}{r} 26 \\ \times\ 24 \\ \hline \end{array}$

22. $\begin{array}{r} 25 \\ \times\ 25 \\ \hline \end{array}$

23. $7\overline{)845}$

24. $9\overline{)1000}$

25. $16.40 ÷ 8

26. $\frac{65}{5}$

27. $\frac{432}{6}$

28. $\begin{array}{r} 2 \\ 4 \\ 6 \\ 8 \\ 6 \\ N \\ 2 \\ 3 \\ +\ 1 \\ \hline 52 \end{array}$

29. $\frac{716}{4}$

30. Draw and shade a circle that shows that $\frac{4}{4}$ equals 1.

LESSON 106 Writing Mixed Numbers as Decimals

> **Speed Test:** 100 Multiplication Facts (Test H in Test Booklet)

The figure at the top of the next page has three squares. Two and three tenths of the squares are shaded. We can describe the shaded parts by using either a mixed number or a decimal number. Both the mixed number and the decimal number are read "two and three tenths."

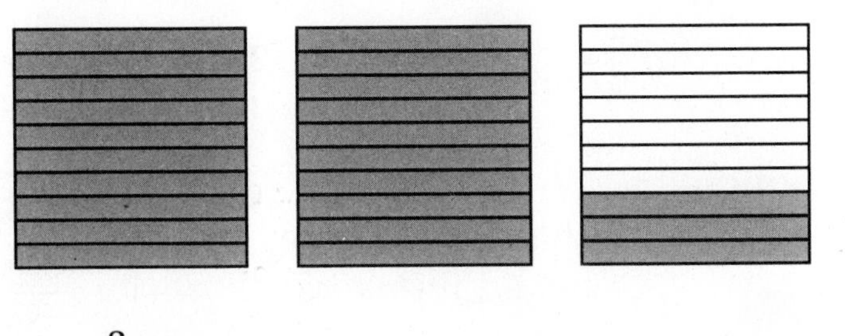

$2\frac{3}{10}$ is shaded or 2.3 is shaded

In the decimal number, the whole number part is to the left of the decimal point and the fraction part is to the right of the decimal point.

In the next figure, one and thirty-three hundredths squares are shaded. We can name one and thirty-three hundredths as a mixed number or as a decimal (number).

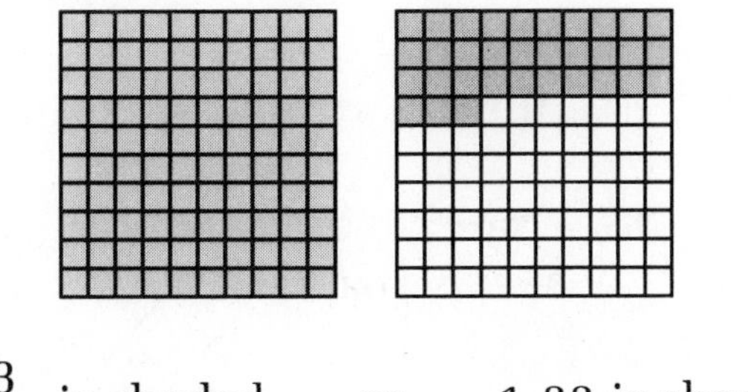

$1\frac{33}{100}$ is shaded or 1.33 is shaded

Both are read "one and thirty-three hundredths."

Example 1 Write each mixed number as a decimal number.

(a) $3\frac{7}{10}$ (b) $16\frac{17}{100}$ (c) $8\frac{9}{100}$

Solution We write the whole number to the left of the decimal point and write the fraction to the right of the decimal point. Tenths are written one place to the right of the decimal point. Hundredths are written two places to the right of the decimal point.

(a) **3.7** (b) **16.17** (c) **8.09**

Example 2 Write each decimal number as a mixed number.

(a) 24.9 (b) 5.23 (c) 4.03

Solution

(a) $\mathbf{24\frac{9}{10}}$ (b) $\mathbf{5\frac{23}{100}}$ (c) $\mathbf{4\frac{3}{100}}$

Practice **a.** Write the number of shaded squares as a mixed number.

b. Write the number of shaded squares as a decimal number.

Write each mixed number as a decimal number.

c. $9\frac{99}{100}$ **d.** $12\frac{3}{10}$ **e.** $10\frac{1}{100}$

Write each decimal number as a mixed number.

f. 60.3 **g.** 6.03 **h.** 5.67

Problem set 106 Use this chart to answer questions 1–4.

Mileage Chart

	Atlanta	Boston	Chicago	Kansas City	Los Angeles	New York	Wash. D.C.
Chicago	730	975		499	2095	840	712
Dallas	805	1819	936	499	1403	1607	1372
Denver	1401	1989	1016	604	1134	1851	1696
Los Angeles	2197	3052	2095	1596		2915	2644
New York	855	216	840	1319	2915		229
St. Louis	558	1178	288	253	1848	966	801

1. The distance from Los Angeles to Boston is how much greater than the distance from Los Angeles to New York?

2. (a) Which two cities are the same distance from Kansas City?

(b) How far apart are the two cities in the answer to (a)?

3. Rebecca is planning a trip from Chicago to Dallas to Los Angeles to Chicago. How many miles will her trip be?

4. There are 3 empty boxes in the chart. What numbers should go in these boxes?

5. Three fourths of the one thousand gold coins were doubloons. How many doubloons were there? Illustrate the problem.

6. Write each mixed number as a decimal.

   (a) $3\frac{5}{10}$ (b) $14\frac{21}{100}$ (c) $9\frac{4}{100}$

7. Estimate the product of 39 and 406.

8. What is the perimeter of this rectangle in millimeters?

9. What is the perimeter of the rectangle in Problem 8 in centimeters?

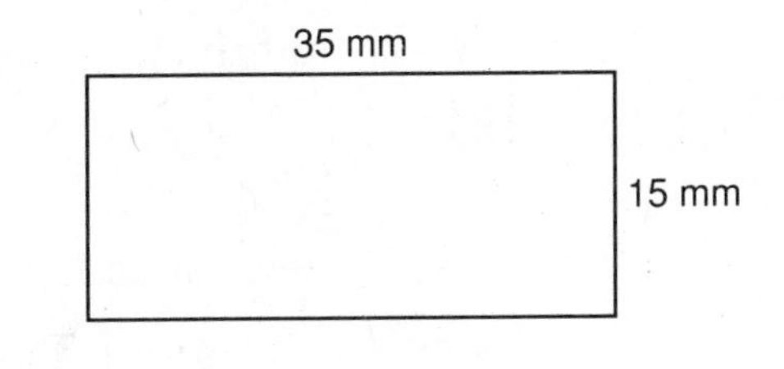

10. James knew that his trip would take 7 hours and 40 minutes. He left at 7 a.m. At what time will he arrive?

11. In the beginning there were quite a few near the edge of town. Just after the big bang fourteen thousand, eleven went away. Then there were eleven thousand, twenty-seven near the edge of town. How many were there in the beginning?

12. Hortense could cap 65 bottles in 5 minutes. How many bottles could she cap in 1 minute?

13. 39,429 + 65,315

14. \$2.74 + \$0.27 + \$6 + 49¢

15. 583,248 − 3,186

16. \$15.75 − (\$2.47 + \$10)

17. 25×40 **18.** 98¢ × 7 **19.** $6 \times 5 \times 3$

20. $\begin{array}{r} 32 \\ \times\ 27 \\ \hline \end{array}$ **21.** $\begin{array}{r} 63 \\ \times\ 41 \\ \hline \end{array}$ **22.** $\begin{array}{r} 35 \\ \times\ 35 \\ \hline \end{array}$ **23.** $\begin{array}{r} 4 \\ 2 \\ 1 \\ 3 \\ 4 \\ 7 \\ 2 \\ 2 \\ 3 \\ 4 \\ +\ X \\ \hline 42 \end{array}$

24. $8\overline{)\$70.00}$ **25.** $9\overline{)4321}$ **26.** $6\overline{)1234}$

27. $800 \div 7$ **28.** $487 \div 3$ **29.** $\dfrac{210}{5}$

30. Draw and shade circles to show that $2\frac{1}{3}$ equals $\frac{7}{3}$.

LESSON 107 Reading Decimal Numbers

Speed Test: 100 Multiplication Facts (Test H in Test Booklet)

In this lesson we will identify decimal place values and write the names of the decimal numbers.

Place value The chart below shows place values from hundreds to hundredths.

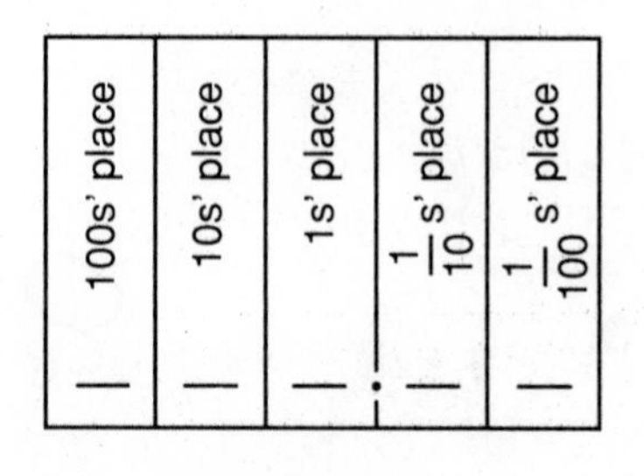

Example 1 Name the place value of the 3 in each number.

(a) 23.4 (b) 2.34 (c) 32.4 (d) 4.23

Solution (a) **ones** (b) **tenths** (c) **tens** (d) **hundredths**

Example 2 Which digit in 23.47 is in the same place as the 5 in 8.5?

Solution In 8.5, the 5 is in the tenths' place. The digit in 23.47 that is in the tenths' place is the **4**.

Naming decimal numbers To name a decimal number, we first name the whole number part. Then we name the decimal part.

To read the decimal number 562.35, we say the whole number part first:

five hundred sixty-two

Next we read the decimal point by saying "and":

five hundred sixty-two **and**

Then we read the number to the right of the decimal point:

five hundred sixty-two **and** thirty-five

Finally we name the place value of the **last digit:**

five hundred sixty-two **and** thirty-five **hundredths**

Example 3 Use words to name each number.

(a) 12.34 (b) 1.3 (c) 6.07

Solution (a) **twelve and thirty-four hundredths**

(b) **one and three tenths**

(c) **six and seven hundredths**

Practice Name the place value of the 7 in these numbers.

a. 17.26 **b.** 61.72

c. 26.17 **d.** 72.61

Which digit is in the tenths' place in these numbers?

e. 43.85 **f.** 438.5 **g.** 4.38 **h.** 1385.4

Use words to name each decimal number.

i. 4.38

j. 43.8

k. 10.6

l. 1.06

m. 36.75

n. 36.7

Problem set 107

Use the facts given to answer questions 1 and 2.

In the Jones family there are 3 children. John is 10 years old. James is 2 years younger than Jill. James is 4 years older than John.

1. How old is James? How old is Jill?

2. When James is 16, how old will Jill be?

3. Denise bought 6 pounds of carrots and an artichoke for \$2.76. If the artichoke cost 84¢, how much was 1 pound of carrots?

4. Write each of these mixed numbers as a decimal number.

   (a) $5\frac{31}{100}$ (b) $16\frac{7}{10}$ (c) $5\frac{7}{100}$

5. Three fifths of the team's 40 points were scored in the first half. How many points did the team score in the first half? Illustrate the problem.

6. Use words to write 7.68.

7. Use words to write 76.8.

8. Estimate the product of 78 and 91.

9. Name the number of shaded squares

   (a) as a mixed number.

   (b) as a decimal.

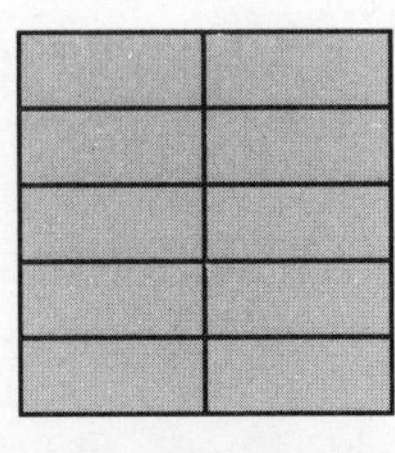

10. In the beginning there were thirty-seven thousand, two hundred sixty-five sitting in the sun. In the end one hundred forty-seven thousand, eight hundred fifty-two more came. How many were sitting in the sun at the end?

11. Sixteen thousand, three hundred thirteen is how much greater than eight thousand, three hundred eighteen?

12. In the beginning it was dark. What time was it 5 hours and 20 minutes later?

Beginning

13. Mr. Toto could bake 27 pizzas in 3 hours. How many could he bake in 1 hour? How many could he bake in 5 hours? (Multiply the first answer by 5.)

14. $15 + $6.15 + $13.85

15. $13.70 − $6.85

16. 26×100 **17.** $9 \times 87¢$ **18.** 14×16

19. $\begin{array}{r} 32 \\ \times\ 15 \\ \hline \end{array}$ **20.** $\begin{array}{r} 64 \\ \times\ 24 \\ \hline \end{array}$ **21.** $\begin{array}{r} 47 \\ \times\ 60 \\ \hline \end{array}$

22. $6\overline{)4248}$ **23.** $7\overline{)1634}$ **24.** $5\overline{)\$49.00}$

25. $3654 \div 8$

28. $2882 \div 9$

29. $\dfrac{456}{6}$

26. $\begin{array}{r} 1 \\ 3 \\ 5 \\ P \\ 7 \\ 3 \\ 2 \\ +\ 3 \\ \hline 44 \end{array}$ **27.** $\begin{array}{r} 8 \\ 4 \\ 5 \\ A \\ 4 \\ 4 \\ 7 \\ +\ 6 \\ \hline 44 \end{array}$

30. Divide a rectangle into parts to show that $\frac{5}{5}$ equals 1.

LESSON 108 Decimal Numbers and Money

Speed Test: 90 Division Facts, Form B (Test J in Test Booklet)

We remember that there are two forms for writing money. In one form the unit is cents. For writing cents we use a cent sign (¢) and no decimal point. When we write

25¢

we mean 25 **whole cents** (or 25 pennies). In the other form of money the unit is dollars. For writing dollars we use a dollar sign ($). We also use a decimal point to show fractions of a dollar. When we write

$.25

we mean 25 **hundredths of a dollar** (a fraction of a dollar).

A penny is 1 cent

A penny is also $\frac{1}{100}$ of a dollar

1¢ = $.01

Sometimes we see notations for money written incorrectly.

Example Something is wrong with this sign. What should be changed so that the sign is correct?

Solution The notation .50¢ is incorrect. We can use the ¢ sign and a whole number to tell how many cents. Or we can use a dollar sign and a decimal point to write the fractional part of a dollar.

50¢ is correct **$0.50 is also correct**

Practice At a vegetable stand Martin saw this incorrect sign. Draw two different signs to show ways to correct this error.

Corn
0.10¢
each
ear

Problem set 108

1. Robin divided his 53 merry men into groups of 6.
   (a) How many groups of 6 could he make?
   (b) How many merry men were left over?
   (c) If the remaining merry men formed a group, how many groups were there in all?

2. Abraham Lincoln was born in 1809 and died in 1865. How long did he live?

3. The parking lot charges $1.25 for the first hour. It charges 50¢ for each additional hour. How much does it cost to park a car in the lot for 3 hours?

4. Two thirds of the team's 45 points were scored in the second half. How many points did the team score in the second half? Draw a diagram to illustrate the problem.

5. Something is wrong with this sign. Draw two different signs to show ways to correct this error.

ICE CREAM
BARS
0.25¢ each

6. What is the value of 3 ten-dollar bills, 4 one-dollar bills, 5 dimes, and 2 pennies?

7. Use words to write 6412.5.

8. Round 5139 to the nearest thousand. Round 6902 to the nearest thousand. Then add the rounded numbers.

9. James opened a gallon bottle of milk and poured out 1 quart. How many quarts of milk were left in the bottle?

10. Find the sum of one hundred forty-seven thousand, three hundred two and three hundred fifty-eight thousand, eleven.

11. Daddy Johnny saw forty-seven thousand, three hundred fifteen on his walk in the woods. Mama Amy only saw sixteen thousand, nine hundred eight. How many more did Daddy Johnny see?

12. Thirteen little people came to the party every hour. After 11 hours, how many little people were at the party?

13. Five full buses held 120 students. How many students were on each bus? How many students could ride on 7 buses? (Multiply the first answer by 7.)

14. $\begin{array}{r} \$68.57 \\ + \$36.49 \\ \hline \end{array}$

15. $\begin{array}{r} \$100.00 \\ - \$\ \ \ 5.43 \\ \hline \end{array}$

16. $\begin{array}{r} 15 \\ 24 \\ 36 \\ 75 \\ 21 \\ 8 \\ 36 \\ + 420 \\ \hline \end{array}$

17. $\begin{array}{r} 47 \\ \times 23 \\ \hline \end{array}$

18. $\begin{array}{r} \$5.08 \\ \times \quad 7 \\ \hline \end{array}$

19. 50×50

20. $\begin{array}{r} 1 \\ 2 \\ 3 \\ 4 \\ P \\ 6 \\ 7 \\ 8 \\ +\ 9 \\ \hline 56 \end{array}$

21. $\begin{array}{r} 1 \\ 2 \\ 4 \\ 6 \\ 8 \\ 6 \\ 4 \\ 3 \\ +\ 3 \\ \hline \end{array}$

22. 49×51

23. 33×25

24. $3\overline{)848}$

25. $9\overline{)6315}$

26. $7\overline{)3641}$

27. $\$5.72 \div 4$

28. $834 \div 5$

29. $\frac{134}{2}$

30. Draw and shade circles to show that $2\frac{2}{3}$ equals $\frac{8}{3}$.

LESSON 109 Circles: Radius and Diameter

Speed Test: 90 Division Facts, Form B (Test J in Test Booklet)

Many of the words we use in mathematics are based on Latin or Greek words. The Latin word for ring was *circus*. The Latin word for the spoke of a wheel was *radius*. The Latin word for through-measure was *diametron*.

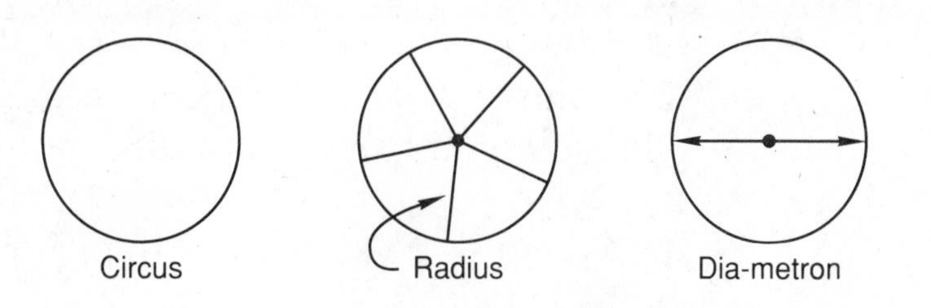

From these Latin words we get the English words **circle, radius,** and **diameter**. The distance from the center to the circle is the radius. The distance across a circle through the center is the diameter.

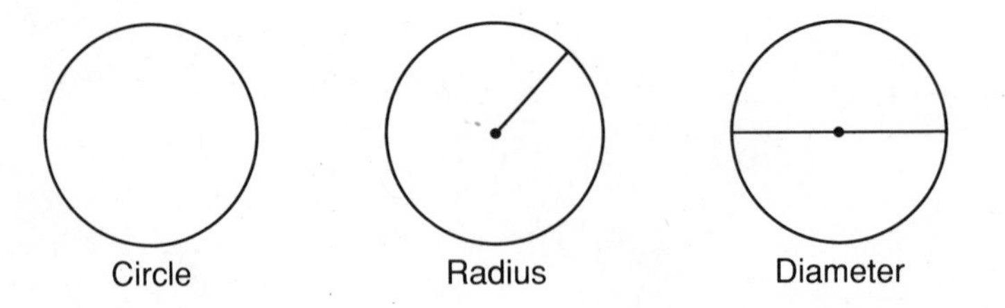

A diameter is twice as long as a radius. A radius is half as long as a diameter.

Example The diameter of this circle is 4 cm. What is its radius?

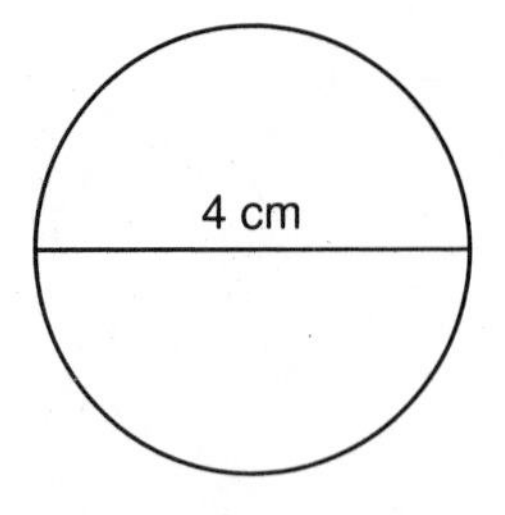

Solution The radius is the distance from the center to the circle. The radius is half the diameter. Since half of 4 cm is 2 cm, the radius is **2 cm**.

Practice

a. What is the name for the distance across the center of a circle?

b. What is the name for the distance from the center to the circle?

c. If the radius of a circle is 4 cm, what is its diameter?

Problem set 109

1. One hundred fifty feet equals how many yards?

2. Tammy gave the clerk $6 to pay for the record. She got back 64¢. How much did the record cost?

3. Sergio is 2 years older than Rebecca. Rebecca is twice as old as Dina. Sergio is 12 years old. How old is Dina?

4. Write each of these decimal numbers as a mixed number.
(a) 3.29 (b) 32.9 (c) 3.09

5. Three fourths of the 84 contestants guessed wrong. How many contestants guessed wrong? Draw a diagram to illustrate the problem.

6. (a) What is the diameter of this circle?
(b) What is the radius of this circle?

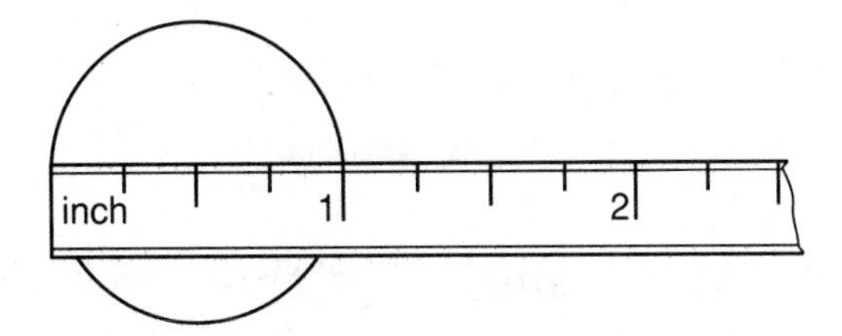

7. Use words to write 8.75.

8. Estimate the sum of 476 and 620.

9. The radius of a circle is 2 cm. How many millimeters is its diameter?

10. Find the sum of four hundred thirty-six, two hundred eighty-seven, and six thousand, three hundred fourteen.

11. The first number was thirteen thousand, one hundred seventy. The second number was only nine thousand, two hundred sixty-one. By how much was the first number greater than the second number?

12. \$43.62 + \$3.60 + 56¢ 13. \$16.25 – (\$6 – 50¢)

14. $5 \times 7 \times 9$ 15. \$7.83 × 6 16. 54×1000

17. $\begin{array}{r} 74 \\ \times\ 16 \\ \hline \end{array}$ 18. $\begin{array}{r} 32 \\ \times\ 40 \\ \hline \end{array}$ 19. $\begin{array}{r} 25 \\ \times\ 24 \\ \hline \end{array}$

20. $6\overline{)3625}$ **21.** $5\overline{)3000}$ **22.** $7\overline{)987}$

23. $461 \div 9$ **24.** $\$13.76 \div 8$ **25.** $\frac{234}{3}$

26. Draw and shade a circle to show that $\frac{8}{8}$ equals 1.

27.
$$\begin{array}{r} 8 \\ 4 \\ 2 \\ 6 \\ 3 \\ N \\ 7 \\ 5 \\ 8 \\ 4 \\ +\ 2 \\ \hline 56 \end{array}$$

28.
$$\begin{array}{r} 9 \\ 3 \\ 8 \\ 1 \\ 7 \\ P \\ 5 \\ 4 \\ 6 \\ +\ 5 \\ \hline 56 \end{array}$$

29.
$$\begin{array}{r} N \\ -\ 421 \\ \hline 325 \end{array}$$

30.
$$\begin{array}{r} 325 \\ +\ \ N \\ \hline 421 \end{array}$$

LESSON 110 Circle Graphs

Speed Test: 90 Division Facts, Form B (Test J in Test Booklet)

We have studied pictographs, bar graphs, and line graphs. Another kind of graph is the **circle graph**. This is often called a **pie graph** because the circle is divided into sections that look like slices of a pie.

Example Use the information in this circle graph to answer these questions.

(a) How many children were there altogether?

(b) Which activity were the most children doing?

(c) How many children were not catching polliwogs?

Solution (a) The title of the graph states that there were **100 children**.

(b) The numbers inside each "slice of pie" show how many children were doing each activity. The largest number were **flying kites**.

(c) Only 19 of the 100 children were catching polliwogs. The rest were not. We subtract 19 from 100 and find that **81 children** were not catching polliwogs.

Practice Use the information from the graph above to answer these questions.

a. How many children were either hiking or skating?

b. How many more children were flying kites than were catching polliwogs?

c. What number should be in the section named "other"?

Problem set 110 Use the information in the circle graph to answer questions 1–3.

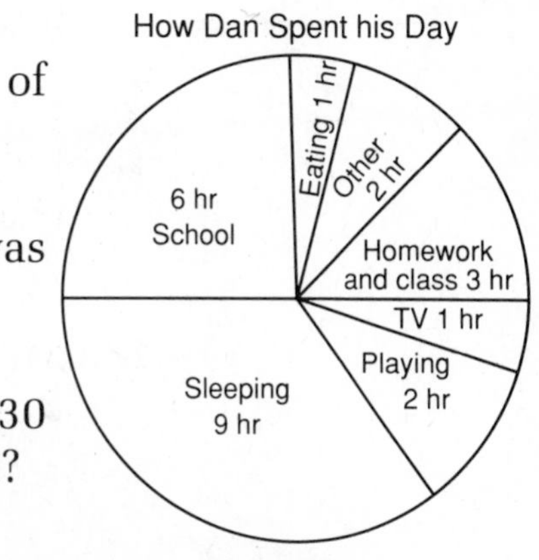

1. What is the total number of hours shown in the graph?

2. What fraction of Dan's day was spent watching TV?

3. If Dan's school starts at 8:30 a.m., at what time does it end?

4. Five sixths of the 288 marchers were out of step. How many marchers were out of step? Illustrate the problem.

5. Something is wrong with this sign. Draw two different signs that show ways to correct this error.

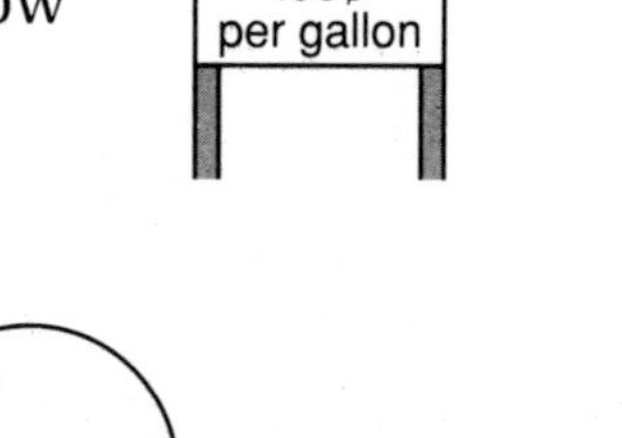

6. (a) What is the radius of this circle?
   (b) What is the diameter of this circle?

7. Use words to write $3126\frac{1}{2}$.

8. Estimate the product of 88 and 22.

9. Apples were priced at 53¢ per pound. What was the cost of 5 pounds of apples?

10. Write the number 403,708 in expanded form. Then use words to write the number.

11. Find the sum of two hundred ninety-three thousand, five hundred seventy-two and seven thousand, nine hundred forty-five.

12. Four pounds of pears cost $1.20. What did 1 pound of pears cost? What did 6 pounds of pears cost?

13. Mike's car would go 150 miles in 3 hours. How far would it go in 1 hour? How far would it go in 8 hours?

14. $\begin{array}{r} \$46.00 \\ -\ \$45.56 \\ \hline \end{array}$

15. $\begin{array}{r} 10{,}165 \\ -\quad 856 \\ \hline \end{array}$

16. $\begin{array}{r} \$\ 0.63 \\ 1.49 \\ 12.24 \\ 0.38 \\ 0.06 \\ 5.00 \\ +\ \ 1.20 \\ \hline \end{array}$

17. 70×70

18. 71×69

19. $\begin{array}{r} 37 \\ \times\ \ 60 \\ \hline \end{array}$

20. $\begin{array}{r} 56 \\ \times\ 42 \\ \hline \end{array}$

21. $\begin{array}{r} \$5.97 \\ \times\quad 8 \\ \hline \end{array}$

22. $4\overline{)\$30.00}$

23. $3\overline{)263}$

24. $5\overline{)4080}$

25. $3156 \div 6$

26. $\begin{array}{r} 1 \\ 2 \\ 4 \\ 3 \\ 5 \\ 7 \\ 8 \\ 2 \\ +\ N \\ \hline 35 \end{array}$

27. $\begin{array}{r} 9 \\ 8 \\ 4 \\ A \\ 6 \\ 2 \\ 9 \\ 4 \\ +\ 3 \\ \hline 49 \end{array}$

28. $890 \div 7$

29. $\dfrac{344}{8}$

30. Draw and shade circles to show that 2 equals $\frac{4}{2}$.

LESSON 111 Decimal Number Line: Tenths

Speed Test: 90 Division Facts, Form B (Test J in Test Booklet)

The arrow is pointing to three and one tenth on this number line. We may write three and one tenth as a mixed number or as a decimal.

2 3 4

$3\frac{1}{10}$ or 3.1

Centimeter scales are often divided into tenths. Since a millimeter is a tenth of a centimeter, each tenth of a centimeter mark on a centimeter scale also marks a millimeter.

Example 1 Find the length of this segment to the nearest tenth of a centimeter.

cm 1 2 3 4 5

Solution The length of the segment is 4 centimeters plus a fraction. The fraction is two tenths. The length of the segment is **4.2 cm.**

Example 2 (a) Find the length of the segment in millimeters.

(b) Find the length of the segment in centimeters.

mm 10 20 30 40 50

cm 1 2 3 4 5

Solution (a) **23 mm** (b) **2.3 cm**

Practice Write the decimal number that is associated with the point to which each arrow is pointing.

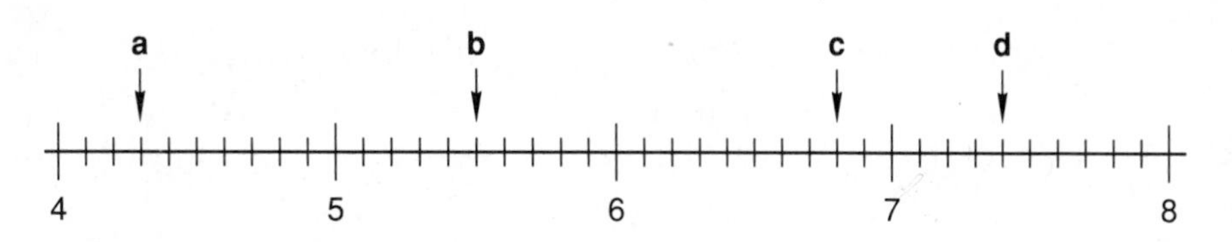

Problem set 111

1. All 110 books must be packed in boxes. Each box will hold 8 books. How many boxes are needed?

2. What number is 5 more than the product of 6 and 7?

3. Gabriel gave the man $7 to pay for the tape. He got back a quarter and two dimes. How much did the tape cost?

4. Four fifths of the 600 gymnasts did back handsprings. How many gymnasts did back handsprings? Illustrate the problem.

5. What is the value of 2 one-hundred-dollar bills, 5 ten-dollar bills, 4 one-dollar bills, 3 dimes, and 1 penny?

6. (a) Find the length of this line segment in millimeters.
(b) Find the length of the segment in centimeters.

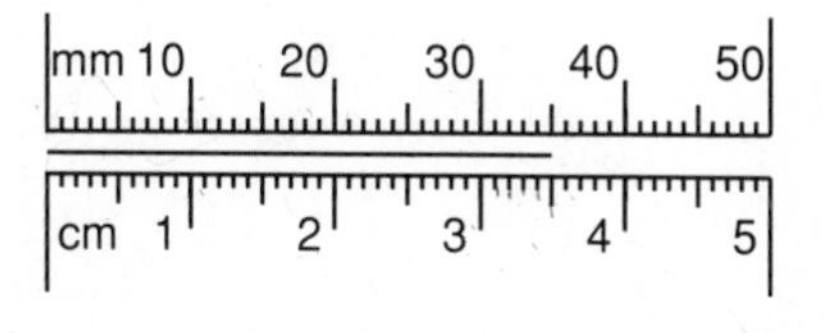

7. Use words to write 12.67.

8. Round 3834 to the nearest thousand.

9. The diameter of the circle is 1 meter. What is the radius of the circle in centimeters?

10. Find the sum of two hundred eighty-six thousand, five hundred fourteen and one hundred thirty-seven thousand, two.

11. At first there were four hundred thirty-one thousand, fifty-nine. Then a whole bunch of them went home. Two hundred fourteen thousand, eleven were left. How many went home?

12. Seven toys cost \$56. What did 1 toy cost? What would 12 toys cost?

13. The big machine could make 248 toys in 8 hours. How many toys could it make in 1 hour? How many toys could it make in 14 hours?

14. \$15 + \$8.75 + \$0.49

15. \$10.00 − (37¢ + \$6)

16. 40×500

17. 41×49

18. $8 \times 5 \times 7$

19. $\begin{array}{r} 32 \\ \times\ 17 \\ \hline \end{array}$

20. $\begin{array}{r} 54 \\ \times\ 63 \\ \hline \end{array}$

21. $\begin{array}{r} 7 \\ 4 \\ 6 \\ 8 \\ 5 \\ 2 \\ 7 \\ 3 \\ +\ K \\ \hline 47 \end{array}$

22. $\begin{array}{r} 9 \\ 1 \\ 7 \\ 6 \\ 5 \\ 2 \\ 8 \\ 4 \\ 3 \\ +\ 2 \\ \hline \end{array}$

23. $\begin{array}{r} 38 \\ \times\ 40 \\ \hline \end{array}$

24. $8\overline{)3616}$

25. $4\overline{)2482}$

26. $7\overline{)3516}$

27. \$4.38 ÷ 6

28. $7162 \div 9$

29. $\frac{1414}{2}$

30. Draw and shade circles to show that 2 equals $\frac{8}{4}$.

LESSON 112

Fractions Equal to 1

Speed Test: 100 Addition Facts (Test A in Test Booklet)

Each of these circles equals 1 whole circle. The fraction below each circle equals the number 1. If the top number and the bottom number of a fraction are the same, the fraction equals 1.

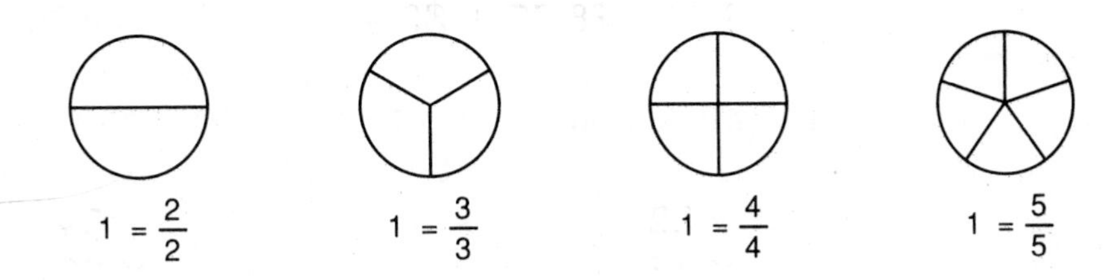

Example 1 Which of these fractions equals 1?

(a) $\frac{1}{6}$ (b) $\frac{6}{6}$ (c) $\frac{7}{6}$

Solution A fraction equals 1 if its top and bottom numbers are the same. The fraction equal to 1 is $\frac{6}{6}$.

Example 2 Write a fraction equal to 1 that has a denominator of 7.

Solution The denominator is the bottom number of the fraction. A fraction equals 1 if its top and bottom numbers are the same. So if the bottom number is 7, the top number must also be 7. We write $\frac{7}{7}$.

Practice **a.** Write a fraction equal to 1 that has a denominator of 6.

b. Which of these fractions equals 1? $\frac{9}{10}$, $\frac{10}{10}$, $\frac{11}{10}$

Which fraction name for 1 is shown by each picture?

c. 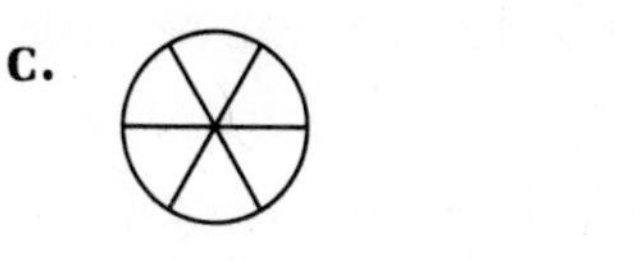

d. 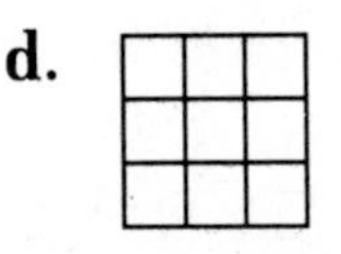

Problem set 112

1. Which even number between 79 and 89 can be divided by 6 without a remainder?

2. How many minutes are in 3 hours?

3. Bill has \$8. Jim has \$2 less than Bill. How much money do they have altogether?

4. Write each fraction or mixed number as a decimal number.

   (a) $\frac{3}{10}$ (b) $4\frac{99}{100}$ (c) $12\frac{1}{100}$

5. Five eighths of the 40 students earned an A on the test. How many students earned an A on the test? Illustrate the problem.

6. (a) What is the diameter of this circle?
   (b) What is the radius of this circle?

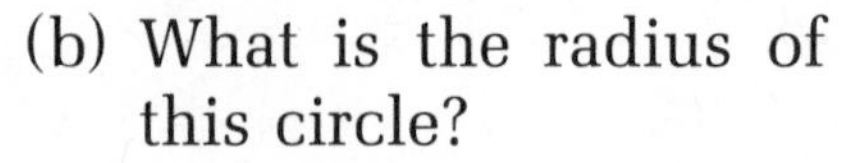

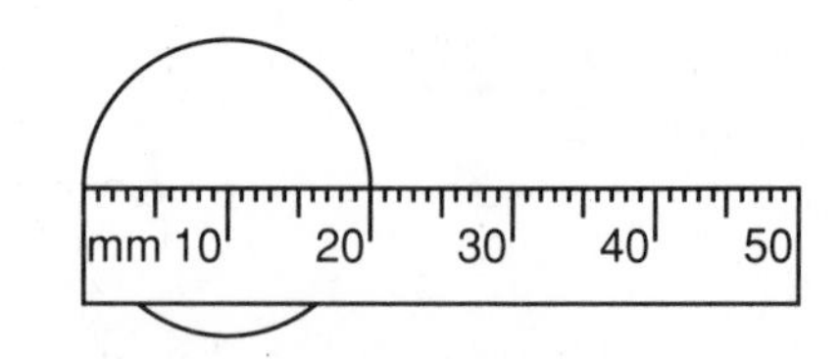

7. Use words to write the number $4325\frac{1}{6}$.

8. Estimate the sum of 498 and 689.

9. Jimbo found that 20 of them would fill 4 containers. How many of them could be put in 1 container? How many of them could he put in 22 containers?

10. Michelle could do 70 jobs in 10 hours. How many jobs could she do in 1 hour? How many jobs could she do in 22 hours?

11. Solve this problem by guessing and then checking your guess. Mike had \$4 more than Sam. Sam had \$56. How much money did Mike have? How much money did both of them have?

12. $\begin{array}{r} 35{,}235 \\ +\ 49{,}609 \\ \hline \end{array}$

13. $\begin{array}{r} \$73.27 \\ +\ \$56.73 \\ \hline \end{array}$

14. $\begin{array}{r} 16 \\ 15 \\ 23 \\ 8 \\ 217 \\ 20 \\ 6 \\ +\ 317 \\ \hline \end{array}$

15. $\begin{array}{r} 47{,}184 \\ -\ 29{,}191 \\ \hline \end{array}$

16. $\begin{array}{r} \$37.50 \\ -\ \$26.35 \\ \hline \end{array}$

17. $\begin{array}{r} \$3.85 \\ \times\quad 7 \\ \hline \end{array}$

18. $\begin{array}{r} 48 \\ \times\ 29 \\ \hline \end{array}$

19. 60×60

20. 59×61

21. $\begin{array}{r} 5 \\ 4 \\ 3 \\ 7 \\ 2 \\ 5 \\ 8 \\ 1 \\ 4 \\ +\ N \\ \hline 45 \end{array}$

22. 37×68

23. $6\overline{)5824}$

24. $9\overline{)\$37.53}$

25. $7\overline{)4205}$

26. $927 \div 3$

27. $8\overline{)3000}$

28. $\dfrac{400}{5}$

29. $\begin{array}{r} 48 \\ -\ N \\ \hline 22 \end{array}$

30. Draw and shade circles to show that $3\frac{3}{4}$ equals $\frac{15}{4}$.

LESSON 113 Changing Improper Fractions to Whole or Mixed Numbers

Speed Test: 100 Addition Facts, (Test A in Test Booklet)

If the top number of a fraction is as large as or larger than the bottom number, the fraction is an **improper fraction**. All of these fractions are improper fractions.

$$\frac{12}{4} \qquad \frac{10}{3} \qquad \frac{9}{4} \qquad \frac{3}{2} \qquad \frac{5}{5}$$

To write an improper fraction as a mixed number, we first need to find out how many wholes the number has. To do this, we divide. The division will tell us how many wholes we have. The remainder is the numerator of the fraction in the mixed number.

Example 1 Write $\frac{13}{5}$ as a mixed number. Draw a picture.

Solution To find the number of wholes, we divide.

$$\begin{array}{r} 2 \\ 5\overline{)13} \\ 10 \\ \hline 3 \end{array} \quad \begin{array}{l} \leftarrow \text{wholes} \\ \\ \\ \leftarrow \text{remainder of 3 \textbf{fifths}} \end{array}$$

This division tells us that $\frac{13}{5}$ equals 2 wholes and 3 fifths left over. We write this $\mathbf{2\frac{3}{5}}$. We can see this if we draw a picture.

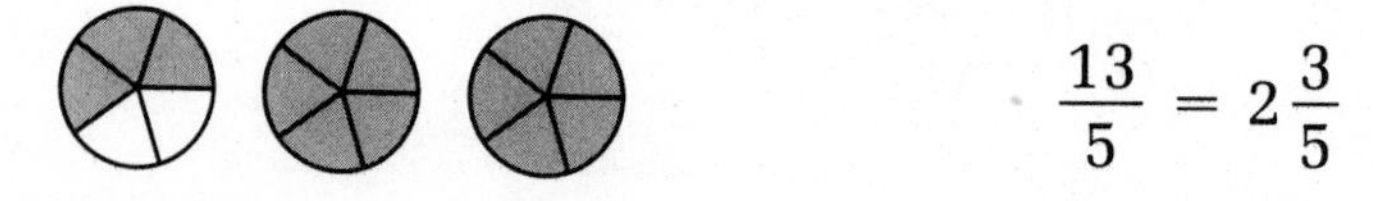

$$\frac{13}{5} = 2\frac{3}{5}$$

Example 2 Write $\frac{10}{3}$ as a mixed number. Then draw a picture.

Solution First we divide.

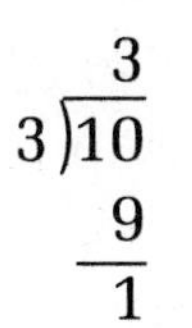

We have 3 wholes and 1 third left over. We write $\mathbf{3\frac{1}{3}}$.

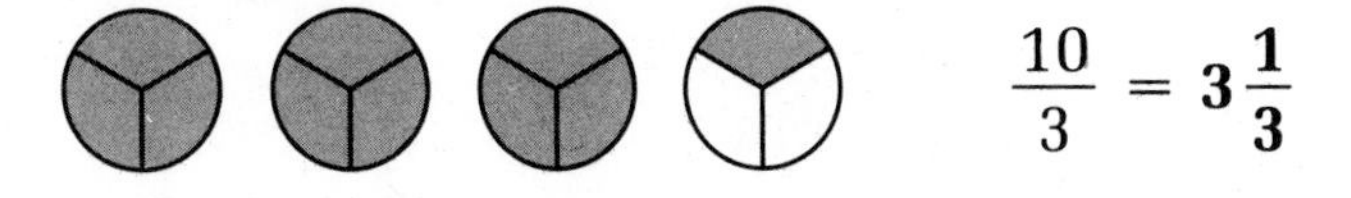

$$\frac{10}{3} = \mathbf{3\frac{1}{3}}$$

Example 3 Write $\frac{12}{4}$ as a whole number. Then draw a picture.

Solution First we divide.

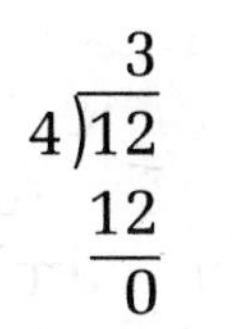

We have 3 wholes and a remainder of zero.

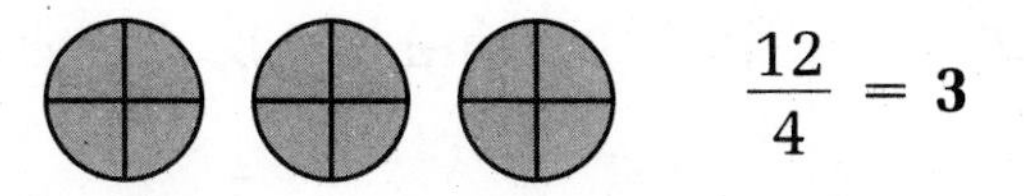

$$\frac{12}{4} = \mathbf{3}$$

Practice Change each improper fraction to a whole number or to a mixed number. Then draw a picture.

a. $\frac{7}{2}$ **b.** $\frac{12}{3}$

c. $\frac{8}{3}$ **d.** $\frac{15}{5}$

e. $\frac{7}{4}$ **f.** $\frac{10}{3}$

Problem set 113

1. If the perimeter of a square is 280 feet, how long is each side of the square?

2. How many full weeks are there in 365 days?

3. Barbara passed out cookies to her 6 friends. Each of her friends got 3 cookies. There were 2 cookies left for Barbara. How many cookies did Barbara have when she began?

4. Three fifths of the 60 leprechauns were less than 2 feet tall. How many leprechauns were less than 2 feet tall? Illustrate the problem.

5. (a) Find the length of this line segment in millimeters.
   (b) Find the length of the line segment in centimeters.

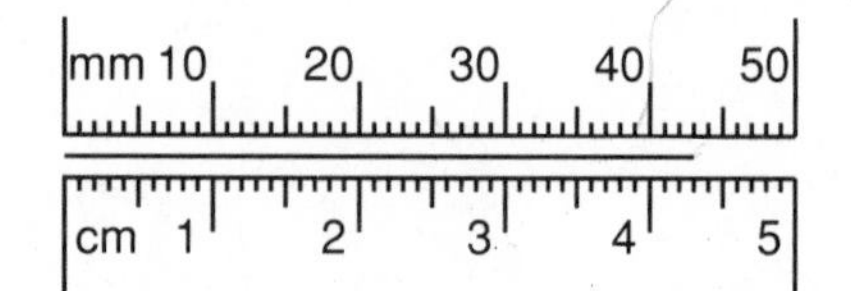

6. What fraction name for 1 is shown by this circle?

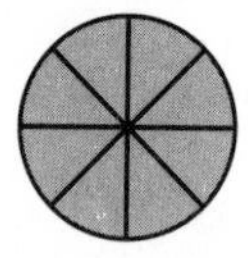

7. Use words to write 432.58.

8. Estimate the product of 87 and 71.

9. Change the improper fraction $\frac{5}{4}$ to a mixed number. Draw a picture to show that the fraction and the mixed number are equal.

10. The wagon train had to travel 1953 miles. The wagons traveled 12 miles per day. How far did the wagons travel in 20 days? How much farther did the wagons have to travel?

11. The cook used about 30 pounds of flour each day to cook pancakes and bread. How many pounds of flour did the cook use in 73 days?

12. The cook found that 132 pounds of potatoes would last 6 days. How many pounds of potatoes were eaten each day? How many pounds of potatoes would be eaten in 73 days?

13. \$6.52 + \$12 + \$1.74 + 26¢

14. 43,217 + 367,490

15. \$80 − (\$63.72 + \$2)

16. 37,614 − 29,148

17. $3 \times 3 \times 3 \times 3$

18. 24×1000

19. 79¢ × 6

20. $\begin{array}{r} 34 \\ \times\ 71 \\ \hline \end{array}$

21. $\begin{array}{r} 86 \\ \times\ 49 \\ \hline \end{array}$

22. $\begin{array}{r} 47 \\ \times\ 63 \\ \hline \end{array}$

23. $4\overline{)2304}$

24. $5\overline{)4815}$

25. $\begin{array}{r} 7 \\ 2 \\ A \\ 6 \\ 5 \\ 7 \\ 3 \\ +\ 2 \\ \hline 57 \end{array}$

26. $6\overline{)3629}$

27. $1436 \div 7$

28. \$18.56 ÷ 8

29. $\dfrac{234}{9}$

30. $\begin{array}{r} N \\ -\ 214 \\ \hline 713 \end{array}$

LESSON 114 Dividing by 10

> **Speed Test:** 100 Addition Facts (Test A in Test Booklet)

We have used a four-step procedure to divide by one-digit numbers. We will use the same four-step procedure to divide by two-digit numbers. In this lesson we will learn how to divide by 10.

Example $10\overline{)432}$

Solution Ten will not divide 4 but will divide 43 four times. In step 1, we are careful to write the 4 above the 3 in 432.

Step 1. We divide $10\overline{)43}$ and write 4.
Step 2. We multiply 4 × 10 and write 40.
Step 3. We subtract 43 − 40 and write 3.
Step 4. We bring down the 2, making 32.

$$\begin{array}{r} \mathbf{43} \\ 10\overline{)432} \\ -\,40 \\ \hline 32 \\ -\,30 \\ \hline \mathbf{r} \rightarrow \mathbf{2} \end{array}$$

Repeat Step 1. We divide $10\overline{)32}$ and write 3.
Step 2. We multiply 3 × 10 and write 30.
Step 3. We subtract 32 − 30 and write 2.
Step 4. There is no number to bring down. The quotient is 43 with a remainder of 2.

Practice **a.** $10\overline{)73}$ **b.** $10\overline{)342}$ **c.** $10\overline{)243}$

d. $10\overline{)720}$ **e.** $10\overline{)561}$ **f.** $10\overline{)380}$

Problem set 114

1. How many 6¢ mints can be bought with 2 quarters?

2. Six times what number is 90?

3. Jason has $8. David has $2 more than Jason. How much money do they have altogether?

4. Three eighths of the 32 elves packed toys on the sleigh. How many elves packed toys on the sleigh? Illustrate the problem.

.5. What is the value of 6 one-hundred-dollar bills, 4 ten-dollar bills, 2 one-dollar bills, 1 dime, and 3 pennies?

6. Write a fraction equal to 1 that has a denominator of 10.

7. Use words to write 8674.3.

8. Estimate the difference we get when 496 is subtracted from 604.

9. Change each improper fraction to a whole number or a mixed number.

(a) $\frac{9}{5}$ (b) $\frac{9}{3}$ (c) $\frac{9}{2}$

10. Gold was discovered in California in 1849. Many people went to California to find gold. If 2400 people came in 10 days, about how many came in 1 day? About how many people came in 1 week?

11. Some miners found about $210 worth of gold each week. What was the approximate value of the gold they found in 1 day? About how much gold did they find in 9 days?

12. One miner bought 6 bags of flour at $4.20 a bag and 8 pounds of salt at 12¢ per pound. How much money did the miner spend?

13. $26.47 + $8.52

14. $\begin{array}{r} 351{,}426 \\ +\ 449{,}576 \\ \hline \end{array}$

15. $\begin{array}{r} 12 \\ 26 \\ 13 \\ 35 \\ 110 \\ 8 \\ +\ 15 \\ \hline \end{array}$

16. 49,249 − 3,755

17. $\begin{array}{r} \$50.00 \\ -\ \$49.49 \\ \hline \end{array}$

18. $\begin{array}{r} \$12.49 \\ \times \quad\ 8 \\ \hline \end{array}$

19. $\begin{array}{r} 73 \\ \times\ 62 \\ \hline \end{array}$

20. 30 × 30 **21.** 28 × 32

24. 54 × 29 **25.** $10\overline{)230}$

26. $7\overline{)2383}$ **27.** $10\overline{)372}$

28. $5.76 ÷ 8 **29.** 412 ÷ 10

30. $\frac{378}{9}$

22.
$$\begin{array}{r} 5 \\ 1 \\ 6 \\ 8 \\ 4 \\ 3 \\ 7 \\ 2 \\ 4 \\ 5 \\ +\,N \\ \hline 62 \end{array}$$

23.
$$\begin{array}{r} 1 \\ 6 \\ 3 \\ 5 \\ K \\ 7 \\ 2 \\ 1 \\ 8 \\ 6 \\ +\,2 \\ \hline 58 \end{array}$$

LESSON 115 Adding Decimal Numbers

Speed Test: 100 Addition Facts (Test A in Test Booklet)

We remember that when we add money, we line up all the decimal points.
We follow this same rule when we add other decimal numbers. When we add decimal numbers, we must be careful to line up the decimal points.

$$\begin{array}{r} \$1.43 \\ +\ \$0.25 \\ \hline \$1.68 \end{array}$$

↑ All decimal points are in line

Example Add:

$$\begin{array}{r} 3.47 \\ 0.36 \\ +\ 1.4 \\ \hline . \end{array}$$

← decimal point in the answer

Solution We add decimals with all the decimal points in line. The decimal point in the answer is also in line. We think of empty places as zeros.

$$\begin{array}{r} 3.47 \\ 0.36 \\ +\ 1.4 \\ \hline \mathbf{5.23} \end{array}$$

Practice Add these decimal numbers. Put the decimal point in each answer before you add.

a. $\begin{array}{r} 2.37 \\ 4.6 \\ +\ 1.25 \\ \hline \end{array}$ **b.** $\begin{array}{r} 1.36 \\ 0.47 \\ +\ 6.2 \\ \hline \end{array}$ **c.** $\begin{array}{r} 3.65 \\ 12.4 \\ +\ 0.57 \\ \hline \end{array}$ **d.** $\begin{array}{r} 3.4 \\ 6.75 \\ +\ 15.1 \\ \hline \end{array}$

Problem set 115

Use the information given to answer questions 1 and 2.

Samantha has 6 cats. Each cat eats $\frac{1}{2}$ can of food each day. Cat food costs 47¢ per can.

1. How many cans of cat food are used each day?

2. How much does Samantha spend on cat food per day?

3. If the perimeter of a square is 240 inches, how long is each side?

4. Math was the favorite class of five sevenths of the 28 students. Math was the favorite class of how many students? Illustrate the problem.

5. Something is wrong with this sign. Draw two different signs to show two ways to correct this error.

Admission .75¢ each

6. (a) What is the radius of this circle?
(b) What is the diameter of this circle?

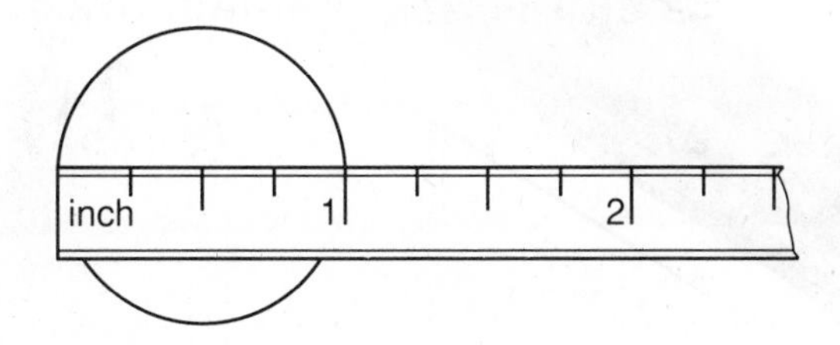

7. Use words to write 417,523.43.

8. Estimate the product of 61 and 397.

9. Change the improper fraction $\frac{4}{4}$ to a whole number.

10. What fraction name for 1 is shown by the squares in this rectangle?

11. Jewell went to the fair. She paid $6.85 for a doll and paid $4.50 for lunch. Then she bought a soft drink for 75¢. How much money did she spend?

12. Mary Jane bought 2 dolls for $7.40 each. She paid the clerk with a $20 bill. How much change did she get back?

13. The big truck that transported the ferris wheel could go only 140 miles in 5 hours. How far could the truck go in 1 hour? How far could the truck go in 23 hours?

14. $\begin{array}{r} 37.49 \\ +\ 6.35 \\ \hline \end{array}$

15. $\begin{array}{r} 5.43 \\ 12.7 \\ +\ 3.2 \\ \hline \end{array}$

16. $\begin{array}{r} 73.48 \\ 5.63 \\ +\ 17.9 \\ \hline \end{array}$

17. $\begin{array}{r} \$65.00 \\ -\ \$29.87 \\ \hline \end{array}$

18. $\begin{array}{r} 24,375 \\ -\ 8,416 \\ \hline \end{array}$

19. $\begin{array}{r} \$3.68 \\ \times\ 9 \\ \hline \end{array}$

20. 90×90

21. 89×91

22. $\begin{array}{r} 2 \\ 5 \\ 7 \\ 4 \\ N \\ 2 \\ 5 \\ 8 \\ 1 \\ +\ 6 \\ \hline 48 \end{array}$

23. 73×36

24. $3\overline{)763}$

25. $10\overline{)430}$

26. $6\overline{)\$57.24}$

27. $765 \div 9$

28. $563 \div 10$

29. $\frac{462}{7}$

30. $\begin{array}{r} 463 \\ -\ N \\ \hline 274 \end{array}$

LESSON 116 Subtracting Decimal Numbers

Speed Test: 100 Addition Facts (Test A in Test Booklet)

When we subtract money, we put all the decimal points in line.

$$\begin{array}{r} \$6.57 \\ -\ \$1.30 \\ \hline \$5.27 \end{array}$$

When we subtract other decimal numbers, we are careful to put all the decimal points in line. The decimal point in the answer is below the other decimal points.

Example Subtract:

$$\begin{array}{r} 6.57 \\ -\ 1.3 \\ \hline . \end{array}$$

← decimal point in the answer

Solution All the decimal points are in line. We think of empty places as zeros. The decimal point in the answer is directly below the other decimal points.

$$\begin{array}{r} 6.57 \\ -\ 1.30 \\ \hline 5.27 \end{array}$$

Practice Subtract.

a. $\begin{array}{r} 26.45 \\ -\ 12.14 \\ \hline \end{array}$ **b.** $\begin{array}{r} 5.46 \\ -\ 2.3 \\ \hline \end{array}$ **c.** $\begin{array}{r} 42.7 \\ -\ 6.4 \\ \hline \end{array}$ **d.** $\begin{array}{r} 15.31 \\ -\ 8.5 \\ \hline \end{array}$

Problem set 116

1. Wendy bought 5 tickets for $2.75 each. She paid for them with a $20 bill. How much money did she get back?

2. Fifty cents is divided equally among 3 friends. There will be some cents left. How many cents will be left?

3. What is the result we get when four hundred nine is subtracted from nine hundred four?

4. Two ninths of the 45 stamps were from Brazil. How many of the stamps were from Brazil? Illustrate the problem.

5. (a) Find the length of this line segment in millimeters.
(b) Find the length of the segment in centimeters.

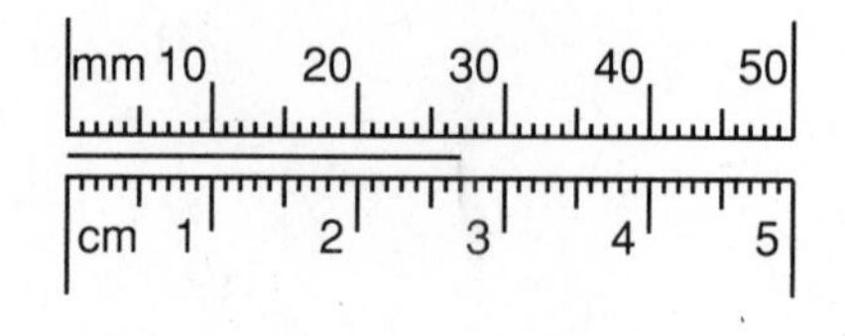

6. The pizza was cut into 10 equal slices. The whole sliced pizza shows what fraction name for 1?

7. Use words to write 3722.91.

8. Round 5167 to the nearest thousand.

9. Change the improper fraction $\frac{9}{4}$ to a mixed number.

10. Which of these fractions is **not** equal to 1?

(a) $\frac{12}{12}$ (b) $\frac{11}{11}$ (c) $\frac{11}{10}$ (d) $\frac{10}{10}$

11. The first group paid \$197 for the trip. There were 7 members in the second group. They paid \$27 each for the trip. Which group paid the most money in all for the trip?

12. In the summer of 1926, there were only 17 stores in the town. Today there are 8 times as many stores in the town. How many stores are there in the town today? How many more stores are there today than there were in 1926?

13. The whole trip was 496 miles. The group traveled for 5 days at 27 miles per day. How much farther did they have to go?

14. $\begin{array}{r} 36.27 \\ +\ 14.7 \\ \hline \end{array}$

15. $\begin{array}{r} 3.62 \\ 12.8 \\ +\ 17.42 \\ \hline \end{array}$

16. $\begin{array}{r} 26.14 \\ 8.9 \\ +\ 13.7 \\ \hline \end{array}$

17. $\begin{array}{r} 36.31 \\ -\ 7.4 \\ \hline \end{array}$

18. $\begin{array}{r} 8.41 \\ -\ 3.26 \\ \hline \end{array}$

19. $\begin{array}{r} 23.73 \\ -\ 9.8 \\ \hline \end{array}$

20. 27×32

21. 62×15

22. 43×49

23. $10\overline{)460}$

24. $9\overline{)\$27.36}$

25. $\begin{array}{r} 9 \\ 8 \\ 7 \\ 2 \\ 4 \\ 6 \\ 5 \\ 4 \\ 3 \\ +\ N \\ \hline 55 \end{array}$

26. $6\overline{)2316}$

27. $1543 \div 7$

28. $532 \div 10$

29. $\dfrac{256}{8}$

30. $\begin{array}{r} 400 \\ -\ XYZ \\ \hline 222 \end{array}$

LESSON 117 Setting Up Decimal Addition Problems

Speed Test: 100 Subtraction Facts (Test B in Test Booklet)

When we add numbers, we add digits that have the same place values. To be sure we add digits with the same place values when we add decimal numbers, we line up the decimal points.

Example Add: $36.4 + 4.26 + 3.1$

Solution We begin by writing the numbers in a column **with the decimal points in line**. This lines up digits with the same place values. **Next we put the decimal point of the answer in line with the other decimal points.** Then we add.

$$\begin{array}{r} 36.4 \\ 4.26 \\ +\ 3.1 \\ \hline . \end{array} \leftarrow \text{decimal point}$$

$$\begin{array}{r} 36.4 \\ 4.26 \\ +\ 3.1 \\ \hline \mathbf{43.76} \end{array}$$

Practice Add these decimal numbers.

a. 4.6 + 12.57

b. 2.36 + 42.8

c. 3.6 + 0.27 + 1.29

d. 14.6 + 36.4 + 16.35

e. 4.65 + 16.3 + 0.84

f. 38.3 + 3.83 + 0.38

Problem set 117

1. Nelson bought 8 pounds of oranges. He gave the storekeeper a $5 bill and got $1.96 back in change. What did 1 pound of oranges cost?

2. Mark had a dozen cookies. He ate two cookies and then gave half of the rest to a friend. How many cookies did he have left?

3. What number is 6 less than the product of 5 and 4?

4. Two thirds of the 12 guitar strings were out of tune. How many of the guitar strings were out of tune? Illustrate the problem.

5. What is the value of 7 ten-dollar bills, 4 one-dollar bills, 3 dimes, and 2 pennies?

6. Write a fraction equal to 1 that has a denominator of 5.

7. Use words to write $397\frac{3}{4}$.

8. Estimate the sum of 4178 and 6899.

9. Change these improper fractions to whole numbers or mixed numbers.

(a) $\frac{7}{3}$ (b) $\frac{8}{4}$ (c) $\frac{9}{5}$

10. The hiking club dues are $15 per year. There were 17 members. The expenses for the year amounted to $270. Did the club have less than enough money or more than enough money? How much more or less?

11. For the first 3 hours they hiked at 3 miles per hour. For the next 2 hours they hiked at 4 miles per hour. If the total trip was 25 miles, how far did they still have to go?

12. Caps for the 17 members cost $20 each. If they had $340 in the bank, how much was left after the caps were paid for?

13. 41.6 + 13.17 + 9.2

14. 45.32 + 4.39 + 16.2

15. 23.4 + 17.8 + 3.27

16. 43.17 + 2.86 + 17.4

17. $\begin{array}{r} \$0.35 \\ \times \quad 8 \\ \hline \end{array}$

18. $\begin{array}{r} 54 \\ \times 12 \\ \hline \end{array}$

19. $\begin{array}{r} 67 \\ \times \quad 30 \\ \hline \end{array}$

20. 80 × 80

21. 79 × 81

22. 73 × 64

23. $4\overline{)\$6.36}$

24. $10\overline{)520}$

25. $\begin{array}{r} 6 \\ 2 \\ 5 \\ 9 \\ 3 \\ 7 \\ 8 \\ 2 \\ 3 \\ 4 \\ + N \\ \hline 57 \end{array}$

26. $7\overline{)4345}$

27. 243 ÷ 10

28. 5448 ÷ 6

29. $\frac{175}{5}$

30. $\begin{array}{r} XYZ \\ + 478 \\ \hline 635 \end{array}$

LESSON 118 Setting Up Decimal Subtraction Problems

Speed Test: 100 Subtraction Facts (Test B in Test Booklet)

When we subtract numbers, we subtract digits that have the same place values. To be sure we subtract digits with the same place values, we line up the decimal points.

Example 1 Subtract: 36.45 − 4.2

Solution We write the first number on top. We write the second number underneath **so that the decimal points are lined up. Then we write the decimal point in the answer so that this decimal point is lined up with the other two decimal points.** Then we subtract. We think of empty places as zeros.

$$\begin{array}{r} 36.45 \\ -\ 4.2 \\ \hline . \end{array} \leftarrow \text{decimal point}$$

$$\begin{array}{r} 36.45 \\ -\ 4.2 \\ \hline \mathbf{32.25} \end{array}$$

Example 2 Subtract:

$$\begin{array}{r} 4.5 \\ -\ 3.27 \\ \hline . \end{array} \leftarrow \text{decimal point in the answer}$$

Solution We subtract the bottom digits from the top digits. We see that we need a digit above the 7. So we write a zero in the empty place. Then we subtract.

$$\begin{array}{r} {\scriptstyle 4\ 1} \\ 4.\not{5}0 \\ -\ 3.27 \\ \hline \mathbf{1.23} \end{array} \leftarrow \text{We write a zero here.}$$

Practice Subtract.

a. $\begin{array}{r} 3.6 \\ -\ 1.43 \\ \hline \end{array}$ b. $\begin{array}{r} 26.2 \\ -\ 12.56 \\ \hline \end{array}$ c. $\begin{array}{r} 8.7 \\ -\ 6.54 \\ \hline \end{array}$ d. $\begin{array}{r} 1.9 \\ -\ 0.19 \\ \hline \end{array}$

e. $3.45 - 2.1$ f. $36.47 - 3.6$ g. $34.6 - 3.46$

Problem set 118 Use the information given to answer questions 1 and 2. Mark kept a tally of the number of vehicles that drove by his house during 1 hour.

Number of Vehicles

Cars	𝍸 𝍸 𝍸 𝍸 IIII
Trucks	𝍸 𝍸
Motorcycles	II
Bicycles	𝍸 II

1. How many more cars than trucks drove by Mark's house?

2. Altogether, how many vehicles drove by Mark's house?

3. What number is 6 less than the sum of 7 and 8?

4. Beth read three tenths of the 180 pages in 1 day. How many pages did she read in 1 day? Illustrate the problem.

5. (a) What is the diameter of the dime?
   (b) What is the radius of the dime?

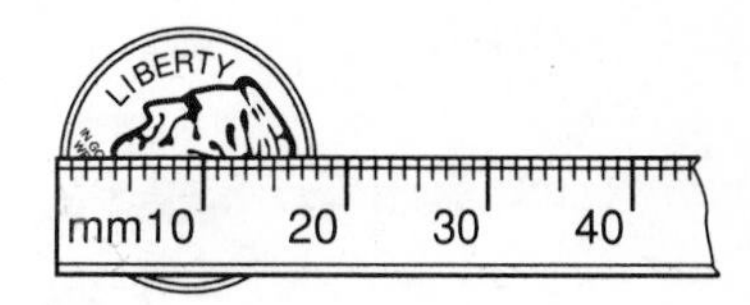

6. The candy bar was broken into 4 equal pieces. The broken candy bar shows what fraction name for 1?

7. Use words to write 17,654.1.

8. Estimate the product of 78 and 32.

9. Change the improper fraction $\frac{5}{2}$ to a mixed number. Draw a picture that shows that the fraction and the mixed number are equal.

10. The camel walked 12 miles the first day. Each of the next 4 days it walked 2 more miles than it walked the first day. How far did the camel walk in 5 days?

11. The group purchased 5 tickets for $630. How much did each ticket cost? What would they have had to pay for 36 tickets?

12. Solve this problem by guessing and then checking your guess. There were 40 red and black checkers. There were 8 more red checkers than black checkers. How many were red and how many were black?

13. $86.34 + 7.96 + 3.1$

14. $29.15 + 4.6 + 16.2$

15. $24.34 - 8.5$

16. $26.4 - 15.18$

17. $4 \times 3 \times 2 \times 1$

18. 26×30

19. 46×12

20. 20×20

21. 19×21

22. 36×64

23. $8\overline{)\$16.48}$

24. $6\overline{)3744}$

25. $$\begin{array}{r} 8 \\ 4 \\ 7 \\ 6 \\ 8 \\ 4 \\ 7 \\ 6 \\ 2 \\ + K \\ \hline 55 \end{array}$$

26. $10\overline{)360}$

27. $3217 \div 5$

28. $423 \div 10$

29. $\dfrac{138}{6}$

30. $$\begin{array}{r} 463 \\ + \quad N \\ \hline 742 \end{array}$$

LESSON 119 Dividing by Multiples of 10, Part 1

Speed Test: 100 Subtraction Facts (Test B in Test Booklet)

We have practiced dividing by 10. In this lesson we will begin dividing by multiples of 10. Multiples of 10 are 20, 30, 40, 50, 60, and so on. Dividing by multiples of 10 is a little harder than dividing by one-digit numbers because we have not memorized the division facts for two-digit numbers. To help us divide by a two-digit number, we may think of dividing by the first digit only.

To help us divide this, → $20\overline{)72}$

we may think this: → $2\overline{)7}$

We use the answer to the easier division for the answer to the more difficult division. Since $2\overline{)7}$ is 3, we use 3 as the division answer. We finish by doing the multiplication and subtraction steps.

$$\begin{array}{r} 3 \\ 20\overline{)72} \\ \underline{60} \\ 12 \end{array} \leftarrow \text{r} \qquad \begin{array}{r} 3 \text{ r } 12 \\ 20\overline{)72} \\ \underline{60} \\ 12 \end{array}$$

← We write the answer this way.

Since we divided 20 into 72, we put the 3 above the 2 in 72.

$$\begin{array}{l} 3 \\ 20\overline{)72} \end{array}$$

This is not correct!
Do not write the 3 above the 7.
This means there are three 20s in 7.

$$\begin{array}{r} 3 \\ 20\overline{)72} \end{array}$$

This is the correct way.
The 3 above the 2 means there are three 20s in 72.

It is important to place the digits in the answer correctly!

Example $30\overline{)127}$

Solution To help us divide this, we may mentally block out the last digit of each number.

We see $30\overline{)127}$, but we think $3\overline{)12}$.

This helps us find that the division answer is 4. Since we are really dividing $30\overline{)127}$, we write the 4 above the 7. Then we multiply 4×30 and write 120. Then we subtract 120 from 127 and write 7. This is the remainder. The answer is **4 r 7.**

$$\begin{array}{r} \mathbf{4\ r\ 7} \\ 30\overline{)127} \\ \underline{120} \\ 7 \end{array}$$

Practice **a.** $30\overline{)72}$ **b.** $20\overline{)87}$ **c.** $40\overline{)95}$

d. $20\overline{)127}$ **e.** $40\overline{)127}$ **f.** $30\overline{)217}$

Problem set 119

1. Eighty students were divided among 3 classrooms as equally as possible. For your answer, name three numbers to show how many students were in each classroom.

2. If the sum of 3 and 4 is subtracted from the product of 3 and 4, what is the difference?

3. Irma is twice as old as her sister and 3 years younger than her brother. Irma's sister is 6 years old. How old is Irma's brother?

4. Four ninths of 513 fans cheered when the touchdown was scored. How many fans cheered? Draw a diagram to illustrate the problem.

5. Something is wrong with this sign. Draw two different signs to show ways to correct this error.

Cash for cans
.85¢ / pound

6. This circle shows what fraction name for 1?

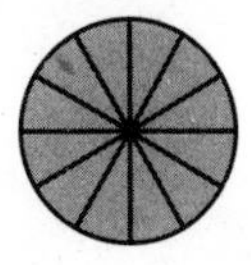

7. Use words to write 1608.72.

8. Estimate the sum of 589 and 398.

9. Change the improper fraction $\frac{6}{2}$ to a whole number. Draw a picture that shows that the fraction and the whole number are equal.

10. Jim could run 7 laps in 42 minutes. How long did each lap take? How long would it take Jim to run 10 laps?

11. Salamona bought 3 scarves for $2.75 each. She paid with a $10 bill. How much change did she receive?

12. Two tickets for the play cost $26. How much would fifteen tickets cost?

13. $7.43 + 6.25 + 12.7$

14. $13.24 + 12.5 + 1.37$

15. $14.36 - 7.5$

16. $28.6 - 2.86$

17. 90×800

18. $8 \times 73¢$

19. $7 \times 6 \times 5 \times 0$

20. 24×30

21. 62×75

22. 57×48

23. $423 \div 7$

24. $5\overline{)\$37.15}$

25.
$$\begin{array}{r} 1 \\ 2 \\ 5 \\ 3 \\ 6 \\ 5 \\ 4 \\ 7 \\ 9 \\ 2 \\ +\ N \\ \hline 48 \end{array}$$

26. $\dfrac{1240}{10}$

27. $60\overline{)240}$

28. $30\overline{)95}$

29. $40\overline{)88}$

30.
$$\begin{array}{r} N \\ -\ 462 \\ \hline 705 \end{array}$$

LESSON 120 Adding Fractions with Like Denominators

Speed Test: 100 Subtraction Facts (Test B in Test Booklet)

Adding fractions is something like adding apples.

1 apple + 1 apple = 2 apples

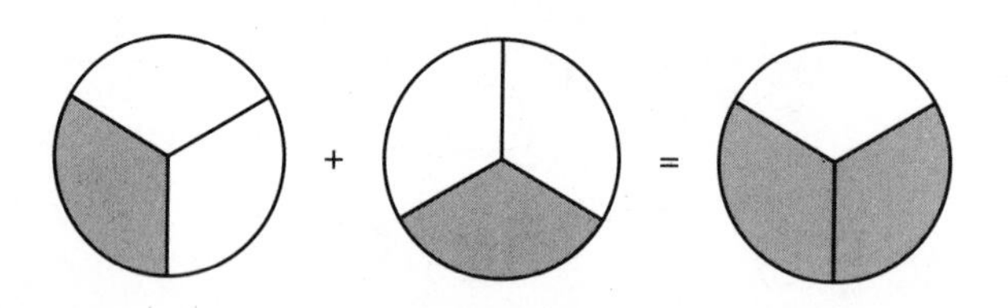

1 third + 1 third = 2 thirds

When we add fractions, we add the top numbers (numerators). We do not add the bottom numbers (denominators).

Example Add: $\frac{3}{5} + \frac{1}{5}$

Solution Three fifths plus one fifth is four fifths.
We add the top numbers only.

Practice Add these fractions.

a. $\frac{1}{3} + \frac{1}{3}$ **b.** $\frac{1}{4} + \frac{2}{4}$ **c.** $\frac{1}{5} + \frac{2}{5}$

d. $\frac{3}{10} + \frac{4}{10}$ **e.** $\frac{5}{12} + \frac{6}{12}$ **f.** $\frac{8}{25} + \frac{9}{25}$

g. How much is three eighths plus four eighths?

Problem set 120 The table below shows how much is charged to ship a package to different parts of the country. Use the information given to answer questions 1 and 2.

WEIGHT IN POUNDS	SHIPPING CHARGES		
	ZONE 1	ZONE 2	ZONE 3
Up to 4 pounds	\$1.75	\$1.93	\$2.36
4 pounds 1 ounce to 7 pounds	\$1.91	\$2.15	\$2.68
7 pounds 1 ounce to 10 pounds	\$2.07	\$2.38	\$3.02
Over 10 pounds	\$2.48	\$2.71	\$3.38

1. How much does it cost to ship an 8-pound package to Zone 2?

2. How much more does it cost to ship a 12-pound package to Zone 3 than to Zone 2?

3. If 2 apples cost 30¢, how much will 4 apples cost?

4. Two fifths of the 70 polliwogs already had back legs. How many of the polliwogs had back legs?

5. (a) Find the length of this line segment in millimeters.
   (b) Find the length of this segment in centimeters.

mm 10 20 30 40 50

cm 1 2 3 4 5

6. The apple was cut into 8 equal slices. The sliced apple shows what fraction name for 1?

7. Use words to write 21,572.4.

8. Round 7842 to the nearest thousand.

9. Change each improper fraction to a whole number or a mixed number.

(a) $\frac{11}{2}$ (b) $\frac{11}{3}$ (c) $\frac{11}{4}$

10. Jeff and Wright found that 11 golf balls would fill each hole. There were 9 holes. How many golf balls did they need to fill all the holes?

11. Nan bought 8 golf balls for $24. How much would she have to pay for 24 golf balls?

12. Solve this problem by guessing and checking your guess. Mike and Carl played the first hole on the golf course. They hit the ball a total of 13 times. Carl hit the ball 3 more times than Mike did. How many times did Carl hit the ball?

13. 9.7 + 14.6 + 0.83

14. 8.79 + 16.4 + 7.3

15. 86.34 − 19.2

16. 13.6 − 4.21

17. $\frac{5}{9} + \frac{2}{9}$

18. $\frac{2}{10} + \frac{3}{10}$

19.
$$\begin{array}{r} 5 \\ 2 \\ 7 \\ 4 \\ 1 \\ 2 \\ 3 \\ 6 \\ 8 \\ +\ N \\ \hline 46 \end{array}$$

20.
$$\begin{array}{r} 25 \\ 4 \\ 13 \\ 7 \\ 32 \\ 5 \\ 47 \\ 8 \\ 6 \\ +\ 13 \\ \hline \end{array}$$

21. $\frac{15}{20} + \frac{2}{20}$

22. 27 × 60

23. 36 × 35

24. 29 × 17

25. $6\overline{)545}$

26. $23.80 ÷ 10

27. $\frac{483}{7}$

28. $30\overline{)153}$ **29.** $20\overline{)84}$ **30.** $40\overline{)367}$

LESSON 121 Subtracting Fractions with Like Denominators

Speed Test: 100 Subtraction Facts (Test B in Test Booklet)

We remember that when we add fractions we add only the top numbers (numerators). We do not add the bottom numbers (denominators).

$$\frac{3}{7} + \frac{2}{7} = \frac{5}{7}$$

When we subtract fractions, we subtract only the top numbers. The bottom number does not change.

$$\frac{5}{7} - \frac{2}{7} = \frac{3}{7}$$

Example 1 Subtract: $\frac{3}{5} - \frac{1}{5}$

Solution **We subtract only the top numbers:** 3 fifths minus 1 fifth is 2 fifths. $\frac{3}{5} - \frac{1}{5} = \mathbf{\frac{2}{5}}$

Example 2 Subtract: $\frac{8}{23} - \frac{5}{23}$

Solution When we subtract fractions, we subtract the top numbers (numerators). We do not change the bottom number.

$$\frac{8}{23} - \frac{5}{23} = \mathbf{\frac{3}{23}}$$

Practice Subtract these fractions.

a. $\frac{2}{3} - \frac{1}{3}$ **b.** $\frac{3}{4} - \frac{2}{4}$ **c.** $\frac{4}{5} - \frac{2}{5}$

d. $\frac{9}{10} - \frac{6}{10}$ e. $\frac{8}{12} - \frac{3}{12}$ f. $\frac{5}{6} - \frac{5}{6}$

Problem set 121 Use the information given in the graph to answer questions 1 and 2.

1. What was Brenda's speeds for the 100-meter dash in Race 3?

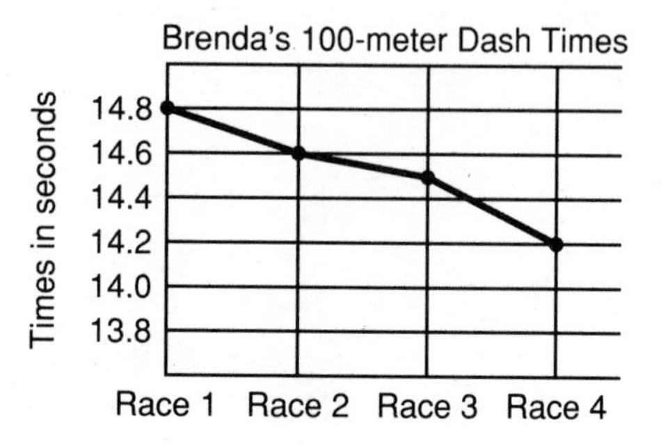

2. Are Brenda's speeds getting faster or slower?

3. Johnny planted two thousand apple seeds. Five hundred twenty grew to be trees. How many seeds did not grow to be trees?

4. Write each decimal number as a mixed number.
(a) 12.41 (b) 24.1 (c) 2.03

5. Seven tenths of the 300 students bought their lunch. How many of the students bought their lunch? Draw a diagram to illustrate the problem.

6. What fraction name for 1 has a denominator of 6?

7. Use words to write 2157.24.

8. Estimate the product of 41 and 396.

9. Change the improper fraction $\frac{8}{3}$ to a mixed number. Draw a picture that shows the two numbers are equal.

10. In Greek mythology, Jason sailed the ship *Argo* in search of the golden fleece. His shipmates were called Argonauts. They sailed 33 miles in 3 days. How far could they sail in 1 day? How far could they sail in 7 days?

11. The Argonauts had fish for supper. There were 8 Argonauts and 50 fish. How many fish were left over after the fish were divided equally among the Argonauts?

12. Solve this problem by guessing and checking your guess. For dessert they had ambrosia. Altogether Jason and Castor ate 33 spoonfuls of ambrosia. Jason ate 5 more spoonfuls than Castor did. How many spoonfuls did each man eat?

13. $3.6 + 4.25 + 16.7$

14. $5.34 + 1.9 + 16.18$

15. $43.27 - 12.6$

16. $16.4 - 1.64$

17. $\frac{4}{10} + \frac{5}{10}$

18. $\frac{5}{11} + \frac{3}{11}$

19. $\frac{3}{5} + \frac{2}{5}$

20. $\frac{8}{9} - \frac{4}{9}$

21. $\frac{5}{7} - \frac{4}{7}$

22.
$$\begin{array}{r} 3 \\ 7 \\ 6 \\ 4 \\ N \\ 2 \\ 8 \\ 5 \\ 4 \\ +\ 7 \\ \hline 51 \end{array}$$

23.
$$\begin{array}{r} 54 \\ 17 \\ 14 \\ 68 \\ 20 \\ 36 \\ 21 \\ 23 \\ 32 \\ +\ 55 \\ \hline \end{array}$$

24. $\frac{1}{3} - \frac{1}{3}$

25.
$$\begin{array}{r} \$4.39 \\ \times \quad 8 \\ \hline \end{array}$$

26.
$$\begin{array}{r} 68 \\ \times\ 27 \\ \hline \end{array}$$

27. $\begin{array}{r} 49 \\ \times\ 73 \\ \hline \end{array}$

28. $40\overline{)91}$

29. $60\overline{)480}$

30. $30\overline{)271}$

LESSON 122 Equivalent Fractions

> **Speed Test:** 64 Multiplication Facts (Test G in Test Booklet)

The same fraction of each circle below has been shaded. We see that different fractions are used to name the shaded parts.

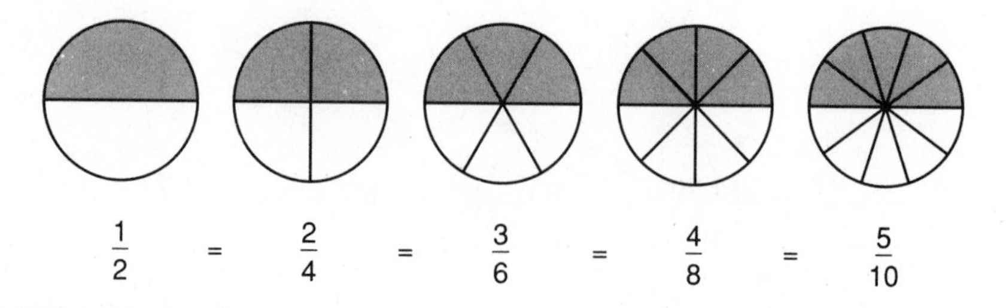

These fractions all name the same amount. Different fractions that name the same amount are called **equivalent fractions.**

Example 1 The rectangle on the left has 3 equal parts. We see that 2 parts are shaded. Two thirds of the figure is shaded.

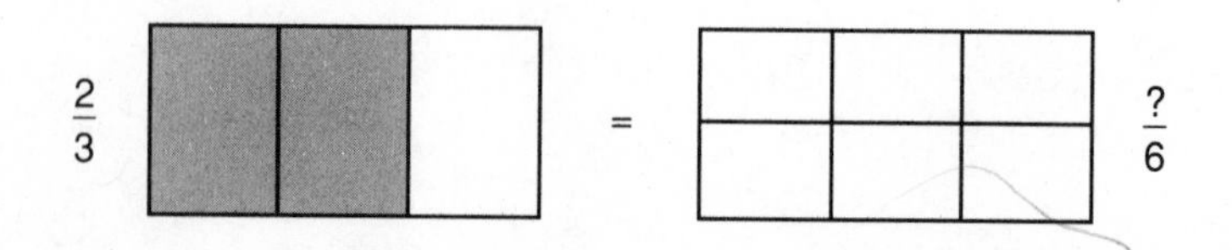

The rectangle on the right has 6 equal parts. How many parts must be shaded so that the same fraction of this rectangle is shaded?

Solution We see that 4 of the 6 parts must be shaded. Two thirds is the same as four sixths.

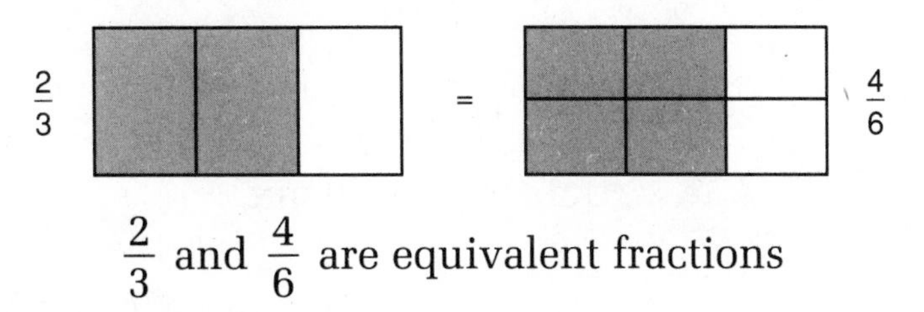

$\frac{2}{3}$ and $\frac{4}{6}$ are equivalent fractions

Example 2 What equivalent fractions are shown?

Solution The same fraction of each rectangle is shaded. The rectangles show that

$$\frac{\mathbf{2}}{\mathbf{8}} = \frac{\mathbf{1}}{\mathbf{4}}$$

Practice Name the equivalent fractions shown by the pictures in (a) and (b).

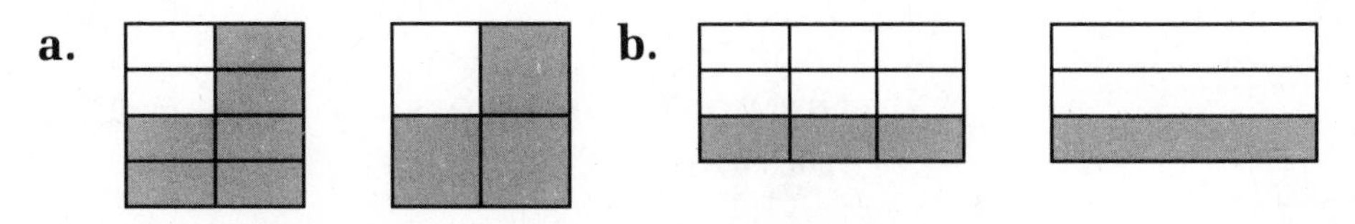

Draw pictures to show that the following pairs of fractions are equivalent.

c. $\frac{2}{4} = \frac{1}{2}$

d. $\frac{4}{6} = \frac{2}{3}$

e. $\frac{2}{8} = \frac{1}{4}$

Problem set 122

1. There are 60 seconds in 1 minute. How many seconds are in 60 minutes? There are 60 minutes in 1 hour. How many seconds are in 1 hour?

2. A can of soup serves 3 people. How many cans of soup are needed to serve 14 people?

3. Mark was 9 years old when his brother was 12. How old will Mark be when his brother is 21 years old?

4. Three tenths of the 2000 trees in the forest were pine trees. How many pine trees were in the forest? Illustrate the problem.

5. Change each improper fraction to a mixed number.

(a) $\frac{19}{3}$ (b) $\frac{19}{4}$ (c) $\frac{19}{5}$

6. Use words to write 19,384.56.

7. What equivalent fractions are shown?

8. Draw a picture to show that $\frac{2}{4}$ and $\frac{1}{2}$ are equivalent fractions.

9. Jamie could drive 135 miles in 3 hours. How far could she drive in 1 hour? How far could she drive in 9 hours?

10. Williken could hop at 2 miles per hour and could run at 5 miles per hour. He hopped for 4 hours and ran for 3 hours. How far did he go?

11. There were 30 seats in each room. There were 17 rooms. How many students could sit in all the rooms?

12. 16.24 − 5.7

13. 32.4 − 1.26

14.
$$\begin{array}{r} 3.4 \\ 2.7 \\ 0.52 \\ 3.6 \\ +\ 1.47 \\ \hline \end{array}$$

15. $\frac{7}{10} + \frac{2}{10}$

16. $\frac{1}{4} + \frac{1}{4} + \frac{1}{4}$

17. $\frac{3}{7} - \frac{1}{7}$

18. $\frac{5}{12} - \frac{4}{12}$

19. $\frac{3}{4} - \frac{3}{4}$

20. $\begin{array}{r} \$4.53 \\ \times \quad 7 \\ \hline \end{array}$

21. $\begin{array}{r} 78 \\ \times\ 36 \\ \hline \end{array}$

22. $\begin{array}{r} 64 \\ \times\ 83 \\ \hline \end{array}$

23. $\$56.40 \div 6$

24. $\begin{array}{r} 7 \\ 4 \\ 8 \\ 5 \\ 3 \\ V \\ 9 \\ 5 \\ +\ 4 \\ \hline 63 \end{array}$

25. $\begin{array}{r} 24 \\ 36 \\ 25 \\ 43 \\ 19 \\ 7 \\ 66 \\ 43 \\ 50 \\ +\ 93 \\ \hline \end{array}$

26. $9\overline{)545}$

27. $\frac{3536}{8}$

28. $260 \div 10$

29. $275 \div 30$

30. $350 \div 50$

LESSON
123

Square Units: An Activity

Speed Test: 64 Multiplication Facts (Test G in Test Booklet)

A square unit is a square with a side of a certain length. Here we show a **square centimeter** and a **square inch** in actual size.

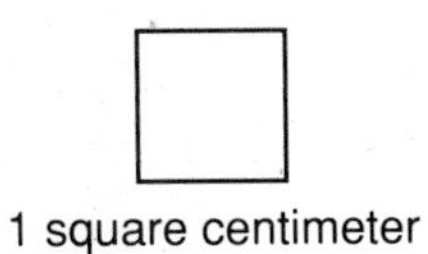
1 square centimeter

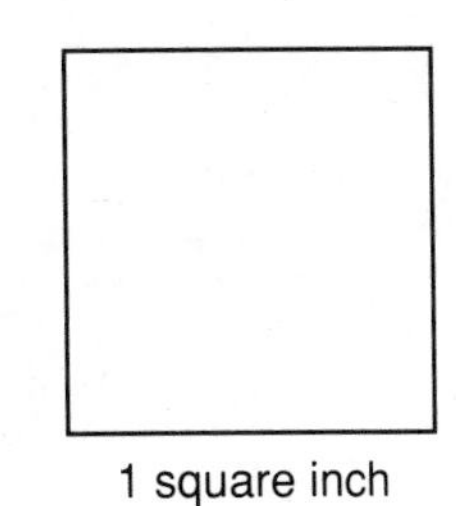
1 square inch

A square centimeter has sides that are 1 centimeter long. A square inch has sides that are 1 inch long.

There are larger square units that will not fit on this page in actual size. A square foot is a square that is a little larger than a page in this book. A square yard is a square with sides 1 yard long. A square mile is a square with sides 1 mile long.

Practice Activity: Cut out some square centimeters and some square inches from construction paper. Be sure they are the same size as the squares on the previous page. Then use the squares you have cut out so that you can answer the questions in each rectangle.

a. How many square inches does it take to cover this rectangle?

b. How many square centimeters does it take to cover this rectangle?

c. On the floor of the room, mark off 1 square foot, 1 square yard, and 1 square meter. Estimate the number of each kind of square it would take to cover the whole floor.

Problem set 123

1. Three hundred seconds is how many minutes?

2. David, Ann, and Chad were playing marbles. Ann had twice as many marbles as David had, and Chad had 5 more marbles than Ann had. David had 9 marbles. How many marbles did Chad have?

3. On each of 5 bookshelves there are 44 books. How many books are on all 5 shelves?

4. Nine tenths of the 30 students remembered their homework. How many students remembered their homework?

5. What fraction name for 1 has a denominator of 3?

6. Use words to write 7462.12.

7. What equivalent fractions are shown?

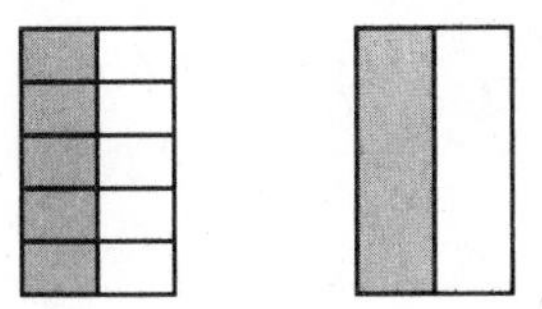

8. Draw a picture to show that $\frac{6}{8}$ and $\frac{3}{4}$ are equivalent fractions.

9. When Susan first looked in the microscope, she could see seven thousand, thirteen microbes. When she looked the next time, she could see only two thousand, seven hundred fifteen microbes. How many more microbes did she see the first time?

10. Sam could do 7 experiments in 42 days. How many experiments could he do in 1 day? How many experiments could he do in 17 days?

11. Sarah had to hurry. The laboratory had to be cleaned by 4:20 p.m. It was already 10:35 a.m. How much time did she have to clean the lab?

12. $4.3 + 12.6 + 3.75$

13. $0.24 + 6.7 + 83.49$

14. $2126.47 - 183.5$

15. $364.1 - 16.41$

16. $\frac{5}{8} + \frac{2}{8}$

17. $\frac{3}{5} + \frac{1}{5}$

18. $\frac{9}{10} - \frac{2}{10}$

19. 59×86

20. 73×48

21. $9 \times 78¢$

22. $9\overline{)2675}$ **23.** $\$35.00 \div 4$ **24.** $\frac{4824}{8}$

25. $60\overline{)540}$

26.
$$\begin{array}{r} 9 \\ 4 \\ 3 \\ 6 \\ N \\ 2 \\ 5 \\ 7 \\ +\ 3 \\ \hline 43 \end{array}$$

27.
$$\begin{array}{r} 4 \\ 23 \\ 5 \\ 47 \\ 1 \\ 58 \\ 6 \\ 44 \\ +\ 2 \\ \hline \end{array}$$

28. $10\overline{)463}$

29. $70\overline{)295}$

30.
$$\begin{array}{r} N \\ +\ 209 \\ \hline 726 \end{array}$$

LESSON 124 Finding the Area of a Rectangle

Speed Test: 64 Multiplication Facts (Test 6 in Test Booklet)

The **area** of a rectangle is the number of squares of a certain size that it takes to cover the rectangle. In the preceding lesson, we covered rectangles with square centimeters and square inches. In this lesson, we will multiply to find the number of squares that are needed to cover a rectangle.

Example 1 Altogether, how many square feet does it take to cover this rectangle?

4 ft
3 ft

Solution Notice that the rectangle is not actual size. Each small square is not 1 square foot. We have drawn the rectangle smaller so it will fit on the page.

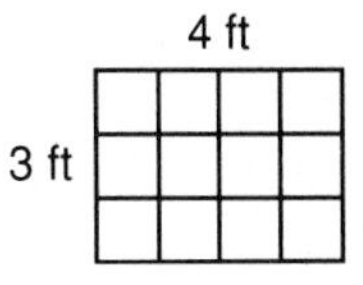

To answer the question, we could copy the rectangle, finish drawing the squares, and count 12 square feet. Another way to answer the question is to multiply the numbers for the length and width of the rectangle. The length equals the number of squares that fit in one row, and the width equals the number of rows. By multiplying 3 rows of 4 squares, we also get **12 square feet.**

Example 2 What is the area of this rectangle?

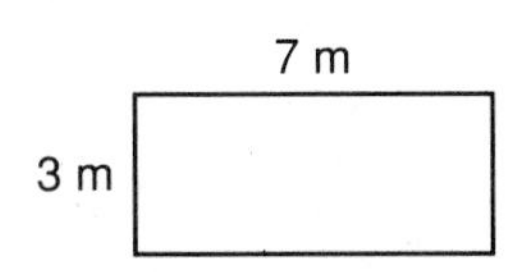

Solution The units used in the drawing are meters. So the area is the number of square meters that will cover the rectangle. The numbers given for the length and width also show the number of squares that will fit along each side. We see that 3 rows of 7 squares will cover the rectangle. We multiply 3×7 and find that the area of the rectangle is **21 square meters.**

Notice that in each of the examples we use "square" units. When stating the area of a shape, we label our answer "square inches," "square meters," "square feet," or whatever square unit is shown by the problem.

Practice Find the area of each rectangle.

a.

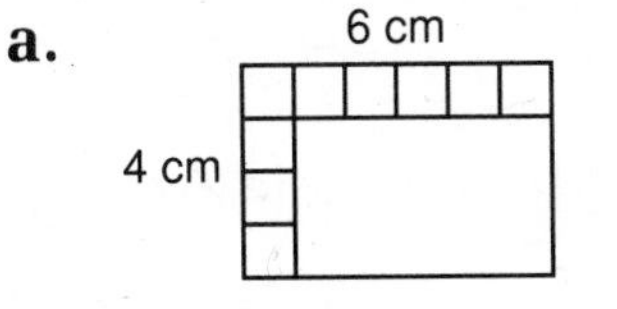

b.

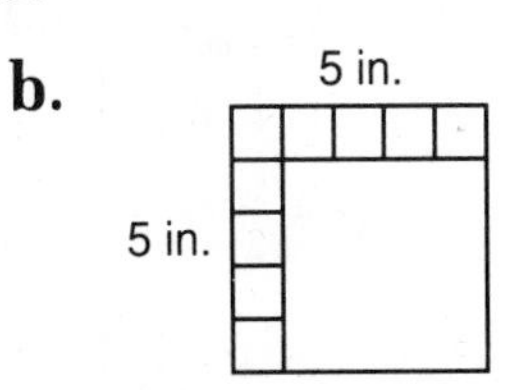

c.

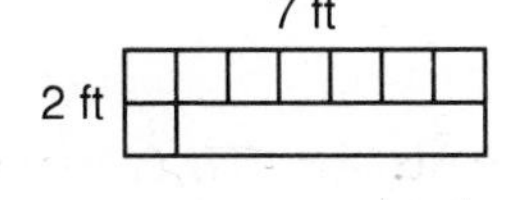

d.

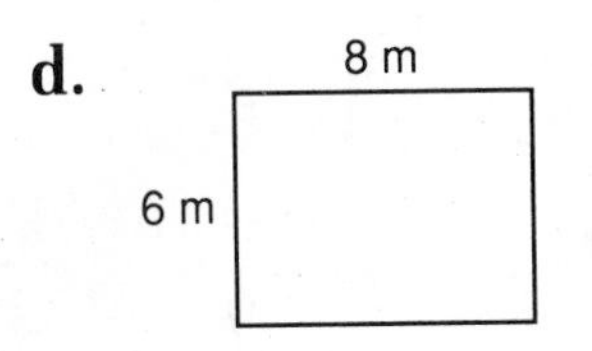

Problem set 124

1. One fence board costs 90¢. It takes 10 boards to build 5 feet of fence. How many boards are needed to build 50 feet of fence? What will the boards cost?

2. Find the area of each rectangle.

3. (a) Find the length of this line segment in millimeters.
   (b) Find the length of the segment in centimeters.

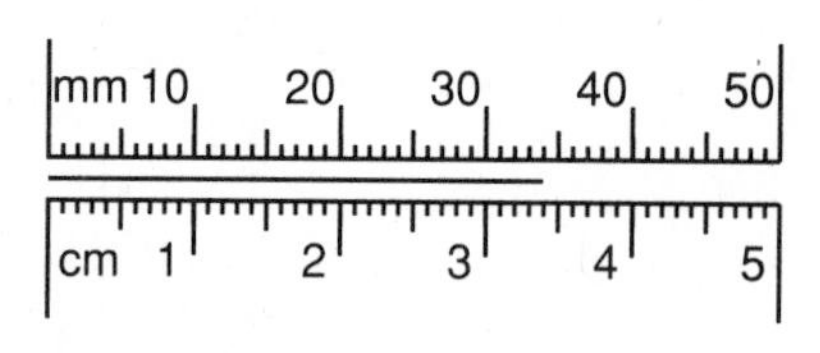

4. Five ninths of the 36 burros were gray. How many of the burros were gray? Illustrate the problem.

5. Change each improper fraction to a whole number or to a mixed number.

   (a) $\frac{15}{2}$ (b) $\frac{15}{3}$ (c) $\frac{15}{4}$

6. Use words to write 493,265.62.

7. What equivalent fractions are shown?

8. Draw a picture to show that $\frac{3}{6}$ and $\frac{1}{2}$ are equivalent fractions.

9. T-shirts cost \$5 each. Mark had \$27. He bought 5 T-shirts. How much money did he have left?

10. Baseballs cost \$1.45 each. Zollie had \$15. She bought 8 baseballs. How much money did she have left?

11. The group bought 12 items for \$1.40 each and 3 items for \$5.20 each. If they began with \$50, how much money did they have left?

12. $25.42 - 24.8$

13. $36.2 - 4.27$

14. $\begin{array}{r} 37.2 \\ 135.7 \\ 10.62 \\ 2.47 \\ +\ 14.0 \\ \hline \end{array}$

15. $\frac{3}{9} + \frac{4}{9}$

16. $\frac{1}{7} + \frac{2}{7} + \frac{3}{7}$

17. $\frac{5}{8} - \frac{5}{8}$

18. $\frac{11}{12} - \frac{10}{12}$

19. $\frac{8}{10} - \frac{5}{10}$

20. $\begin{array}{r} 48 \\ \times\ 36 \\ \hline \end{array}$

21. $\begin{array}{r} 72 \\ \times\ 58 \\ \hline \end{array}$

22. $\begin{array}{r} \$4.08 \\ \times\ \ \ 7 \\ \hline \end{array}$

23. $\frac{4716}{6}$

24. $7\overline{)2549}$

25. $\$19.40 \div 5$

26. $490 \div 10$

27. $90 \div 20$

28. $171 \div 40$

29. $\begin{array}{r} N \\ +\ 462 \\ \hline 523 \end{array}$

30. $\begin{array}{r} N \\ -\ 462 \\ \hline 743 \end{array}$

LESSON 125 Dividing by Multiples of 10, Part 2

Speed Test: 64 Multiplication Facts (Test 6 in Test Booklet)

When we divide by multiples of 10, we continue to follow the four-step division method. We divide, multiply, subtract, and bring down. We have practiced division problems with one-digit answers. In this lesson, we will practice division problems with two-digit answers.

Example 1 $30\overline{)963}$

Solution Step 1. We begin by breaking the problem into the smaller division problem, $30\overline{)96}$, and write 3.

Step 2. We multiply 3 × 30 and write 90.

Step 3. We subtract 96 − 90 and write 6.

Step 4. We bring down the 3, making 63.

$$\begin{array}{r} \mathbf{32}\ \mathbf{r}\ \mathbf{3} \\ 30\overline{)963} \\ \underline{90} \\ 63 \\ \underline{60} \\ 3 \end{array}$$

Repeat Step 1. We divide $30\overline{)63}$ and write 2.

Step 2. We multiply 2 × 30 and write 60.

Step 3. We subtract 63 − 60 and write 3.

Step 4. There are no more numbers to bring down. The quotient is 32 with a remainder of 3.

Example 2 $40\overline{)2531}$

Solution Step 1. We begin by breaking the problem into a smaller division problem. Since $40\overline{)25}$ does not give an answer of 1 or more, we start with the division $40\overline{)253}$ and write 6 above the 3.

$$\begin{array}{r} \mathbf{63} \text{ r } \mathbf{11} \\ 40\overline{)2531} \\ \underline{240} \\ 131 \\ \underline{120} \\ 11 \end{array}$$

Step 2. We multiply 6 × 40 and write 240.
Step 3. We subtract 253 − 240 and write 13.
Step 4. We bring down the 1, making 131.

Repeat Step 1. We divide $40\overline{)131}$ and write 3.
Step 2. We multiply 3 × 40 and write 120.
Step 3. We subtract 131 − 120 and write 11.
Step 4. There are no more numbers to bring down. The quotient is 63 with a remainder of 11.

Practice Divide.

a. $20\overline{)862}$ **b.** $30\overline{)870}$ **c.** $40\overline{)888}$

d. $50\overline{)2300}$ **e.** $60\overline{)2400}$ **f.** $70\overline{)2453}$

Problem set 125

1. Admission to the zoo was $3.25 for adults and $1.50 for children. Gary bought tickets for 2 adults and 3 children. Altogether, how much did the tickets cost?

2. Find the area of each rectangle.

3. Ten quarters is the same as how many nickels?

4. Something is wrong with this sign. Draw two different signs to show ways to correct the error.

Lemonade
.35¢
per glass

5. What fraction of the letters in AARDVARK are A's?

6. What fraction name for 1 has a denominator of 4?

7. Use words to write 36,497.8.

8. Draw a picture to show that $\frac{1}{3}$ and $\frac{2}{6}$ are equivalent fractions.

9. The champion bike rider could ride 58 miles in 2 hours. At the same speed, how far could the champion ride in 11 hours?

10. The bikers bought 11 caps at \$5.40 each and 7 straps at \$1.20 each. They paid for the items with a \$100 bill. How much change did they receive?

11. Forty-nine students stood patiently in line. Eight of them could ride in each vehicle. There were only 5 vehicles. How many students did not get a ride?

12. 38.42 + 5.71 + 14.3

13. \$6 + 6¢ + \$3.45

14. 43.0 − 7.49

15. 125.46 − 24.9

16. $\frac{1}{9} + \frac{1}{9}$

17. $\frac{4}{11} + \frac{5}{11}$

18. $\frac{8}{9} - \frac{1}{9}$

19. $\frac{8}{12} - \frac{1}{12}$

20. 58 × 26

21. 8 × \$3.75

22. 47 × 36

23. $\$36.28 \div 4$

24.
$$\begin{array}{r} 4 \\ 3 \\ 8 \\ 5 \\ 2 \\ 7 \\ 9 \\ 9 \\ 4 \\ + N \\ \hline 73 \end{array}$$

25.
$$\begin{array}{r} 17 \\ 22 \\ 33 \\ 42 \\ 17 \\ 85 \\ 21 \\ 37 \\ 46 \\ 13 \\ + 20 \\ \hline \end{array}$$

26. $9\overline{)4321}$

27. $\dfrac{456}{6}$

28. $30\overline{)480}$

29. $40\overline{)512}$

30. $20\overline{)1240}$

LESSON 126 Multiplying a Three-Digit Number by a Two-Digit Number

Speed Test: 64 Multiplication Facts (Test G in Test Booklet)

We have learned to multiply a two-digit number by another two-digit number. In this lesson we will learn to multiply a three-digit number by a two-digit number.

Example 1 Multiply: 364×24

Solution We write the three-digit number on top. We write the two-digit number underneath so that the last digits are lined up. We multiply 364 by 4. Next we multiply 364 by 2. We write the last digit of this product in the tens' place, which is under the 2 in 24. Then we add and find that the product is **8736**.

$$\begin{array}{r} {\scriptstyle 1} \\ {\scriptstyle 21} \\ 364 \\ \times \quad 24 \\ \hline 1456 \\ 728 \\ \hline 8736 \end{array}$$

Example 2 Multiply:

$$\begin{array}{r} 407 \\ \times\ 38 \\ \hline \end{array}$$

Solution First we multiply 407 by 8. Then we multiply 407 by 3. We write the last digit of this product in the tens' place, which is under the 3 in 38. We add and find the product is **15,466**.

$$\begin{array}{r} \scriptstyle 5 \\ 407 \\ \times\ 38 \\ \hline 3256 \\ 1221\ \\ \hline 15,466 \end{array}$$

Practice Multiply.

a. 235 × 24 **b.** 14 × 430 **c.** 125 × 28

d. $\begin{array}{r} 406 \\ \times\ 35 \\ \hline \end{array}$ **e.** $\begin{array}{r} 620 \\ \times\ 31 \\ \hline \end{array}$ **f.** $\begin{array}{r} 562 \\ \times\ 47 \\ \hline \end{array}$

Problem set 126

1. Carrie's cousin lives 3000 miles away. If Carrie drove 638 miles the first day, 456 miles the second day, and 589 miles the third day, how much farther does she need to drive to get to her cousin's house?

2. Find the area of each rectangle.

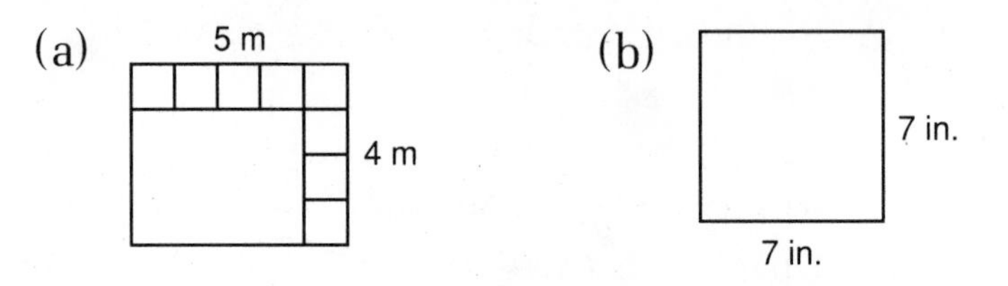

3. If the perimeter of a square is 2 meters, each side is how many centimeters long?

4. Gracie found 35 pines cones. Four sevenths of the pine cones still had seeds. How many of the pine cones still had seeds? Illustrate the problem.

5. Round 6843 to the nearest thousand.

6. What fraction name for 1 has a denominator of 5?

7. Use words to write 55,374.25.

8. Draw a picture to show that $\frac{1}{2}$ and $\frac{4}{8}$ are equivalent fractions.

9. Dwight found twenty-three thousand, five hundred seventy-two on the first day. He found nine hundred forty-four on the second day. How many more did Dwight find on the first day?

10. Daniel Boone could furnish the settlers with 750 pounds of meat in 5 days. How much meat could he furnish in 1 day? How much meat could he furnish in 13 days?

11. The explorer Zebulon Pike estimated that the mountain was eight thousand, seven hundred forty-two feet high. This was five thousand, three hundred sixty-eight feet less than the actual height. Today, we call this mountain Pike's Peak. How high is Pike's Peak?

12. 30.07 − 3.7

13. 46.0 − 12.46

14.
$$\begin{array}{r} 37.15 \\ 6.84 \\ 1.29 \\ 29.1 \\ +\ 3.6 \\ \hline \end{array}$$

15. $\frac{8}{15} + \frac{6}{15}$

16. $\frac{3}{10} + \frac{3}{10} + \frac{3}{10}$

17. $\frac{5}{15} - \frac{3}{15}$

18. $\frac{7}{8} - \frac{7}{8}$

19. $\frac{5}{6} - \frac{4}{6}$

20.
$$\begin{array}{r} 320 \\ \times\ \ 46 \\ \hline \end{array}$$

21.
$$\begin{array}{r} 4 \\ 3 \\ 2 \\ 2 \\ 1 \\ 3 \\ 5 \\ N \\ +\ 4 \\ \hline 32 \end{array}$$

22.
$$\begin{array}{r} 25 \\ 31 \\ 42 \\ 22 \\ 13 \\ 17 \\ 91 \\ 84 \\ +\ 23 \\ \hline \end{array}$$

23.
$$\begin{array}{r} 142 \\ \times\ 23 \\ \hline \end{array}$$

24.
$$\begin{array}{r} 307 \\ \times\ 25 \\ \hline \end{array}$$

25. $6\overline{)4837}$

26. $\dfrac{1372}{4}$

27. $\$30.00 \div 8$

28. $40\overline{)960}$

29. $30\overline{)725}$

30. $20\overline{)1360}$

LESSON 127 Finding Equivalent Fractions by Multiplying

Speed Test: 90 Division Facts, Form A (Test I in Test Booklet)

We remember that when we multiply a number by 1 the answer equals the number.

$$2 \times 1 = 2 \qquad 2000 \times 1 = 2000 \qquad \frac{1}{2} \times 1 = \frac{1}{2}$$

We also remember that there are many ways to write 1.

$$1 = \frac{2}{2} = \frac{3}{3} = \frac{4}{4} = \frac{5}{5} = \frac{6}{6} = \cdots$$

We can use these two facts to find equivalent fractions. If we multiply a fraction by a fraction name for 1, the result is an equivalent fraction.

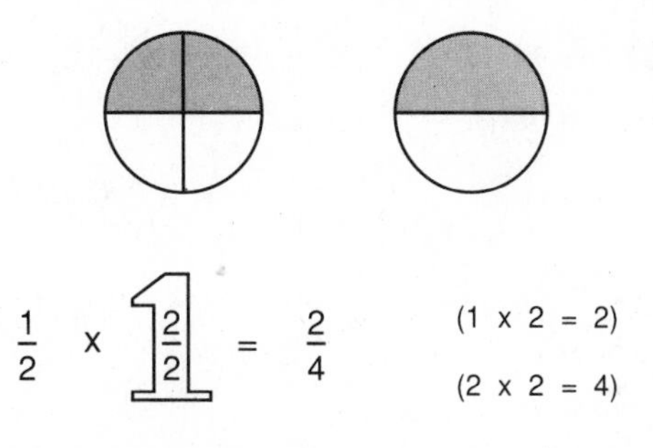

We found that $\frac{1}{2}$ is equivalent to $\frac{2}{4}$ by multiplying $\frac{1}{2}$ by $\frac{2}{2}$, which is a fraction name for 1. We can find other fractions equivalent to $\frac{1}{2}$ by multiplying by other fraction names for 1.

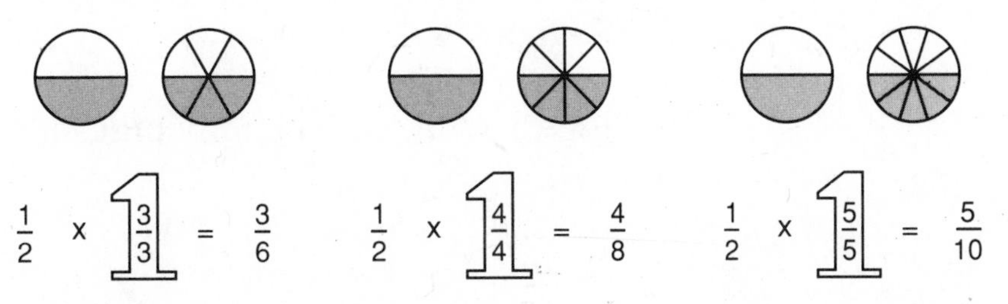

Example Find four fractions equivalent to $\frac{1}{3}$ by multiplying $\frac{1}{3}$ by $\frac{2}{2}$, $\frac{3}{3}$, $\frac{4}{4}$, and $\frac{5}{5}$.

Solution

$$\frac{1}{3} \times \frac{2}{2} = \frac{2}{6} \qquad \frac{1}{3} \times \frac{3}{3} = \frac{3}{9}$$

$$\frac{1}{3} \times \frac{4}{4} = \frac{4}{12} \qquad \frac{1}{3} \times \frac{5}{5} = \frac{5}{15}$$

We find that four fractions equivalent to $\frac{1}{3}$ are

$$\mathbf{\frac{2}{6}, \frac{3}{9}, \frac{4}{12}, \frac{5}{15}}$$

Practice Find four equivalent fractions for each of these fractions by multiplying by $\frac{2}{2}$, $\frac{3}{3}$, $\frac{4}{4}$, and $\frac{5}{5}$.

a. $\frac{1}{4}$

b. $\frac{5}{6}$

c. $\frac{2}{5}$

d. $\frac{1}{10}$

Problem set 127 Use the information in the table to answer questions 1 and 2.

Lengths of Rivers

	MILES	KILOMETERS
Nile	4145	6671
Amazon	3915	6300
Yangtze	3900	6276
Congo	2718	4374
Mississippi	2348	3778

1. The Nile is how many kilometers longer than the Yangtze?

2. The Amazon is how many miles longer than the Congo?

3. How many feet are in 10 yards?

4. The squirrel found 1236 acorns during the summer. It saved $\frac{2}{3}$ of the acorns for winter. How many acorns did the squirrel save for winter?

5. (a) Find the perimeter of this rectangle in centimeters.

(b) Find the perimeter of the rectangle in millimeters.

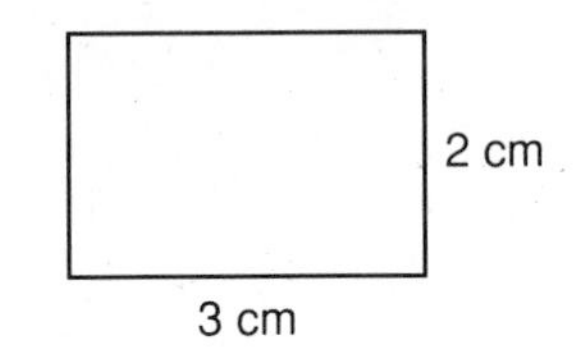

6. Estimate the sum of 627 and 288.

7. Use words to name $3578\frac{2}{3}$.

8. Find three fractions equivalent to $\frac{1}{2}$ by multiplying by $\frac{3}{3}$, $\frac{6}{6}$, and $\frac{10}{10}$.

9. On his way to discovering the Pacific Ocean, Vasco de Balboa could travel only 42 miles in 7 days. How far could he travel in 1 day? How far could he travel in 13 days?

10. Solve this problem by guessing and checking your answer. In one week Balboa saw 77 buffalo and deer. He saw 13 more buffalo than deer. How many deer did he see?

11. Solve this problem by guessing and then checking your guess. DeSoto's men carried a total of 30 lances and pikes. If they had 4 more lances than pikes, how many lances did they have?

12. $4.63 + 26.7 + 0.34$

13. $8.4 + 23.57 + 16.9$

14. $23.91 - 9.3$

15. $60.4 - 0.64$

16. $\frac{8}{12} + \frac{3}{12}$

17. $\frac{2}{7} + \frac{3}{7}$

18. $\frac{15}{16} - \frac{6}{16}$

19. $\frac{3}{8} - \frac{2}{8}$

20. 234×42

21. 36×470

22.
$$\begin{array}{r} 7 \\ 4 \\ 2 \\ 1 \\ 4 \\ N \\ 2 \\ 7 \\ 4 \\ 8 \\ \hline 68 \end{array}$$

23.
$$\begin{array}{r} 42 \\ 23 \\ 35 \\ 47 \\ 74 \\ 53 \\ 32 \\ 24 \\ 21 \\ 5 \\ +\ 7 \\ \hline \end{array}$$

24. 408×29

25. $9\overline{)981}$

26. $\$25.00 \div 4$

27. $\frac{427}{7}$

28. $50\overline{)750}$

29. $40\overline{)1280}$

30. $30\overline{)843}$

LESSON 128 Reducing Fractions

Speed Test: 90 Division Facts, Form A (Test I in Test Booklet)

When we **reduce** a fraction, we find an equivalent fraction written with **smaller** numbers. The picture below shows $\frac{4}{6}$ reduced to $\frac{2}{3}$.

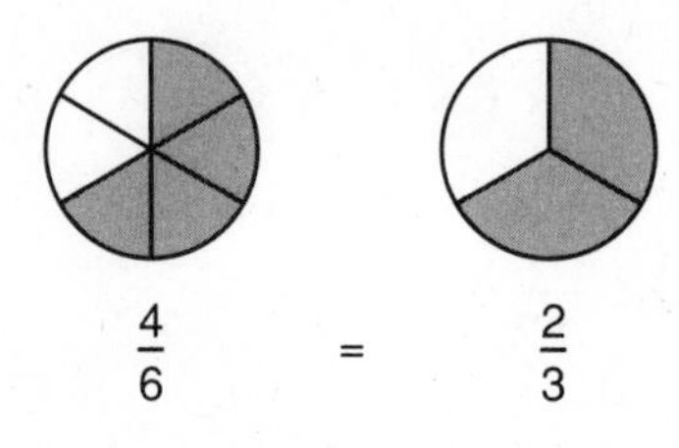

$$\frac{4}{6} = \frac{2}{3}$$

Not all fractions can be reduced. Only a fraction whose top and bottom numbers can be divided by the same number can be reduced. Since the top and bottom numbers of $\frac{4}{6}$ can both be divided by 2, we can reduce the fraction $\frac{4}{6}$.

To reduce a fraction, we divide the fraction by a name for 1. We reduce $\frac{4}{6}$ by dividing 4 by 2 and dividing 6 by 2.

$$\frac{4}{6} \div \frac{2}{2} = \frac{2}{3} \qquad \begin{matrix}(2 \div 2 = 2)\\(6 \div 2 = 3)\end{matrix}$$

Example Write the reduced form of each fraction.

(a) $\frac{6}{8}$ (b) $\frac{6}{9}$ (c) $\frac{6}{7}$

Solution To reduce a fraction, we divide both parts of the fraction by a number that divides both parts without a remainder.

(a) The only numbers that divide 6 and 8 evenly are 1 and 2. Dividing by $\frac{1}{1}$ does not change the fraction, so we divide 6 by 2 and divide 8 by 2.

$$\frac{6}{8} \div \frac{2}{2} = \mathbf{\frac{3}{4}}$$

(b) In problem (b), both 6 and 9 can be divided by 3, so we divide 6 by 3 and divide 9 by 3.

$$\frac{6}{9} \div \frac{3}{3} = \frac{2}{3}$$

(c) In problem (c), the fraction $\frac{6}{7}$ cannot be reduced. The only number that divides both 6 and 7 is 1, and dividing by $\frac{1}{1}$ does not change the fraction.

$$\frac{6}{7} \div \frac{1}{1} = \frac{6}{7}$$

The reduced form of $\frac{6}{7}$ is $\mathbf{\frac{6}{7}}$.

Practice Write the reduced form of each fraction.

a. $\frac{2}{4}$ **b.** $\frac{2}{6}$ **c.** $\frac{3}{9}$ **d.** $\frac{3}{8}$

e. $\frac{2}{10}$ **f.** $\frac{6}{10}$ **g.** $\frac{9}{12}$ **h.** $\frac{9}{10}$

Problem set 128

1. Stagecoaches were pulled by 4-horse teams, and covered wagons were pulled by 6-horse teams. Altogether, how many horses were needed to pull 3 stagecoaches and 7 covered wagons?

2. (a) Find the perimeter of this rectangle.
(b) Find the area of the rectangle.

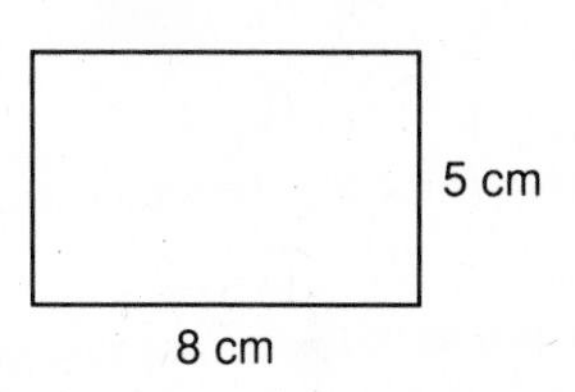

3. Sixteen quarters equals how many dimes?

4. Gilbert watched TV for 60 minutes. He found that commercials were on for two tenths of the time. Commercials were on for how many minutes? Draw a diagram to illustrate the problem.

5. (a) What is the radius of this circle?
(b) What is the diameter of the circle?

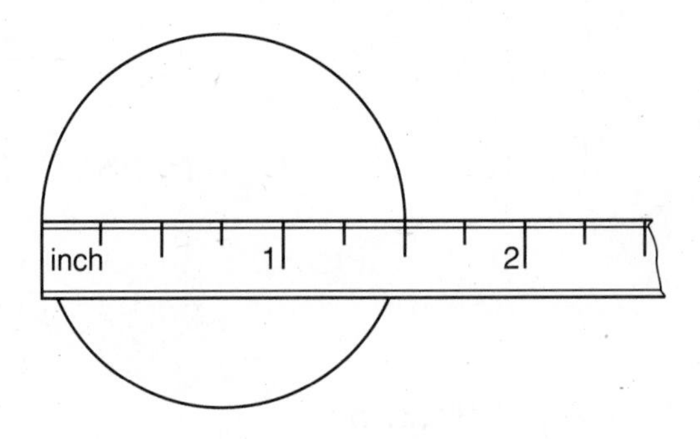

6. Write the reduced form of each fraction.

(a) $\frac{4}{6}$ (b) $\frac{5}{10}$ (c) $\frac{6}{7}$

7. Draw a picture to show that $\frac{4}{6}$ and $\frac{2}{3}$ are equivalent fractions.

8. Find three fractions equivalent to $\frac{2}{3}$ by multiplying by $\frac{2}{2}$, $\frac{3}{3}$, and $\frac{4}{4}$.

9. Maria could bake 65 cookies in 5 batches. How many cookies were in 1 batch? How many cookies would be in 22 batches?

10. James bought 15 apples at 23¢ each, 13 oranges at 41¢ each, and 3 cans of juice at 63¢ each. How much money did he spend?

11. Ninety-three thousand, eleven is how much less than ninety-seven thousand, three hundred sixty-three?

12. 7.6 − 0.76

13. 26.04 − 25.7

14.
$$\begin{array}{r} 4.5 \\ 0.35 \\ 12.1 \\ 32.47 \\ +\ 8.9 \\ \hline \end{array}$$

15. $\frac{1}{10} + \frac{2}{10}$

16. $\frac{5}{13} + \frac{4}{13} + \frac{3}{13}$

17. $\frac{4}{9} - \frac{3}{9}$ **18.** $\frac{6}{7} - \frac{3}{7}$ **19.** $\frac{4}{33} - \frac{4}{33}$

20. $\begin{array}{r} 502 \\ \times \ 52 \\ \hline \end{array}$ **21.** $\begin{array}{r} 327 \\ \times \ 26 \\ \hline \end{array}$ **22.** $\begin{array}{r} 490 \\ \times \ 57 \\ \hline \end{array}$

23. $1000 \div 5$ **24.** $\$42.72 \div 6$ **25.** $\frac{3627}{9}$

26. $60\overline{)1320}$ **27.** $30\overline{)2220}$ **28.** $40\overline{)963}$

29. $\begin{array}{r} 91,325 \\ - \quad N \\ \hline 7,463 \end{array}$ **30.** $\begin{array}{r} 91,325 \\ + \quad N \\ \hline 97,815 \end{array}$

LESSON 129 Dividing by Two-Digit Numbers, Part 1

Speed Test: 90 Division Facts, Form A (Test I in Test Booklet)

We have divided by two-digit numbers that are multiples of 10. In this lesson we will begin dividing by other two-digit numbers. When we divide by two-digit numbers, sometimes we accidentally choose an answer to a division problem that is too large. If this happens, we start over and try a smaller number for the answer.

Example 1 $31\overline{)95}$

Solution Step 1. To help us divide $31\overline{)95}$, we think $3\overline{)9}$ and try that answer. We write 3.

Step 2. We multiply 3×31 and write 93.

Step 3. We subtract 93 from 95 and write 2.

Step 4. There are no digits to bring down. The quotient is **3 with a remainder of 2.**

$$\begin{array}{r} 3 \text{ r } 2 \\ 31\overline{)95} \\ 93 \\ \hline 2 \end{array}$$

Example 2 $43\overline{)246}$

Solution Step 1. To help us divide $43\overline{)246}$, we may think $4\overline{)24}$ and try that answer. We write 6.

Step 2. We multiply 6 × 43 and write 258. We see that 258 is greater than 246. We tried 6, but 6 was too large.

$$\begin{array}{r} 6 \\ 43\overline{)246} \\ \underline{258} \end{array} \leftarrow \text{too large}$$

Start over Step 1. This time we try 5 as the division answer.

Step 2. We multiply 5 × 43 and write 215.

Step 3. We subtract and write 31.

Step 4. There are no digits to bring down. The quotient is **5 with a remainder of 31**.

$$\begin{array}{r} 5 \text{ r } 31 \\ 43\overline{)246} \\ \underline{215} \\ 31 \end{array}$$

Practice Divide.

a. $32\overline{)128}$ **b.** $21\overline{)90}$ **c.** $25\overline{)68}$

d. $42\overline{)250}$ **e.** $46\overline{)164}$ **f.** $31\overline{)225}$

Problem set 129

1. There are 60 minutes in 1 hour and 24 hours in 1 day. How many minutes are in 1 day?

2. Davey had 56 baseball cards. He gave half of them to Janeen. Janeen gave half of those to Jason. How many cards did Jason get?

3. (a) What is the perimeter of this square?
(b) What is the area of the square?

3 ft
3 ft

4. If 2 apples cost 60¢, how much money would 6 apples cost?

5. (a) What is the diameter of the dime?
(b) What is the radius of the dime?

6. Use words to write 27,384.6.

7. Write the reduced form of each fraction.
(a) $\frac{3}{6}$ (b) $\frac{3}{12}$ (c) $\frac{8}{10}$

8. Find three fractions equivalent to $\frac{3}{4}$ by multiplying $\frac{3}{4}$ by $\frac{2}{2}$, $\frac{3}{3}$, and $\frac{4}{4}$.

9. Guess and check your guess to solve this problem. There were 14 rocks and bricks in all. There were 2 more rocks than bricks. How many rocks were there?

10. There were 41 sticks and stones. There were 13 more sticks than stones. How many sticks were there?

11. Able, Baker, and Charlie carried weights. Able could carry twice as much as Baker. Baker could carry 10 pounds. Charlie could carry the same amount as Able. How many pounds could all three boys carry?

12. $39.42 + 9.5 + 16.1$

13. $6.34 + 0.17 + 19.8$

14. $37.1 - 9.46$

15. $4.91 - 4.91$

16. $\frac{3}{10} + \frac{6}{10}$

17. $\frac{1}{8} + \frac{4}{8}$

18. $\frac{5}{7} - \frac{1}{7}$

19. $\frac{15}{16} - \frac{8}{16}$

20. 476×14

21. 240×68

22. 34×904

23. $8\overline{)4325}$

26. $47.16 ÷ 9

27. $\dfrac{1221}{3}$

28. $40\overline{)3210}$

29. $32\overline{)98}$

30. $34\overline{)184}$

24.
$$\begin{array}{r} 3 \\ 4 \\ 2 \\ 1 \\ 7 \\ 3 \\ 2 \\ 6 \\ N \\ +\ 2 \\ \hline 45 \end{array}$$

25.
$$\begin{array}{r} 14 \\ 27 \\ 36 \\ 42 \\ 81 \\ 18 \\ 27 \\ 36 \\ 28 \\ +\ 28 \\ \hline \end{array}$$

LESSON 130 Adding and Subtracting Mixed Numbers

Speed Test: 90 Division Facts, Form A (Test I in Test Booklet)

Adding mixed numbers We remember that a mixed number is a number with a whole number part and a fraction part. To add mixed numbers, we add the fraction parts together. Then we add the whole number parts together.

Example 1 Add: $2\frac{3}{5} + 3\frac{1}{5}$

Solution We may write the numbers one above the other. We add the fractions and get $\frac{4}{5}$. We add the whole numbers and get 5. The sum of the mixed numbers is $\mathbf{5\frac{4}{5}}$.

$$\begin{array}{r} 2\frac{3}{5} \\ +\ 3\frac{1}{5} \\ \hline 5\frac{4}{5} \end{array}$$

Example 2 Add: $3 + 1\frac{1}{3}$

Solution The number 3 is a whole number, not a mixed number. Therefore, there is no fraction to add to $\frac{1}{3}$. We add the whole numbers and get $\mathbf{4\frac{1}{3}}$.

$$\begin{array}{r} 3 \\ +\ 1\frac{1}{3} \\ \hline 4\frac{1}{3} \end{array}$$

Subtracting mixed numbers To subtract mixed numbers, we subtract the fraction. Then we subtract the whole numbers.

Example 3 Subtract: $5\frac{2}{3} - 1\frac{1}{3}$

Solution We must subtract the second number from the first number. To do this, it is helpful to write the first number above the second number. Then we subtract the fractions and get $\frac{1}{3}$. Then we subtract the whole numbers and get 4. The difference is $\mathbf{4\frac{1}{3}}$.

$$\begin{array}{r} 5\frac{2}{3} \\ -\ 1\frac{1}{3} \\ \hline 4\frac{1}{3} \end{array}$$

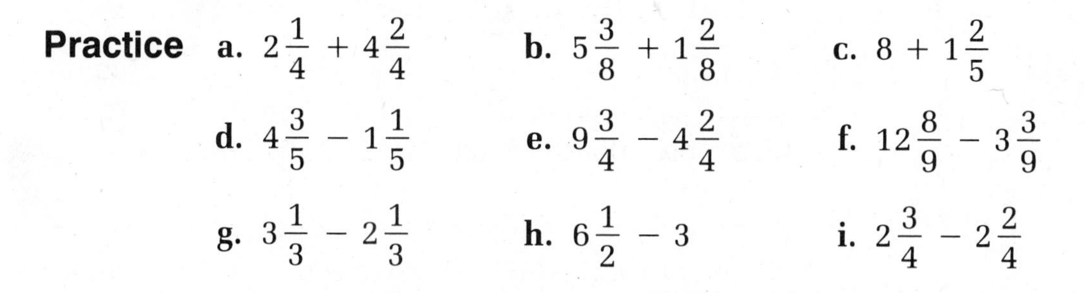

Practice

a. $2\frac{1}{4} + 4\frac{2}{4}$ **b.** $5\frac{3}{8} + 1\frac{2}{8}$ **c.** $8 + 1\frac{2}{5}$

d. $4\frac{3}{5} - 1\frac{1}{5}$ **e.** $9\frac{3}{4} - 4\frac{2}{4}$ **f.** $12\frac{8}{9} - 3\frac{3}{9}$

g. $3\frac{1}{3} - 2\frac{1}{3}$ **h.** $6\frac{1}{2} - 3$ **i.** $2\frac{3}{4} - 2\frac{2}{4}$

Problem set 130

1. If Thorton can do 1 problem in 3 seconds, how many **minutes** will it take him to do 100 problems?

2. (a) What is the area of this rectangle?
(b) What is the perimeter of this rectangle?

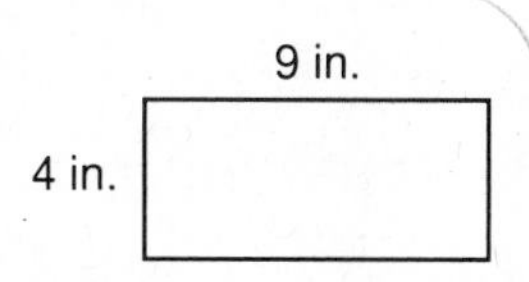

3. Gilbert earned \$40. He saved $\frac{5}{8}$ of what he earned. How much money did he save?

4. After spending \$2.50 on a ticket and \$1.25 on popcorn, Jill still had 75¢. How much money did she start with?

5. Round 3096 to the nearest thousand.

6. Use words to write 14,723.8.

7. Write the reduced form of each fraction.

(a) $\frac{2}{8}$ (b) $\frac{4}{10}$ (c) $\frac{4}{9}$

8. Draw a picture to show that $\frac{5}{10}$ and $\frac{1}{2}$ are equivalent fractions.

9. The first weight was twice the second weight. The third weight was 3 times the second weight. The second weight was 42 pounds. What was the sum of the three weights?

10. The machine made 144 washers in 9 time periods. How many washers could it make in 1 time period? How long would it take to make 32 washers?

11. Roger ran 42 laps every week. How many laps would he run in 73 weeks?

12. $36.4 + 6.75 + 7.3$

13. $6.3 - (4.21 - 3.6)$

14. $3 + 1\frac{1}{4}$

15. $3\frac{2}{5} + 1\frac{1}{5}$

16. $12\frac{1}{3} + 1\frac{1}{3}$

17. $5\frac{4}{5} - 3\frac{1}{5}$

18. $3\frac{1}{2} - 1\frac{1}{2}$

19. $6\frac{2}{3} - 4$

20. $\begin{array}{r} 473 \\ \times \quad 62 \\ \hline \end{array}$

21. $\begin{array}{r} 590 \\ \times \quad 37 \\ \hline \end{array}$

22. $\begin{array}{r} \$12.57 \\ \times \quad 6 \\ \hline \end{array}$

23. \$2.92 ÷ 4

24.
$$\begin{array}{r} 4 \\ 7 \\ N \\ 3 \\ 2 \\ 1 \\ 6 \\ 4 \\ +\ 2 \\ \hline 35 \end{array}$$

25.
$$\begin{array}{r} 14 \\ 21 \\ 43 \\ 88 \\ 77 \\ 66 \\ 31 \\ 25 \\ +\ 43 \\ \hline \end{array}$$

26. $7\overline{)3565}$

27. $\dfrac{6876}{9}$

28. $60\overline{)4900}$

29. $21\overline{)89}$

30. $42\overline{)244}$

LESSON 131 Simplifying Fraction Answers

Speed Test: 90 Division Facts, Form A (Test I in Test Booklet)

It is customary to write math answers in the simplest form possible. If an answer contains a fraction, there are two procedures that we usually follow.

1. Improper fractions are written as mixed numbers (or whole numbers).
2. Fractions are reduced when possible.

Example 1 Add: $\dfrac{2}{3} + \dfrac{2}{3}$

Solution We add to get the sum $\frac{4}{3}$. We notice that $\frac{4}{3}$ is an improper fraction. We take an extra step and change $\frac{4}{3}$ to the mixed number $\mathbf{1\frac{1}{3}}$.

$$\frac{2}{3} + \frac{2}{3} = \frac{4}{3}$$

$$\frac{4}{3} = \mathbf{1\frac{1}{3}}$$

Example 2 Subtract: $\frac{3}{4} - \frac{1}{4}$

Solution We subtract and get the difference $\frac{2}{4}$. We notice that $\frac{2}{4}$ can be reduced. We take an extra step and reduce $\frac{2}{4}$ to $\frac{1}{2}$.

$$\frac{3}{4} - \frac{1}{4} = \frac{2}{4}$$

$$\frac{2}{4} = \mathbf{\frac{1}{2}}$$

Example 3 Add: $3\frac{1}{3} + 4\frac{2}{3}$

Solution We add and get the sum $7\frac{3}{3}$. We notice that $\frac{3}{3}$ is an improper fraction that equals 1. So $7\frac{3}{3} = 7 + 1$, which is **8**.

$$3\frac{1}{3} + 4\frac{2}{3} = 7\frac{3}{3}$$

$$7\frac{3}{3} = \mathbf{8}$$

Example 4 Add: $5\frac{3}{5} + 6\frac{4}{5}$

Solution We add and get $11\frac{7}{5}$. We notice that $\frac{7}{5}$ is an improper fraction. We change $\frac{7}{5}$ to $1\frac{2}{5}$. So $11\frac{7}{5}$ equals $11 + 1\frac{2}{5}$, which is $\mathbf{12\frac{2}{5}}$.

$$5\frac{3}{5} + 6\frac{4}{5} = 11\frac{7}{5}$$

$$11\frac{7}{5} = \mathbf{12\frac{2}{5}}$$

Example 5 Subtract: $6\frac{5}{8} - 1\frac{3}{8}$

Solution We subtract and get $5\frac{2}{8}$. We notice that $\frac{2}{8}$ can be reduced. We reduce $\frac{2}{8}$ to $\frac{1}{4}$. So $5\frac{2}{8}$ equals $\mathbf{5\frac{1}{4}}$.

$$6\frac{5}{8} - 1\frac{3}{8} = 5\frac{2}{8}$$

$$5\frac{2}{8} = \mathbf{5\frac{1}{4}}$$

Practice Simplify each answer.

a. $\frac{4}{5} + \frac{4}{5}$ **b.** $\frac{5}{6} - \frac{1}{6}$ **c.** $3\frac{2}{3} + 1\frac{2}{3}$

d. $5\frac{1}{4} + 6\frac{3}{4}$ **e.** $7\frac{7}{8} - 1\frac{1}{8}$ **f.** $5\frac{3}{5} + 1\frac{3}{5}$

Problem set 131

1. Sharon made 78 photocopies. If she paid 6¢ per copy, how much change did she get back from a $5 bill?

2. (a) What is the area of this square?

(b) What is the perimeter of this square?

6 cm

3. Danny has \$9. David has twice as much money as Danny. Chris has \$4 more than David. How much money does Chris have?

4. There are 40 quarters in a roll of quarters. What is the value of 2 rolls of quarters?

5. Estimate the product of 29 and 312.

6. Use words to write 136,402.35.

7. Write the reduced form of each fraction.

(a) $\frac{2}{12}$ (b) $\frac{6}{8}$ (c) $\frac{3}{9}$

8. Find three fractions equivalent to $\frac{1}{6}$ by multiplying by $\frac{2}{2}$, $\frac{3}{3}$, and $\frac{4}{4}$.

9. Draw diagrams to help work this problem. Their noses were black, red, and green. Green came just before black and red was not last. Who came first?

10. Pencils cost 22¢ each. Erasers cost 41¢ each. Pens cost \$1.41 each. Gracie bought 13 pencils and 14 erasers. How much money did she spend?

11. Jamie ran at 5 miles per hour for 3 hours and at 3 miles per hour for 8 hours. How far did he run?

12. $4.62 + 16.7 + 9.8$

13. $14.62 - (6.3 - 2.37)$

14. $\frac{3}{5} + \frac{4}{5}$

15. $16 + 3\frac{3}{4}$

16. $1\frac{2}{3} + 3\frac{1}{3}$

17. $\frac{2}{5} + \frac{3}{5}$

18. $7\frac{4}{5} + 7\frac{1}{5}$

19. $6\frac{2}{3} + 3\frac{2}{3}$

20. 372×39

21. 47×142

22. 360×24

23. $8\overline{)4834}$

24. $\begin{array}{r} 2 \\ 5 \\ 1 \\ 3 \\ 2 \\ 3 \\ 7 \\ 8 \\ 2 \\ + N \\ \hline 48 \end{array}$

25. $\begin{array}{r} 22 \\ 35 \\ 46 \\ 7 \\ 2 \\ 90 \\ 51 \\ 73 \\ + 64 \\ \hline \end{array}$

26. $\frac{2940}{7}$

27. $\$36.00 \div 5$

28. $30\overline{)963}$

29. $51\overline{)310}$

30. $24\overline{)124}$

LESSON 132 Renaming Fractions

Speed Test: 100 Multiplication Facts (Test H in Test Booklet)

We remember that if we multiply a fraction by a fraction that equals 1, the result is an equivalent fraction. If we multiply by $\frac{1}{2}$ by $\frac{2}{2}$, we get $\frac{2}{4}$. The fractions $\frac{1}{2}$ and $\frac{2}{4}$ are equivalent fractions because they have the same value.

$$\frac{1}{2} \times \frac{2}{2} = \frac{2}{4}$$

Sometimes we must choose a particular multiplier that is equal to 1.

Example 1 Find the equivalent fraction for $\frac{1}{4}$ whose denominator is 12.

Solution To change 4 to 12, we must multiply by 3. So we multiply $\frac{1}{4}$ by $\frac{3}{3}$.

$$\frac{1}{4} \times \frac{3}{3} = \frac{3}{12}$$

The fraction $\frac{1}{4}$ is equivalent to $\mathbf{\frac{3}{12}}$.

Example 2 Complete the equivalent fraction: $\frac{2}{3} = \frac{?}{15}$

Solution The bottom number changed from 3 to 15. The bottom number was multiplied by 5. The correct multiplier is $\frac{5}{5}$.

$$\frac{2}{3} \times \frac{5}{5} = \frac{10}{15}$$

The fraction $\frac{2}{3}$ and the fraction $\mathbf{\frac{10}{15}}$ are equivalent fractions.

Practice Find the top numbers so the pairs of the fractions will be equivalent fractions.

a. $\frac{1}{4} = \frac{?}{12}$ **b.** $\frac{2}{3} = \frac{?}{12}$ **c.** $\frac{5}{6} = \frac{?}{12}$

d. $\frac{3}{5} = \frac{?}{10}$ **e.** $\frac{2}{3} = \frac{?}{9}$ **f.** $\frac{3}{4} = \frac{?}{8}$

Problem set 132

1. If a can of soup costs $1.29 and serves 3 people, how much would it cost to serve soup to 12 people?

2. (a) What is the perimeter of this rectangle?
(b) What is the area of this rectangle?

3. What number is 8 less than the product of 9 and 10?

4. The woodpecker found 3069 bugs in the tree and ate $\frac{2}{3}$ of them. How many bugs were left in the tree? Illustrate the problem.

5. (a) Find the length of this line segment in centimeters.
(b) Find the length of the segment in millimeters.

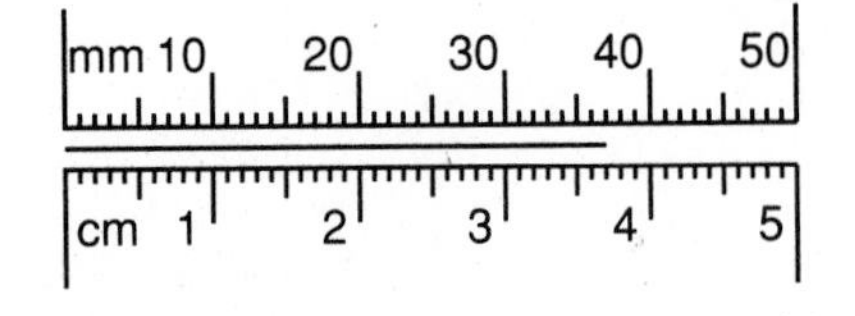

6. Use words to write 356,420.

7. Complete each equivalent fraction.

(a) $\frac{1}{2} = \frac{?}{6}$ (b) $\frac{1}{3} = \frac{?}{6}$ (c) $\frac{2}{3} = \frac{?}{6}$

8. Write the reduced form of each fraction.

(a) $\frac{2}{6}$ (b) $\frac{6}{9}$ (c) $\frac{9}{16}$

9. There were 40 workers on the job. Of those workers, 10 had worked overtime. What fraction of the workers had worked overtime?

10. Bill got $10 for his tenth birthday. Each year after that he got $1 dollar more than he got on his previous birthday. He saved all his birthday money. How much money did he have when he was 17 years old?

11. The big machine could make 32 figures in 16 minutes. How many figures could it make in 1 minute? How long would it take to make 100 figures?

12. $9.36 - (4.37 + 3.8)$ **13.** $24.32 - (8.61 + 12.5)$

14. $\frac{4}{5} + \frac{4}{5}$ **15.** $5\frac{5}{8} + 3\frac{3}{8}$ **16.** $6\frac{3}{10} + \frac{2}{10}$

17. $8\frac{2}{3} - 5\frac{1}{3}$

18. $4\frac{3}{4} - 2\frac{1}{4}$

19. $7\frac{1}{2} - 5$

20. 60×400

21. 536×27

22. $12 \times \$1.50$

23. $6\overline{)3642}$

24. $\$125 \div 5$

25. $\frac{4380}{10}$

26. $40\overline{)645}$

27. $41\overline{)165}$

28. $\begin{array}{r} 26 \\ 9 \\ 14 \\ 11 \\ 2 \\ +\ 34 \\ \hline \end{array}$

29. $33\overline{)93}$

30. $\begin{array}{r} AB,CDE \\ -\ 14,962 \\ \hline 27,853 \end{array}$

LESSON 133 Naming Geometric Solids

Speed Test: 100 Multiplication Facts (Test H in Test Booklet)

We have practiced naming shapes such as triangles, rectangles, and circles. These are flat shapes. They do not take up **space**. Objects that take up space are things like cars, basketballs, desks, houses, and people.

Geometric shapes that take up space are called **geometric solids**. The chart below shows the names of some geometric solids.

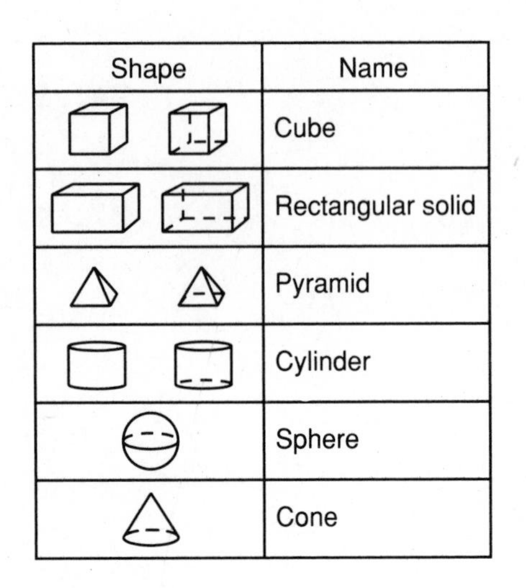

Shape	Name
	Cube
	Rectangular solid
	Pyramid
	Cylinder
	Sphere
	Cone

Example 1 Name each shape.

(a) (b) (c)

Solution If we check each shape with the chart, we see that (a) is a **sphere**, (b) is a **cube**, and (c) is a **cone**.

Example 2 What is the shape of a soup can?

Solution A soup can has the shape of a **cylinder**.

Example 3 This rectangular solid is made of how many small cubes?

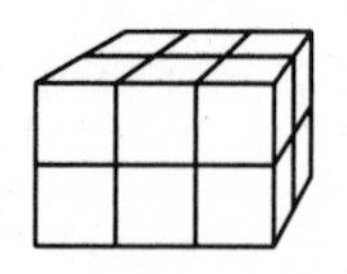

Solution We see that the rectangular solid is made of 2 layers of cubes with 6 cubes on each layer. $2 \times 6 = 12$. The rectangular solid is made of **12 cubes**.

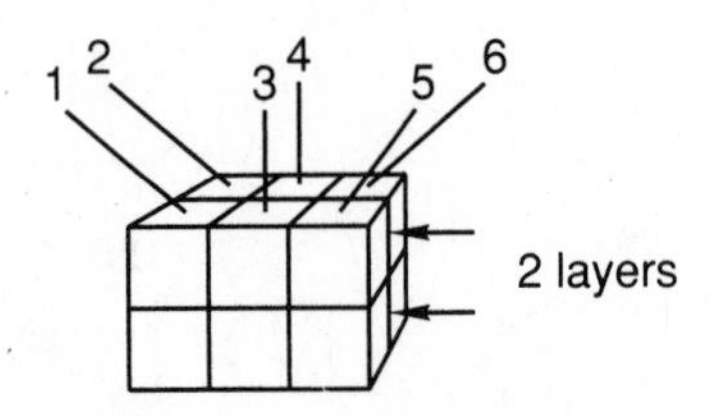

Practice Name the shape of each object listed.

a. A basketball

b. A shoebox

c. An ice cream cone

d. What is the shape of this Egyptian landmark?

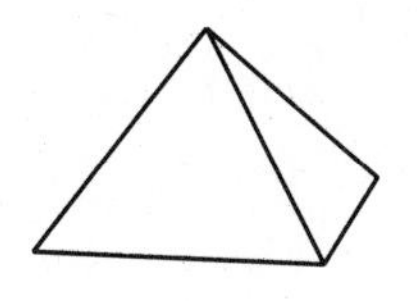

e. This rectangular solid is made of how many small cubes?

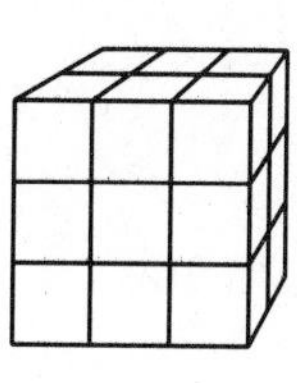

Problem set 133 Use the information given in the graph to answer questions 1–3.

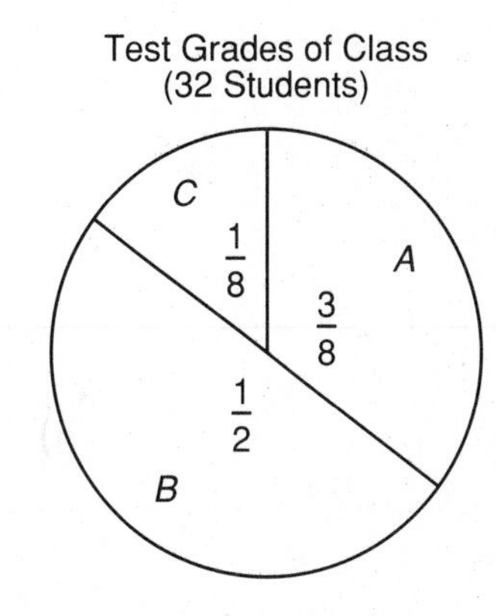

1. Of the 32 students in the class, how many earned a C on the test?

2. How many students earned an A?

3. Altogether, how many students earned an A or a B?

4. If the perimeter of a square is 3 feet, each side is how many inches long?

5. This rectangular solid is made of how many small cubes?

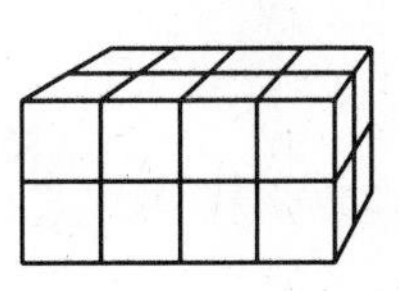

6. Name each shape.

(a) (b) 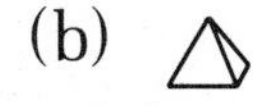 (c)

7. Complete each equivalent fraction.

(a) $\frac{1}{2} = \frac{?}{8}$ (b) $\frac{1}{4} = \frac{?}{8}$ (c) $\frac{3}{4} = \frac{?}{8}$

8. Write the reduced form of each fraction.

(a) $\frac{9}{10}$ (b) $\frac{10}{15}$ (c) $\frac{4}{8}$

9. Draw diagrams to help solve this problem. Wilbur was not first. Roger came just before Sarah. Sarah was not last. Who was in the middle?

10. What number is four hundred ninety-seven greater than fourteen thousand, seventeen?

11. There were 7 containers for 37 items. If the items were divided as evenly as possible between the containers, how many items would be in each container?

12. $9.43 - (6.43 + 1.7)$

13. $16.1 - (1.23 + 12.3)$

14. $\frac{2}{3} + \frac{2}{3} + \frac{2}{3}$

15. $3\frac{1}{2} + 4\frac{1}{2}$

16. $5\frac{1}{6} + 1\frac{1}{6}$

17. $3\frac{1}{3} - 1\frac{1}{3}$

18. $4\frac{3}{8} - 1\frac{1}{8}$

19. $4\frac{1}{4} - 2$

20. $20 \times 20 \times 20$

21. $25 \times \$6.40$

22. 68×47

23. $6\overline{)4564}$

24. $\$12.40 \div 10$

25. $\frac{4624}{8}$

26. $30\overline{)246}$

27. $41\overline{)246}$

28. $32\overline{)246}$

29. $\begin{array}{r} 246 \\ -\ N \\ \hline 172 \end{array}$

30. $\begin{array}{r} 246 \\ +\ N \\ \hline 372 \end{array}$

LESSON 134 Roman Numerals through 39

Speed Test: 100 Multiplication Facts (Test H in Test Booklet)

Roman numerals are letters used by the ancient Romans to write numbers. Roman numerals are still used today on clocks and buildings and are used to number book chapters or Super Bowl games.

Some Roman numerals and their values are:

I which equals 1
V which equals 5
X which equals 10

Roman numerals do not use place value. Instead, the value of the digits are added or sometimes subtracted.

II means 2 ones, which is 2, not 11

Below are listed the Roman numerals for the numbers 1–20. Notice the I before V means **1 less** than 5. The I before the X means **1 less** than 10. Study the patterns in the list below.

I =	1	XI =	11
II =	2	XII =	12
III =	3	XIII =	13
IV =	4	XIV =	14
V =	5	XV =	15
VI =	6	XVI =	16
VII =	7	XVII =	17
VIII =	8	XVIII =	18
IX =	9	XIX =	19
X =	10	XX =	20

Example (a) Write XXVII in our number system.
(b) Write 34 in Roman numerals.

Solution (a) We add 2 tens plus 5 plus 2 ones and get 27.

$$\begin{array}{ccccc} XX & & V & & II \\ 20 & + & 5 & + & 2 = \mathbf{27} \end{array}$$

(b) We think of 34 as 30 plus 4. Thirty is 3 tens, which is XXX. 4 is 1 less than 5, which is IV. So 34 is written **XXXIV**.

Practice Write the Roman numerals in order from 1–39.

Problem set 134

1. Grampa bought meat for \$4.70, potatoes for \$1.59, bananas for \$1.65, and lettuce for 87¢. He gave the grocer \$10. How much did he get back?

2. What is the area of this rectangle?

15 mm
10 mm

3. Bill can blow up 1 balloon in 1 minute. At that rate, how many balloons can Bill blow up in ten hours?

4. This rectangular solid is made of how many small cubes?

5. Four dresses cost \$96. How much did 1 dress cost? What would 15 dresses cost?

6. Draw a diagram to help with this problem. You have to drive through Subligna to get to Dalton from Chattanooga. It is 25 miles from Chattanooga to Subligna. It is 78 miles from Chattanooga to Dalton. How far is it from Subligna to Dalton?

7. At first glance Wilba could see seven thousand. At second glance she could see only six thousand, seventy-seven. How many more did she see at first glance?

8. (a) Write XXIV in our number system.
(b) Write 36 in Roman numerals.

9. Name the shape of each object.
(a) a marble (b) a cereal box

10. Complete each equivalent fraction.
(a) $\frac{1}{2} = \frac{?}{10}$ (b) $\frac{1}{5} = \frac{?}{10}$ (c) $\frac{4}{5} = \frac{?}{10}$

11. Write the reduced form of each fraction.
(a) $\frac{2}{10}$ (b) $\frac{4}{6}$ (c) $\frac{4}{12}$

12. $4.76 + 12.8 + 6.7$

13. $24.3 - (12.14 - 3.2)$

14. $\frac{5}{8} + \frac{6}{8}$

15. $4\frac{1}{4} + 2\frac{3}{4}$

16. $3 + 4\frac{5}{6}$

17. $5\frac{3}{5} + 3\frac{1}{5}$

18. $8\frac{1}{4} + 3\frac{1}{4}$

19. $\frac{7}{8} + \frac{1}{8}$

20. 24 × 15¢

21.
$$\begin{array}{r} 1 \\ 6 \\ 3 \\ N \\ 4 \\ 8 \\ 9 \\ 7 \\ +\ 6 \\ \hline 63 \end{array}$$

22.
$$\begin{array}{r} 24 \\ 32 \\ 81 \\ 18 \\ 27 \\ 72 \\ 38 \\ 15 \\ +\ 24 \\ \hline \end{array}$$

23. 472 × 60

24. 54 × 23

25. $6.72 ÷ 8

26. $10\overline{)737}$

27. $\frac{4935}{7}$

28. $60\overline{)4270}$

29. $41\overline{)210}$

30. $31\overline{)210}$

LESSON 135 Common Denominators

Speed Test: 100 Multiplication Facts (Test H in Test Booklet)

Two fractions have common denominators if the denominators are equal.

$\frac{3}{8}$ $\frac{5}{8}$

These two fractions have common denominators.

$\frac{3}{8}$ $\frac{5}{9}$

These two fractions do **not** have common denominators.

Fractions that do not have common denominators can be renamed so that they do have common denominators.

Example 1 Rename $\frac{2}{3}$ and $\frac{3}{4}$ so that they have a common denominator of 12.

Solution To rename a fraction, we multiply the fraction by a name for 1. To change the denominator of $\frac{2}{3}$ to 12, we multiply by $\frac{4}{4}$. To change the denominator of $\frac{3}{4}$ to 12, we multiply by $\frac{3}{3}$.

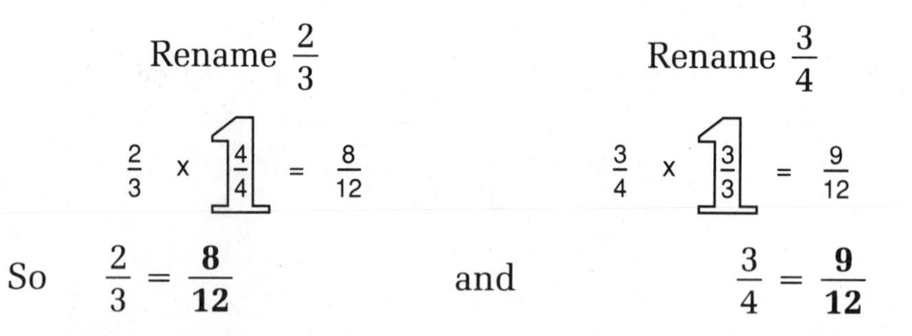

So $\frac{2}{3} = \mathbf{\frac{8}{12}}$ and $\frac{3}{4} = \mathbf{\frac{9}{12}}$

Example 2 Rename $\frac{1}{2}$ and $\frac{1}{3}$ so that they have a common denominator.

Solution This time we need to figure out a common denominator before we can rename the fractions. The denominators are 2 and 3. The common denominator is a multiple of both 2 and 3. Multiples of 2 are 2, 4, 6, 8, and so on. Multiples of 3 are 3, 6, 9, 12, and so on. We see that 6 is a multiple of both 2 and 3. So we will use 6 as the common denominator.

Rename $\frac{1}{2}$ Rename $\frac{1}{3}$

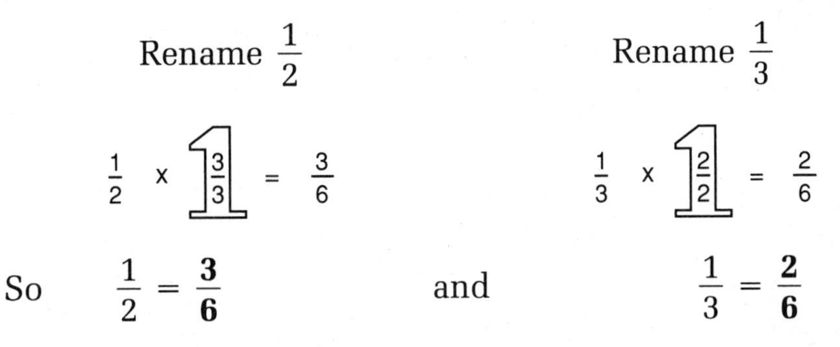

So $\frac{1}{2} = \frac{\mathbf{3}}{\mathbf{6}}$ and $\frac{1}{3} = \frac{\mathbf{2}}{\mathbf{6}}$

Practice a. Rename $\frac{1}{2}$ and $\frac{1}{5}$ so that they have a common denominator of 10.

b. Rename $\frac{3}{4}$ and $\frac{5}{6}$ so that they have a common denominator of 12.

Rename each pair of fractions in (c)–(f) so that they have common denominators.

c. $\frac{1}{2}$ and $\frac{2}{3}$

d. $\frac{1}{3}$ and $\frac{1}{4}$

e. $\frac{1}{2}$ and $\frac{3}{5}$

f. $\frac{2}{3}$ and $\frac{2}{5}$

Problem set 135

1. Tim caught 24 polliwogs. If he let one fourth of them go, how many did he keep? Illustrate the problem.

2. Rectangular Park was 2 miles long and 1 mile wide. Gordon ran around the park twice. How many miles did he run?

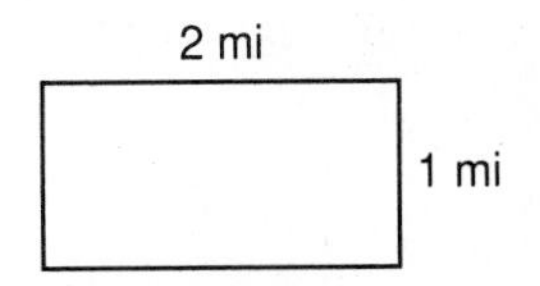

3. If 2 oranges cost 42¢, how much would 8 oranges cost?

4. Three fourths of the 64 baseball cards showed players from the American League. How many of the baseball cards showed American League players? Illustrate the problem.

5. (a) Write XIX in our number system.
(b) Write 28 in Roman numerals.

6. Complete each equivalent fraction.
(a) $\frac{1}{2} = \frac{?}{12}$ (b) $\frac{1}{3} = \frac{?}{12}$ (c) $\frac{1}{4} = \frac{?}{12}$

7. Write the reduced form of each fraction.
(a) $\frac{5}{10}$ (b) $\frac{8}{15}$ (c) $\frac{6}{12}$

8. Randy paid 42¢ for 6 clips and 64¢ for 8 erasers. Find the cost for 1 clip and for 1 eraser. What would be the cost of 14 clips and 20 erasers?

9. There were 14 the first year, 16 the second year, and 18 the third year. If they continued to increase at 2 per year, how many would there be in the tenth year?

10. Ricky's father drove at 40 miles per hour for 3 hours and at 50 miles per hour for 7 hours. How far did he drive in all?

11. (a) Rename $\frac{1}{4}$ and $\frac{2}{3}$ so that they have a common denominator of 12.
(b) Rename $\frac{1}{3}$ and $\frac{3}{4}$ so that they have a common denominator.

12. $47.14 - (3.63 + 36.3)$ **13.** $50.1 + (6.4 - 1.46)$

14. $\frac{3}{4} + \frac{3}{4} + \frac{3}{4}$ **15.** $4\frac{1}{6} + 1\frac{1}{6}$ **16.** $5\frac{3}{5} + 1\frac{2}{5}$

17. $\frac{5}{6} + \frac{1}{6}$

18. $12\frac{3}{4} + 3\frac{1}{4}$

19.
$$\begin{array}{r} 4 \\ 1 \\ 2 \\ 3 \\ N \\ 5 \\ 5 \\ 7 \\ 7 \\ +\ 2 \\ \hline 56 \end{array}$$

20. $6\frac{1}{5} + 1\frac{1}{5}$

21. 340×15

22. 26×307

23. 70×250

24. $\frac{3550}{5}$

25. $\$20.00 \div 8$

26. $9\overline{)5784}$

27. $432 \div 30$

28. $342 \div 42$

29. $243 \div 43$

30.
$$\begin{array}{r} 42{,}186 \\ -\ \ 6{,}497 \\ \hline \end{array}$$

LESSON 136 Place Value through Hundred Millions

> **Speed Test:** 100 Multiplication Facts (Test H in Test Booklet)

The table below names the values of the first nine whole number places.

hundred millions	ten millions	millions	hundred thousands	ten thousands	thousands	hundreds	tens	ones	decimal point
_	_	_	, _	_	_	, _	_	_	•

Example 1 Which digit is in the ten-millions' place in 156,374,289?

Solution We see that the ten-millions' place is eight places from the right. We count eight places in 156,374,289 and find that the digit in the ten-millions' place is **5**.

Example 2 The 3 is in what place in 23,485,179?

Solution The 3 is seven places from the right. We see that seven places from the right is the **millions' place**.

Practice The 9 is in what place in each number?

a. 93,000,000

b. 379,286,142

c. 1,987,364

d. 28,398,167

e. 986,452,137

f. 57,349,286

Problem set 136

1. Eighteen soldiers guarded the castle. If one third of them were sleeping, how many were awake? Illustrate the problem.

2. Stephanie bought a juice bar and gave the clerk a dollar. She got back two quarters, a dime, and a nickel. How much did the juice bar cost?

3. If the perimeter of a square is 6 centimeters, each side is how many millimeters long?

4. This cube is made of how many smaller cubes?

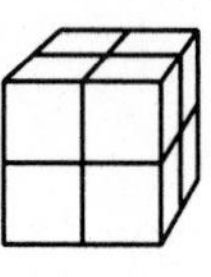

5. (a) Write XVIII in our number system.
(b) Write 33 in Roman numerals.

6. Miguel paid $63 for 7 balls and paid $80 for 10 bats. How much would 1 bat cost? How much would 1 ball cost? What would he have to pay for 13 balls and 15 bats?

7. Carey could dig 15 holes in 3 hours. How long would it take her to dig 40 holes?

8. There were 7 in the first year. There were 12 in the second year. There were 17 in the third year. Find the pattern and tell how many there would be in the eighth year.

9. Complete each equivalent fraction.

(a) $\frac{2}{3} = \frac{?}{12}$ (b) $\frac{3}{4} = \frac{?}{12}$ (c) $\frac{5}{6} = \frac{?}{12}$

10. (a) Rename $\frac{1}{2}$ and $\frac{3}{5}$ so that they have a common denominator of 10.
(b) Rename $\frac{1}{2}$ and $\frac{2}{3}$ so that they have common denominators.

11. Which digit is in the millions' place in 316,254,798?

12. 35.4 – (7.6 + 12.42)

13. 79.17 – (3.6 + 17.4)

14. $4\frac{2}{4} + 3\frac{3}{4}$

15. $5\frac{1}{6} + 3\frac{2}{6}$

16. $\frac{5}{12} + \frac{4}{12}$

17. $\frac{7}{10} + \frac{2}{10}$

18. $4\frac{7}{8} + 1\frac{1}{8}$

19. $4\frac{5}{9} + 2\frac{5}{9}$

20. 400×500

21. $\$6.37 \times 9$

22. $$\begin{array}{r} 4 \\ 2 \\ 1 \\ 7 \\ 3 \\ 6 \\ N \\ 8 \\ 4 \\ +\ 7 \\ \hline 43 \end{array}$$

23. 57×630

24. $\$48.00 \div 10$

25. $\dfrac{6416}{8}$

26. $40\overline{)484}$

27. $12\overline{)96}$

28. $21\overline{)124}$

29. $$\begin{array}{r} 47{,}816 \\ -\ 3{,}724 \\ \hline \end{array}$$

30. $$\begin{array}{r} 47{,}816 \\ +\ 14{,}726 \\ \hline \end{array}$$

LESSON 137 Naming Numbers through Hundred Millions

Speed Test: 90 Divisions Facts, Form B (Test J in Test Booklet)

To read whole numbers of more than six digits, we use two commas. To read the number

496874235

we place the thousands' comma three places from the right end. Then we place the millions' comma three places to the left of the thousands' comma.

496 , 874 , 235

↑ millions' comma ↑ thousands' comma

To write this number in words, we read the number to the left of the millions' comma.

four hundred ninety-six

Now we read the millions' comma by saying "million," and we put a comma after the word million.

four hundred ninety-six million,

Then we read the rest of the number.

four hundred ninety-six million, eight hundred seventy-four thousand, two hundred thirty-five

We remembered to write a comma after the word "thousand" just as we did after the word "million."

Example Use words to write the number 374196285.

Solution We use commas to separate the number into three sections, which we read one section at a time. When we get to the first comma, we write "million." When we get to the second comma, we write "thousand."

374 , 196 , 285
million thousand

three hundred seventy-four million, one hundred ninety-six thousand, two hundred eighty-five

Practice Use words to name each of these numbers. Insert commas as necessary.

a. 27196384

b. 176245398

c. 3541286

d. 19574328

Problem set 137

1. Robin separated his 45 merry men as equally as possible into 4 groups. How many merry men were in the largest group?

2. (a) What is the area of this rectangle?
(b) What is the perimeter of this rectangle?

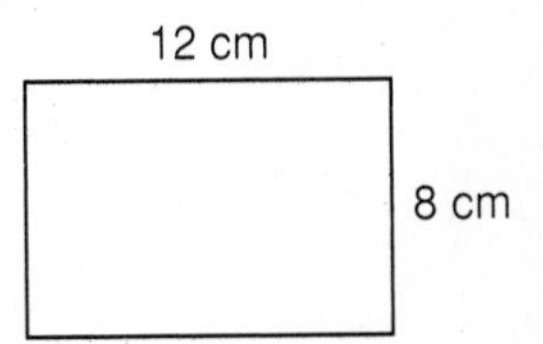

3. Tom answered $\frac{5}{6}$ of the 90 questions correctly. How many questions did Tom answer correctly? Illustrate the problem.

4. Name the shape of each object.
   (a) a roll of paper towels (b) a baseball (c) dice

5. Write the reduced form of each fraction.
   (a) $\frac{3}{6}$ (b) $\frac{5}{15}$ (c) $\frac{8}{12}$

6. Rename $\frac{3}{4}$ and $\frac{5}{6}$ so that they have a common denominator.

7. Which digit is in the ten-millions' place in 328,496,175?

8. Draw a picture to help you solve this problem. Winder is on the road from Atlanta to Athens. It is 77 miles from Athens to Atlanta. It is 31 miles from Winder to Athens. How far is it from Winder to Atlanta?

9. Uniforms were expensive. Blouses cost $13 each and pants cost $17 each. The coach bought 12 blouses and 12 pairs of pants. How much did the coach spend?

10. Use words to name 2534167.

11. $4.36 + 12.7 + 10.72$
12. $8.54 - (4.2 - 2.17)$
13. $\frac{5}{9} + \frac{5}{9}$
14. $3\frac{2}{3} + 1\frac{2}{3}$
15. $4\frac{5}{8} + 1$
16. $7\frac{2}{3} + 1\frac{2}{3}$
17. $4\frac{4}{9} + 1\frac{1}{9}$
18. $\frac{11}{12} + \frac{1}{12}$
19. 570×64
20. 382×31
21. 54×18
22. $\frac{3731}{7}$
23. $9\overline{)5432}$
24. $\$20.80 \div 8$

25. $60\overline{)548}$ **26.** $13\overline{)50}$ **27.** $72\overline{)297}$

28.
$$\begin{array}{r} 87{,}654 \\ -\ 39{,}207 \\ \hline \end{array}$$

29.
$$\begin{array}{r} 87{,}654 \\ -\quad N \\ \hline 39{,}207 \end{array}$$

30. Write the Roman numerals between XX and XXX.

LESSON 138 Dividing Two-Digit Numbers, Part 2

Speed Test: 90 Division Facts, Form B (Test J in Test Booklet)

We have used four steps to do two-digit division problems that have one-digit answers. In this lesson we will practice two-digit division problems that have two-digit answers.

We will continue to follow the four steps: divide, multiply, subtract, and bring down.

Example $21\overline{)487}$

Solution Step 1. We break the problem into a smaller division problem, $21\overline{)48}$, and write 2.

Step 2. We multiply 2 × 21 and write 42.

Step 3. We subtract 42 from 48 and write 6.

Step 4. We bring down the 7, making 67.

$$\begin{array}{r} 23 \text{ r } 4 \\ 21\overline{)487} \\ \underline{42} \\ 67 \end{array}$$

Repeat Step 1. We divide $21\overline{)67}$ and write 3.

Step 2. We multiply 3 × 21 and write 63.

Step 3. We subtract 63 from 67 and write 4.

Step 4. There are no digits to bring down. The answer is **23 with a remainder of 4**.

$$\begin{array}{r} 23 \text{ r } 4 \\ 21\overline{)487} \\ \underline{42} \\ 67 \\ \underline{63} \\ 4 \end{array}$$

Practice a. $21\overline{)672}$ b. $12\overline{)384}$ c. $32\overline{)675}$

d. $23\overline{)510}$ e. $41\overline{)880}$ f. $11\overline{)555}$

Problem set 138 Use the information in the graph to answer questions 1–3.

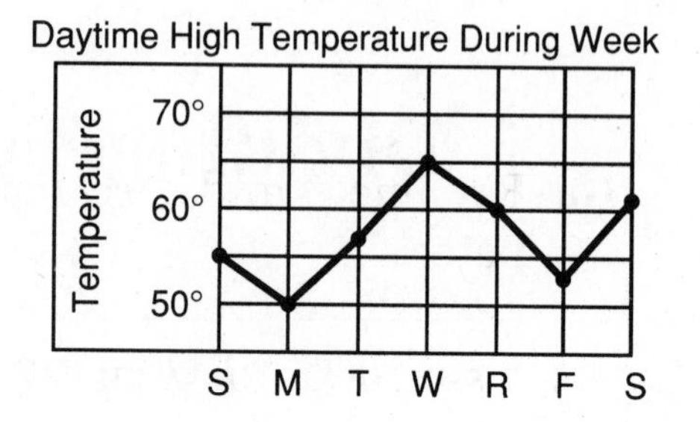

1. On what day was the temperature the highest?

2. What was the temperature on Tuesday?

3. From Monday to Wednesday, the high temperature went up how many degrees?

4. (a) What is the perimeter of this rectangle?
   (b) What is the area of this rectangle?

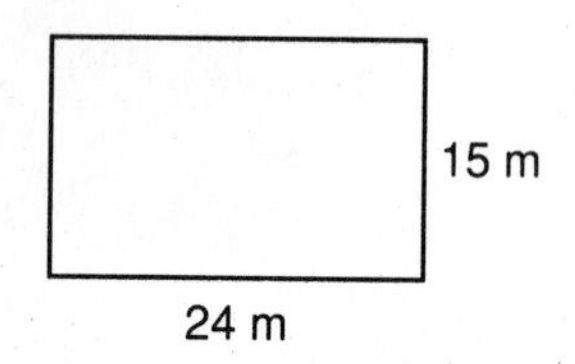

5. The big buses would hold 40 players. The small buses would hold 13 players. One hundred forty-six players were in the line. The coach had 3 big buses. How many small buses did the coach need?

6. Rosanne bought 3 small ones for \$1.27 each, 7 middle-sized ones for \$2.33 each, and 5 big ones for \$15.04 each. How much money did she spend?

7. Roger drove 140 miles at 20 miles per hour. Then he drove 100 miles at 25 miles per hour. How long did the trip take?

8. Name each shape.

(a) (b) (c)

9. Write the reduced form of each fraction.

(a) $\frac{6}{8}$ (b) $\frac{4}{9}$ (c) $\frac{4}{16}$

10. Rename $\frac{2}{3}$ and $\frac{3}{4}$ so that they have a common denominator.

11. Use words to name 27386415.

12. $4.75 + 16.14 + 10.9$

13. $18.4 - (4.32 - 2.6)$

14. $4\frac{4}{5} + 3\frac{3}{5}$

15. $5\frac{1}{6} + 1\frac{2}{6}$

16. $7\frac{3}{4} + \frac{1}{4}$

17. $5\frac{3}{8} + 5\frac{1}{8}$

18.
$$\begin{array}{r} 5 \\ 2 \\ 4 \\ 7 \\ 3 \\ 5 \\ 3 \\ 6 \\ 5 \\ N \\ +\ 4 \\ \hline 55 \end{array}$$

19.
$$\begin{array}{r} 28 \\ 47 \\ 74 \\ 36 \\ 91 \\ 87 \\ 21 \\ 12 \\ +\ 14 \\ \hline \end{array}$$

20. $4\frac{1}{6} + 2\frac{1}{6}$

21. 720×36

22. 147×54

23. $8\overline{)5766}$

24. $\$36.30 \div 10$

25. $\frac{4735}{5}$

26. $21\overline{)441}$

27. $32\overline{)960}$

28. $12\overline{)518}$

29.
$$\begin{array}{r} 714 \\ -\ N \\ \hline 516 \end{array}$$

30.
$$\begin{array}{r} N \\ -\ 21,765 \\ \hline 81,426 \end{array}$$

LESSON 139 Adding and Subtracting Fractions with Unlike Denominators

Speed Test: 90 Division Facts, Form B (Test J in Test Booklet)

We have practiced adding and subtracting fractions that have the same denominators. In this lesson, we will begin adding and subtracting fractions that have different denominators.

Same denominator	Different denominators
$\frac{1}{5} \leftrightarrow \frac{3}{5}$	$\frac{1}{2} \leftrightarrow \frac{3}{4}$

To add or subtract fractions that have different denominators, we must first change the name of the fractions so that they will have the same denominators.

Recall that we change the name of a fraction by multiplying the fraction by a fraction name for 1.

Example 1 Add: $\frac{1}{4} + \frac{3}{8}$

Solution The denominators are not the same. We multiply $\frac{1}{4}$ by $\frac{2}{2}$ and get $\frac{2}{8}$. Now we can add.

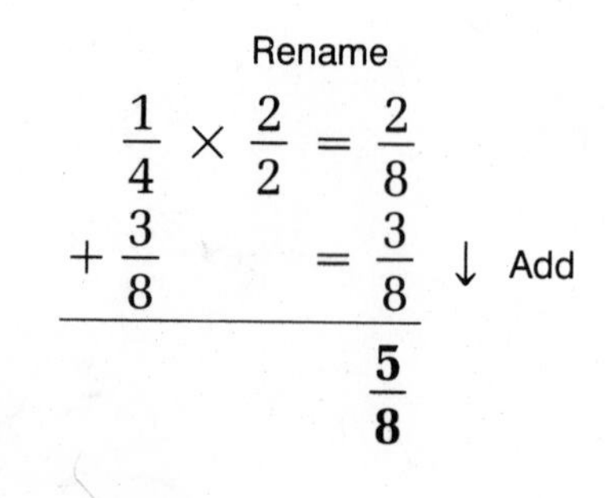

$$\frac{1}{4} \times \frac{2}{2} = \frac{2}{8}$$
$$+\frac{3}{8} = \frac{3}{8}$$
$$\mathbf{\frac{5}{8}}$$

Example 2 Subtract: $\frac{5}{6} - \frac{1}{2}$

Solution The denominators are not the same. We change $\frac{1}{2}$ to a fraction whose denominator is 6. Then we subtract. We reduce the answer.

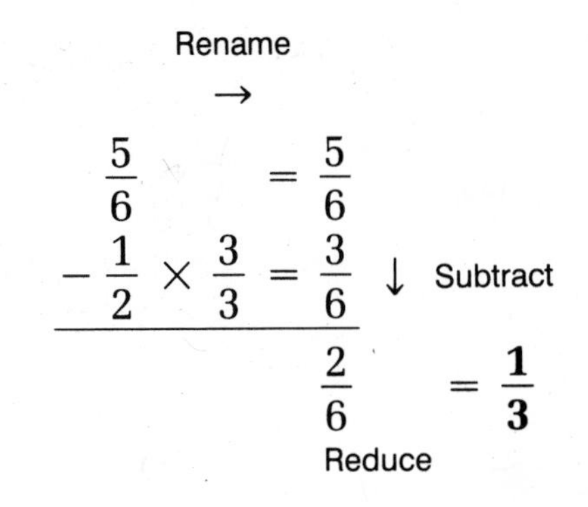

$$\begin{array}{r} \frac{5}{6} \qquad\quad = \frac{5}{6} \\ -\frac{1}{2} \times \frac{3}{3} = \frac{3}{6} \\ \hline \frac{2}{6} = \mathbf{\frac{1}{3}} \end{array}$$

Practice Add or subtract. Reduce when possible.

a. $\frac{1}{2} + \frac{2}{6}$ **b.** $\frac{1}{3} + \frac{1}{9}$ **c.** $\frac{1}{8} + \frac{1}{2}$

d. $\frac{3}{8} - \frac{1}{4}$ **e.** $\frac{2}{3} - \frac{2}{9}$ **f.** $\frac{7}{8} - \frac{1}{2}$

Problem set 139

1. Clotilda laid 1-foot-square floor tiles in a room 15 feet long and 12 feet wide. How many floor tiles did she use?

2. What is the perimeter of this triangle?

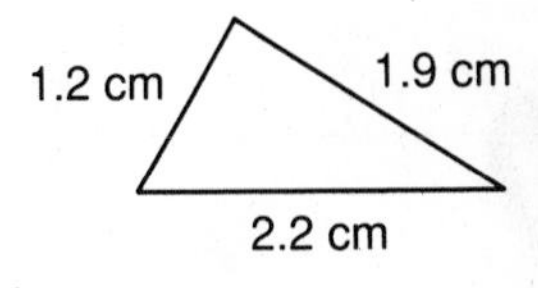

3. Tim found that $\frac{3}{8}$ of the 32 pencils in the room had no erasers. How many pencils had no erasers? Illustrate the problem.

4. Seventy-two eggs is how many dozen eggs?

5. This cube is made of how many smaller cubes?

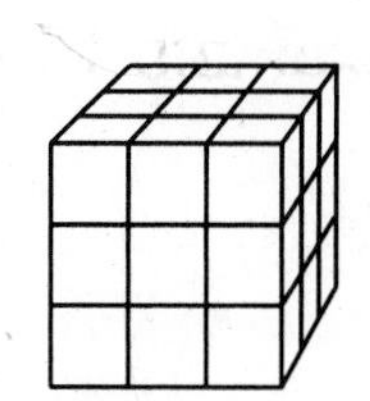

6. The big containers held 120 items each. The small containers held 15 items each. Jim had 900 items. He had 7 big containers. How many small containers did he need?

7. Farmica bought 7 tapes at $4.23 each, 6 records at $2.47 each, and 8 compact disks at $10.95 each. How much money did she need?

8. Roger drove 85 miles in 5 hours. What was his average speed? How far could he drive in 13 hours?

9. (a) Write XXIX in our number system.
(b) Write 24 Roman numerals.

10. Write the reduced form of each fraction.

(a) $\frac{8}{10}$ (b) $\frac{6}{15}$ (c) $\frac{8}{16}$

11. Use words to name the number 123415720.

12. $8.3 + 4.72 + 0.6 + 12.1$

13. $17.42 - (6.7 - 1.23)$

14. $3\frac{3}{8} + 3\frac{3}{8}$

15. $\frac{1}{4} + \frac{1}{8}$

16. $\frac{1}{2} + \frac{1}{6}$

17. $5\frac{5}{6} - 1\frac{1}{6}$

18. $\frac{1}{4} - \frac{1}{8}$

19. $\frac{1}{2} - \frac{1}{6}$

20. 87×16

21. 49×340

22. 504×30

23. $\$35.40 \div 6$

24. $\frac{4784}{4}$

25. $7\overline{)2385}$

26. $30\overline{)450}$

27. $32\overline{)450}$

28. $15\overline{)450}$

29. $\begin{array}{r} 71{,}432 \\ -\ 56{,}666 \\ \hline \end{array}$

30. $\begin{array}{r} 8124 \\ -\quad N \\ \hline 735 \end{array}$

LESSON 140 Writing Numbers through Hundred Millions

Speed Test: 90 Division Facts, Form B (Test J in Test Booklet)

For a few lessons, we have practiced naming numbers through hundred millions. In this lesson we will begin writing these numbers.

Example Use digits to write forty-six million, eight hundred ninety-five thousand, two hundred seventy.

Solution It is a good idea to read the entire number before we begin writing it. We see millions and thousands, so we know the number will have this form:

___, ___, ___

We go back and read "forty-six million" and write 46 and a comma. We read "eight hundred ninety-five thousand" and write 895 and a comma. We read "two hundred seventy" and write 270.

46,895,270

Practice Use digits to write each number.

a. Nine million, two hundred sixty-three thousand, five hundred twelve

b. Eleven million, one hundred twenty-three thousand, four hundred

c. Nine hundred eighty-seven million, six hundred fifty-four thousand, three hundred twenty-one

d. Sixteen million, five hundred forty-three thousand

e. Ninety-three million

f. Seven hundred fifty-three million, eight hundred seventy-two thousand, one hundred ninety

Problem set 140

1. The Martins drank 11 gallons of milk each week. How many quarts of milk did they drink each week?

2. Sixty fleas leaped on Rover as he ran through the field. If one fourth of them perished from flea powder, how many survived? Illustrate the problem.

3. (a) What is the area of this square? 10 mm

   (b) What is the perimeter of this square?

4. Mark is 7 inches taller than Jim. Jim is 5 inches taller than Jan. Mark is 60 inches tall. How many inches tall is Jan?

5. Five students were in the first line. Three times that number were in the second line. The third line held 6 times the number that were in the first line. What was the total number of students in all three lines?

6. Mayville is on the road from Altoona to Watson. It is 47 miles from Mayville to Altoona. It is 24 miles from Mayville to Watson. How far is it from Altoona to Watson?

7. How much greater is seven hundred forty-seven thousand, six hundred two than two hundred eleven thousand, seven hundred seventy-seven?

8. (a) Write XXXIII in our number system.
   (b) Write 39 in Roman numerals.

9. Name each shape.

   (a) 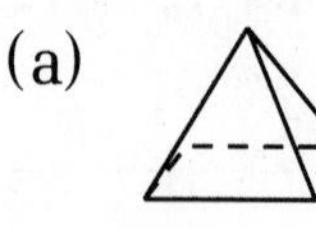 (b) 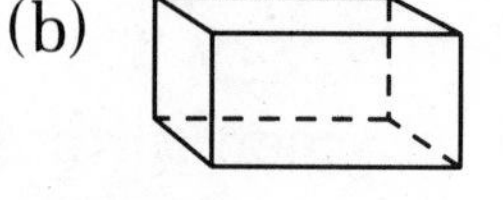 (c) 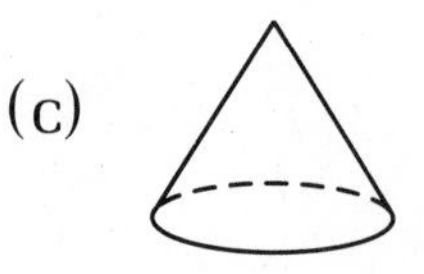

10. Write the reduced form of each fraction.

    (a) $\frac{9}{15}$ (b) $\frac{10}{12}$ (c) $\frac{12}{16}$

11. Use digits to write one hundred nineteen million, two hundred forty-seven thousand, nine hundred eighty-four.

12. $14.94 - (3.6 + 4.7)$

13. $6.8 - (1.37 + 2.2)$

14. $3\frac{2}{5} + 1\frac{4}{5}$

15. $\frac{5}{8} + \frac{1}{4}$

16. $\frac{1}{3} + \frac{1}{6}$

17. $5\frac{9}{10} - 1\frac{1}{10}$

18. $\frac{5}{8} - \frac{1}{4}$

19. $\frac{1}{3} - \frac{1}{6}$

20. 38×217

21. 173×60

22. 90×500

23. $\frac{6752}{4}$

24.
$$\begin{array}{r} 7 \\ 5 \\ 8 \\ 6 \\ 9 \\ N \\ 8 \\ 7 \\ 6 \\ +\ 4 \\ \hline 75 \end{array}$$

25.
$$\begin{array}{r} 27 \\ 99 \\ 77 \\ 86 \\ 68 \\ 54 \\ 45 \\ 74 \\ 85 \\ +\ 27 \\ \hline \end{array}$$

26. $7\overline{)2942}$

27. $\$80.01 \div 9$

28. $10\overline{)453}$

29. $11\overline{)453}$

30. $22\overline{)453}$

LESSON 141 Reading Roman Numerals through Thousands

Speed Test: 90 Division Facts, Form B (Test J in Test Booklet)

We have practiced using these Roman numerals:

I V X

In this lesson, we will practice using more Roman numerals. These are listed in the table.

NUMERAL	VALUE
I	1
V	5
X	10
L	50
C	100
D	500
M	1000

Example Write each Roman numeral in our number system.

(a) LXX (b) DCCL (c) XLIV (d) MMI

Solution (a) LXX is 50 + 10 + 10, which is **70**.

(b) DCCL is 500 + 100 + 100 + 50, which is **750**.

(c) XLIV is 10 less than 50 and 1 less than 5. That is, 40 + 4 = **44**.

(d) MMI is 1000 + 1000 + 1, which is **2001**.

Practice Write each Roman numeral in our number system.

a. CCCLXII **b.** CCLXXXV **c.** CD

d. XLVII **e.** MMMCCLVI **f.** MCMXCIX

Problem set 141

1. Beth is reading a 210-page book. If she has read one third of the book, how many pages does she still have to read? Illustrate the problem.

2. Iceland covers an area of 39,768 square miles. The area of Virginia is 40,767 square miles. How much greater is the area of Virginia than the area of Iceland?

3. Molly Pitcher carried 21 buckets of water to the thirsty troops. If each bucket held 3 gallons, how many **quarts** of water did she deliver?

4. What is the perimeter of this triangle?

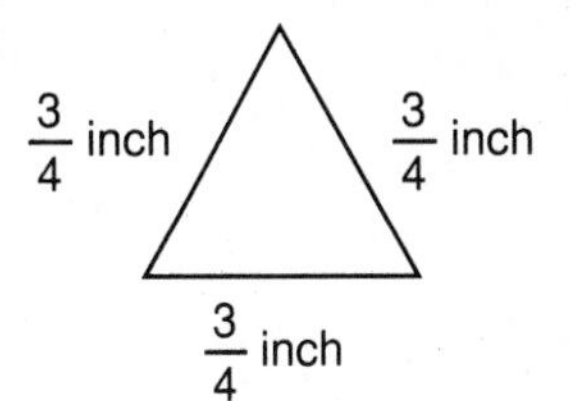

5. Write each Roman numeral in our number system.
(a) MDCLXVI (b) CCXXIV (c) XCIX

6. Write the reduced form of each fraction.
(a) $\frac{9}{12}$ (b) $\frac{12}{15}$ (c) $\frac{10}{20}$

7. This cube is made of how many smaller cubes?

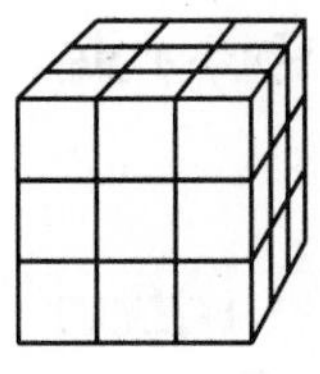

8. Use digits to write ninety-three million, one hundred seventy-five thousand, two hundred twelve.

9. Rita flipped a coin 21 times. It came up heads 17 times. What fraction tells the number of times it did not come up heads?

10. Five red shirts cost \$100. Six green jackets cost \$480. What did each red shirt and each green jacket cost? What would be the cost of 14 red shirts and 13 green jackets?

11. Harry drove 4 hours at 30 miles per hour and 7 hours at 40 miles per hour. How far did he drive?

12. $59.34 - (12.6 + 7.5)$ **13.** $47.6 - (3.7 + 12.23)$

14. $8\frac{1}{4} + 1\frac{3}{4}$ **15.** $\frac{1}{2} + \frac{3}{10}$ **16.** $\frac{2}{3} + \frac{1}{6}$

17. $7\frac{3}{8} - 1\frac{1}{8}$ **18.** $\frac{1}{2} - \frac{3}{10}$ **19.** $\frac{2}{3} - \frac{1}{6}$

20. 47×31

21.
$$\begin{array}{r} 4 \\ 7 \\ 3 \\ 2 \\ 5 \\ 8 \\ 6 \\ N \\ +\ 5 \\ \hline 54 \end{array}$$

22.
$$\begin{array}{r} 27 \\ 32 \\ 11 \\ 22 \\ 33 \\ 41 \\ 14 \\ 26 \\ +\ 48 \\ \hline \end{array}$$

23. 430×26

24. 48×250

25. $9\overline{)7227}$

26. $\$37.00 \div 4$

27. $\dfrac{810}{6}$

28. $70\overline{)6440}$

29. $32\overline{)650}$

30. $24\overline{)560}$

LESSON 142 Average · Probability

Speed Test: 90 Division Facts, Form B (Test J in Test Booklet)

Average Here we show three stacks of pancakes.

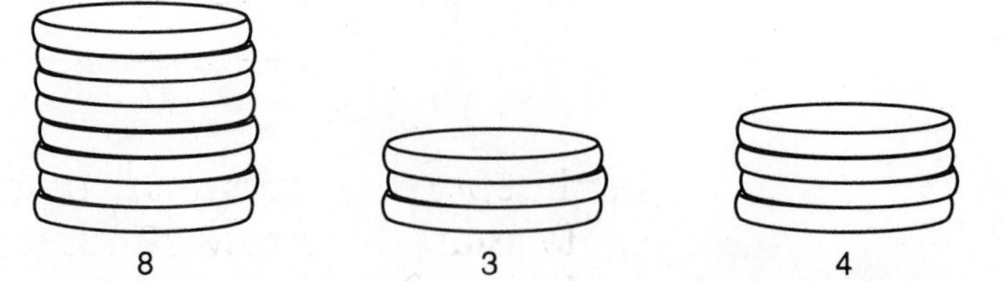

There are 15 pancakes in all. If we put an equal number in each stack, we get 5 in each stack.

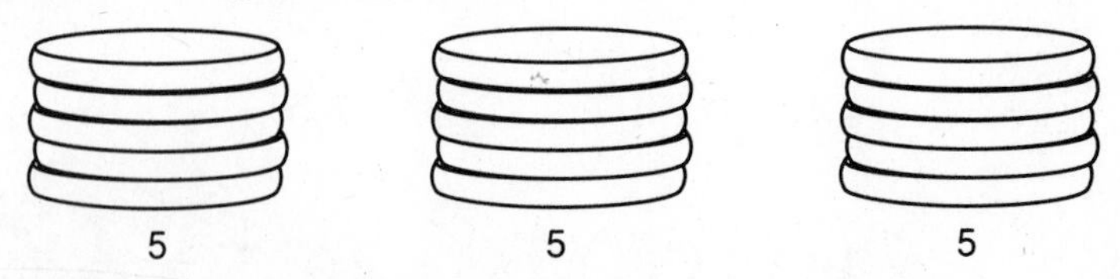

We say that the **average** of the three numbers 8, 3, and 4 is the number 5. The average is the number in each group if the groups are leveled out. To find the average, we add the numbers and divide by the number of numbers.

$$\text{Average} = \frac{8 + 3 + 4}{3} = \frac{15}{3} = 5$$

Example Find the average of 5, 4, 2, and 9.

Solution There are 4 numbers. We add the numbers and divide the sum by 4.

$$\text{Average} = \frac{5 + 4 + 2 + 9}{4} = \frac{20}{4} = \mathbf{5}$$

Probability The probability of an event tells how likely the event is to happen. If the spinner shown here is spun many times, it will probably stop on A about $\frac{1}{4}$ of the time.
It will stop on each of the other letters about $\frac{1}{4}$ of the time. We say about $\frac{1}{4}$ of the time because probability is not exact. A good way to show this is to make a graph of the outcome of a number of spins.

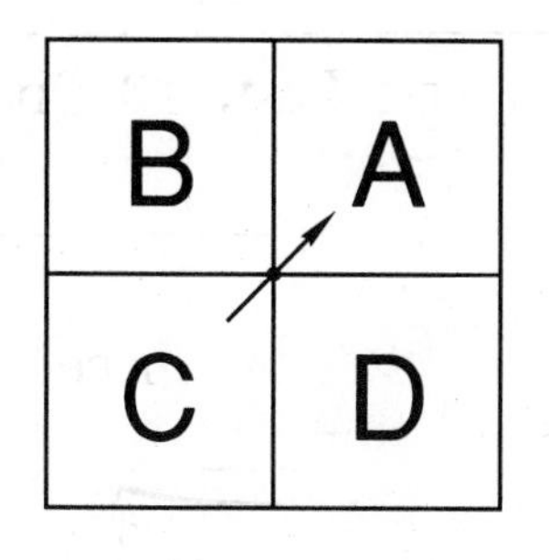

Jamie spun the arrow 10 times and it stopped on A twice. He spun the arrow 20 times and it stopped on A 6 times. He spun the arrow 40 times and it stopped on A 8 times. If we graph these numbers, we get

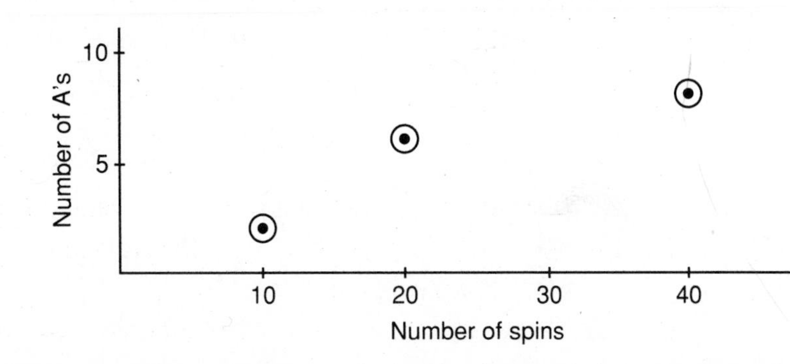

If we estimate a line through the points, we can predict about how many times the spinner will stop on A for any number of spins.

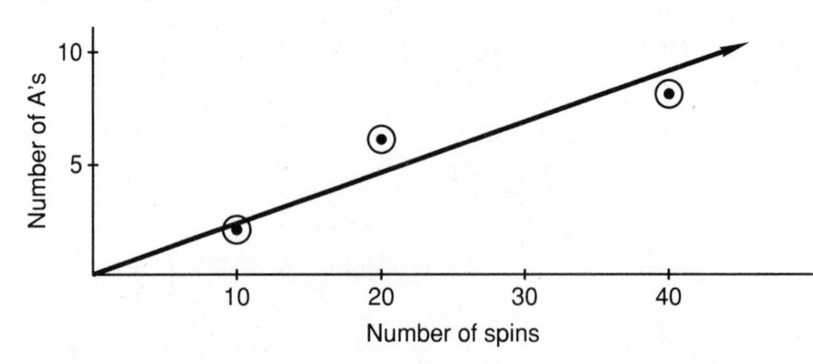

Practice a. Find the average of 14, 7, and 3.

b. Spin a spinner like the one shown in the lesson. Spin it 10 times, 30 times, and 50 times. Make a graph of the number of times it stops on B. Use the graph to help you guess how many times the spinner will stop on B if you spin it 20 times. Discuss what you have done.

Problem set 142

1. There are 108 students in the big room. One fourth of the students wore red hats. How many students did not wear red hats?

2. Jamie piled seventy-two thousand, four hundred forty of them in the first container. In the second container he was able to pile eighty-seven thousand, two. How many more were in the second container?

3. There are 1000 milliliters in a liter. Janie's bucket held 4500 milliliters. How many liters did Janie's bucket hold?

4. A square has one side that is $7\frac{1}{2}$ inches long. What is the perimeter of the square?

5. Dogs at the animal supply store cost \$40 each and parrots cost \$120 each. Sonny's father bought 5 dogs and 7 parrots for his pet store. How much did he spend?

6. Mike drove 6 hours at 48 miles per hour and 3 hours at 58 miles per hour. How far did he drive?

7. Carey drove 126 miles in 3 hours. At the same speed how far could she drive in 13 hours?

8. Write each Roman numeral in our number system.
(a) MCLXVII (b) CCCXXIX (c) CXII

9. Write the reduced form of each fraction.
(a) $\frac{12}{15}$ (b) $\frac{8}{10}$ (c) $\frac{40}{80}$

10. How many small cubes are in this rectangular solid?

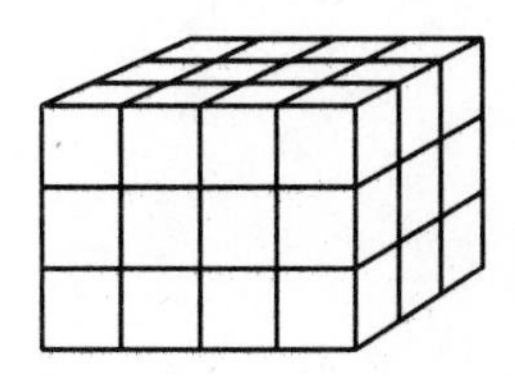

11. Use digits to write seventy-two million, two hundred forty thousand, thirteen.

12. $46.24 - (14.2 + 3.6)$

13. $49.7 - (4.6 + 14.37)$

14. $8\frac{1}{5} + 3\frac{3}{20}$

15. $\frac{1}{4} + \frac{3}{8}$

16. $\frac{2}{3} + \frac{5}{9}$

17. $7\frac{3}{8} - 5\frac{3}{8}$

18. $\frac{1}{2} - \frac{5}{10}$

19. $\frac{5}{6} - \frac{2}{3}$

20. 48×32

21. 420×43

22. 49×210

23. $9\overline{)2529}$

24. $\$43.00 \div 4$

25. $\frac{852}{6}$

26. $70\overline{)6160}$

27. $32\overline{)640}$

28. $26\overline{)560}$

29. $\begin{array}{r} N \\ -\ 4168 \\ \hline 2736 \end{array}$

30. $\begin{array}{r} N \\ +\ 4168 \\ \hline 5436 \end{array}$

Appendix A

Multiplication by 10's, 11's, and 12's

We know that when we multiply by 10 the answer has a zero at the end.

$1 \times 10 = 10$	$5 \times 10 = 50$	$9 \times 10 = 90$
$2 \times 10 = 20$	$6 \times 10 = 60$	$10 \times 10 = 100$
$3 \times 10 = 30$	$7 \times 10 = 70$	$11 \times 10 = 110$
$4 \times 10 = 40$	$8 \times 10 = 80$	$12 \times 10 = 120$

Multiplying by 11 is easy also because every answer listed below except the last three has double digits that are the same.

$1 \times 11 = 11$	$5 \times 11 = 55$	$9 \times 11 = 99$
$2 \times 11 = 22$	$6 \times 11 = 66$	$10 \times 11 = 110$
$3 \times 11 = 33$	$7 \times 11 = 77$	$11 \times 11 = 121$
$4 \times 11 = 44$	$8 \times 11 = 88$	$12 \times 11 = 132$

Multiplying by 12 is important because many items are sold in sets of 12. A set of 12 things is called a **dozen**. Memorizing this 12's table requires practice in saying the table aloud.

$1 \times 12 = 12$	$5 \times 12 = 60$	$9 \times 12 = 108$
$2 \times 12 = 24$	$6 \times 12 = 72$	$10 \times 12 = 120$
$3 \times 12 = 36$	$7 \times 12 = 84$	$11 \times 12 = 132$
$4 \times 12 = 48$	$8 \times 12 = 96$	$12 \times 12 = 144$

Appendix B

Supplemental Practice Problems for Selected Lessons

This appendix contains additional practice problems for concepts presented in selected lessons. It is very important that no problems in the regular problem sets be omitted to make room for these problems. This book is designed to produce long-term retention of concepts, and long-term practice of all the concepts is necessary. The practice problems in the problem sets provide enough initial exposure to concepts for most students. If a student continues to have difficulty with certain concepts, some of these problems can be assigned as remedial exercises.

Supplemental Practice for Lesson 18

1. $\begin{array}{r} 42 \\ -\ 23 \\ \hline \end{array}$ **2.** $\begin{array}{r} 30 \\ -\ 16 \\ \hline \end{array}$ **3.** $\begin{array}{r} 24 \\ -\ 17 \\ \hline \end{array}$ **4.** $\begin{array}{r} 54 \\ -\ 27 \\ \hline \end{array}$

5. $\begin{array}{r} 31 \\ -\ 24 \\ \hline \end{array}$ **6.** $\begin{array}{r} 60 \\ -\ 36 \\ \hline \end{array}$ **7.** $\begin{array}{r} 23 \\ -\ AB \\ \hline 6 \end{array}$ **8.** $\begin{array}{r} CD \\ -\ 19 \\ \hline 21 \end{array}$

9. $\begin{array}{r} 57 \\ -\ EF \\ \hline 29 \end{array}$ **10.** $\begin{array}{r} GH \\ -\ 36 \\ \hline 34 \end{array}$ **11.** $\begin{array}{r} 42 \\ -\ JK \\ \hline 5 \end{array}$ **12.** $\begin{array}{r} LM \\ -\ 47 \\ \hline 27 \end{array}$

13. $36 - 18$ **14.** $24 - 17$ **15.** $40 - 23$

16. $60 - RS = 33$ **17.** $PQ - 39 = 18$ **18.** $72 - 64$

19. $TV - 46 = 28$ **20.** $35 - WX = 7$

Supplemental Practice for Lesson 20

1. $\begin{array}{r} 12 \\ 8 \\ 15 \\ +\ 7 \\ \hline \end{array}$ **2.** $\begin{array}{r} 36 \\ 8 \\ 24 \\ +\ 16 \\ \hline \end{array}$ **3.** $\begin{array}{r} 12 \\ 23 \\ 24 \\ +\ 20 \\ \hline \end{array}$ **4.** $\begin{array}{r} 16 \\ 36 \\ 54 \\ +\ 32 \\ \hline \end{array}$

5. $\begin{array}{r} 74 \\ 37 \\ 60 \\ +\ 46 \\ \hline \end{array}$ **6.** $\begin{array}{r} 57 \\ 24 \\ 38 \\ +\ 83 \\ \hline \end{array}$ **7.** $\begin{array}{r} 95 \\ 9 \\ 78 \\ +\ 35 \\ \hline \end{array}$ **8.** $\begin{array}{r} 47 \\ 58 \\ 62 \\ +\ 55 \\ \hline \end{array}$

9. $\begin{array}{r} 34 \\ 27 \\ 8 \\ +\ 27 \\ \hline \end{array}$ **10.** $\begin{array}{r} 67 \\ 15 \\ 436 \\ +\ 25 \\ \hline \end{array}$ **11.** $\begin{array}{r} 314 \\ 28 \\ 116 \\ +\ 42 \\ \hline \end{array}$ **12.** $\begin{array}{r} 9 \\ 32 \\ 154 \\ +\ 97 \\ \hline \end{array}$

13. $\begin{array}{r} 374 \\ 257 \\ 38 \\ +\ 146 \\ \hline \end{array}$ **14.** $\begin{array}{r} 66 \\ 207 \\ 84 \\ +\ 259 \\ \hline \end{array}$ **15.** $\begin{array}{r} 360 \\ 45 \\ 179 \\ +\ 78 \\ \hline \end{array}$ **16.** $\begin{array}{r} 40 \\ 95 \\ 379 \\ +\ 86 \\ \hline \end{array}$

17. $\begin{array}{r} 36 \\ 275 \\ 175 \\ +\ 384 \\ \hline \end{array}$ 18. $\begin{array}{r} 436 \\ 39 \\ 147 \\ +\ 88 \\ \hline \end{array}$ 19. $\begin{array}{r} 363 \\ 247 \\ 152 \\ +\ 148 \\ \hline \end{array}$ 20. $\begin{array}{r} 273 \\ 54 \\ 106 \\ +\ 50 \\ \hline \end{array}$

Supplemental Practice for Lesson 36

1. $\begin{array}{r} 263 \\ -\ 147 \\ \hline \end{array}$ 2. $\begin{array}{r} 432 \\ -\ 141 \\ \hline \end{array}$ 3. $\begin{array}{r} 520 \\ -\ 336 \\ \hline \end{array}$ 4. $\begin{array}{r} 287 \\ -\ 179 \\ \hline \end{array}$

5. $\begin{array}{r} 196 \\ -\ 57 \\ \hline \end{array}$ 6. $\begin{array}{r} 479 \\ -\ 286 \\ \hline \end{array}$ 7. $\begin{array}{r} 360 \\ -\ 134 \\ \hline \end{array}$ 8. $\begin{array}{r} 424 \\ -\ 254 \\ \hline \end{array}$

9. $\begin{array}{r} 316 \\ -\ 79 \\ \hline \end{array}$ 10. $\begin{array}{r} 260 \\ -\ 146 \\ \hline \end{array}$ 11. $\begin{array}{r} 415 \\ -\ 387 \\ \hline \end{array}$

12. 247 − 79 13. 163 − 127 14. 459 − 367

15. 770 − 287 16. 612 − 78 17. 340 − 149

18. 210 − 86 19. 436 − 156 20. 520 − 417

Supplemental Practice for Lesson 42

Use words to write each of these numbers.

1. 363 2. 1,246

3. 12,280 4. 25,362

5. 123,570 6. 253,500

7. 112,060 8. 220,405

9. 204,050 10. 546,325

Use digits to write each of these numbers.

11. one thousand, two hundred seventy-eight

12. eleven thousand, five hundred forty-four

13. twenty-two thousand, four hundred thirty
14. fifty-seven thousand, nine hundred
15. one hundred seventy-one thousand, two hundred thirty
16. two hundred ten thousand, nine hundred
17. five hundred sixty-three thousand, fifty-eight
18. nine hundred eighty-seven thousand, six hundred fifty-four
19. one hundred five thousand, seventy
20. six hundred fifty thousand, four hundred three

Supplemental Practice for Lesson 45

Use a fraction or mixed number to name every point marked with an arrow on these number lines.

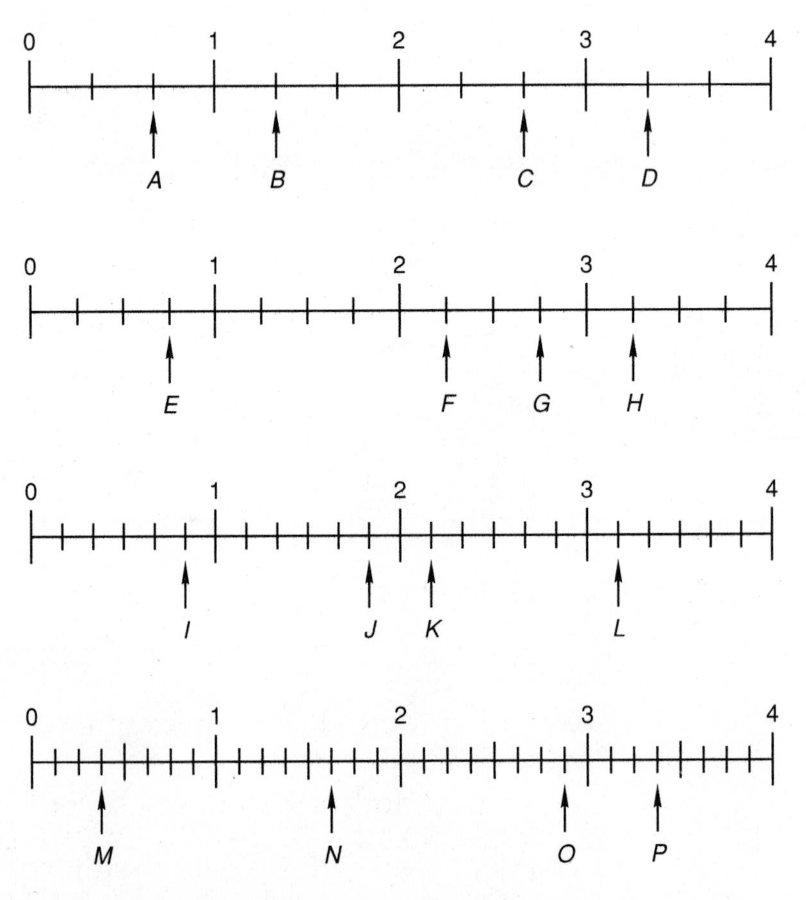

0 1 2 3 4

Q R S T

Supplemental Practice for Lesson 50

1. $\begin{array}{r} 300 \\ -136 \\ \hline \end{array}$
2. $\begin{array}{r} 403 \\ -257 \\ \hline \end{array}$
3. $\begin{array}{r} 200 \\ -143 \\ \hline \end{array}$
4. $\begin{array}{r} 306 \\ -157 \\ \hline \end{array}$
5. $\begin{array}{r} 600 \\ -249 \\ \hline \end{array}$
6. $\begin{array}{r} 201 \\ -165 \\ \hline \end{array}$
7. $\begin{array}{r} 100 \\ -36 \\ \hline \end{array}$
8. $\begin{array}{r} 405 \\ -229 \\ \hline \end{array}$
9. $\begin{array}{r} 400 \\ -349 \\ \hline \end{array}$
10. $\begin{array}{r} 101 \\ -94 \\ \hline \end{array}$
11. $\begin{array}{r} 700 \\ -436 \\ \hline \end{array}$
12. $\begin{array}{r} 501 \\ -127 \\ \hline \end{array}$
13. $\begin{array}{r} 100 \\ -79 \\ \hline \end{array}$
14. $\begin{array}{r} 907 \\ -65 \\ \hline \end{array}$
15. $\begin{array}{r} 500 \\ -249 \\ \hline \end{array}$
16. $\begin{array}{r} 804 \\ -756 \\ \hline \end{array}$
17. $\begin{array}{r} 800 \\ -48 \\ \hline \end{array}$
18. $\begin{array}{r} 602 \\ -575 \\ \hline \end{array}$
19. $\begin{array}{r} 900 \\ -617 \\ \hline \end{array}$
20. $\begin{array}{r} 107 \\ -28 \\ \hline \end{array}$

Supplemental Practice for Lesson 52

1. $6.35 + $4
2. $4.84 − $3
3. 48¢ + 67¢
4. $0.49 − 15¢
5. $0.36 + 85¢
6. $1.25 − 8¢
7. $2.45 + 6¢
8. 75¢ − $0.67
9. 98¢ + $12
10. $1.00 − 95¢
11. $3.46 + $2 + 49¢
12. 50¢ − $0.07
13. 36¢ + $0.12 + $2.14
14. $2 − $1.37

15. $\$15 + \$1.50 + 15¢$

16. $\$1 - 37¢$

17. $36¢ + 24¢ + 78¢$

18. $\$5 - \$4.63$

19. $\$5 + \$0.36 + 9¢$

20. $87¢ - 78¢$

Supplemental Practice for Lesson 57

1. $\begin{array}{r} 36 \\ \times\ 2 \\ \hline \end{array}$

2. $\begin{array}{r} 43 \\ \times\ 5 \\ \hline \end{array}$

3. $\begin{array}{r} 57 \\ \times\ 3 \\ \hline \end{array}$

4. $\begin{array}{r} 24 \\ \times\ 6 \\ \hline \end{array}$

5. $\begin{array}{r} 38 \\ \times\ 4 \\ \hline \end{array}$

6. $\begin{array}{r} 52 \\ \times\ 7 \\ \hline \end{array}$

7. $\begin{array}{r} 27 \\ \times\ 5 \\ \hline \end{array}$

8. $\begin{array}{r} 19 \\ \times\ 8 \\ \hline \end{array}$

9. 28×6

10. 9×56

11. 47×7

12. 4×89

13. 64×8

14. 5×75

15. 63×9

16. 4×76

17. 97×6

18. 2×68

19. 54×7

20. 3×45

Supplemental Practice for Lesson 61

1. $\begin{array}{r} 3486 \\ -\ 1687 \\ \hline \end{array}$

2. $\begin{array}{r} 2175 \\ -\ 1346 \\ \hline \end{array}$

3. $\begin{array}{r} 3747 \\ -\ 1654 \\ \hline \end{array}$

4. $\begin{array}{r} 4403 \\ -\ 1475 \\ \hline \end{array}$

5. $\begin{array}{r} 6300 \\ -\ 3149 \\ \hline \end{array}$

6. $\begin{array}{r} 2048 \\ -\ 1951 \\ \hline \end{array}$

7. $\begin{array}{r} 3000 \\ -\ 1346 \\ \hline \end{array}$

8. $\begin{array}{r} 4005 \\ -\ 2418 \\ \hline \end{array}$

9. $\begin{array}{r} 3040 \\ -\ 1535 \\ \hline \end{array}$

10. $\begin{array}{r} 6000 \\ -\ 2164 \\ \hline \end{array}$

11. $\begin{array}{r} 8010 \\ -\ 7825 \\ \hline \end{array}$

12. $\begin{array}{r} 5007 \\ -\ 1838 \\ \hline \end{array}$

13. $\begin{array}{r} 36{,}247 \\ -\ 1{,}456 \\ \hline \end{array}$

14. $\begin{array}{r} 30{,}148 \\ -\ 23{,}109 \\ \hline \end{array}$

15. $\begin{array}{r} 40{,}015 \\ -\ 16{,}438 \\ \hline \end{array}$

16. $\begin{array}{r} 30{,}000 \\ -\ 256 \\ \hline \end{array}$

17. $\begin{array}{r} 30{,}604 \\ -\ 1{,}915 \\ \hline \end{array}$

18. $\begin{array}{r} 90{,}040 \\ -\ 37{,}478 \\ \hline \end{array}$

19. $\begin{array}{r} 376{,}142 \\ -\ 36{,}174 \\ \hline \end{array}$

20. $\begin{array}{r} 403{,}700 \\ -\ 394{,}672 \\ \hline \end{array}$

Supplemental Practice for Lesson 62

1. $2\overline{)15}$
2. $5\overline{)23}$
3. $3\overline{)25}$
4. $26 \div 6$
5. $31 \div 4$
6. $50 \div 7$
7. $5\overline{)37}$
8. $8\overline{)35}$
9. $6\overline{)43}$
10. $30 \div 9$
11. $45 \div 7$
12. $17 \div 2$
13. $8\overline{)49}$
14. $3\overline{)25}$
15. $9\overline{)60}$
16. $27 \div 4$
17. $15 \div 8$
18. $32 \div 5$
19. $3\overline{)20}$
20. $6\overline{)34}$

Supplemental Practice for Lesson 67

1. $\begin{array}{r} 136 \\ \times\ 2 \\ \hline \end{array}$

2. $\begin{array}{r} 235 \\ \times\ 5 \\ \hline \end{array}$

3. $\begin{array}{r} 430 \\ \times\ 3 \\ \hline \end{array}$

4. 216×6

5. 450×4

6. 7×642

7. $\begin{array}{r} 307 \\ \times\ 5 \\ \hline \end{array}$

8. $\begin{array}{r} 458 \\ \times\ 8 \\ \hline \end{array}$

9. $\begin{array}{r} 740 \\ \times\ 6 \\ \hline \end{array}$

10. 368×7

11. 9×403

12. 490×8

13. $\begin{array}{r} 609 \\ \times\ 2 \\ \hline \end{array}$

14. $\begin{array}{r} 470 \\ \times\ 9 \\ \hline \end{array}$

15. $\begin{array}{r} 518 \\ \times\ 3 \\ \hline \end{array}$

16. 2×296 **17.** 708×4 **18.** 3×430

19. 275×5 **20.** 4×308

Supplemental Practice for Lesson 73

1. $3\overline{)48}$ **2.** $2\overline{)56}$ **3.** $4\overline{)72}$ **4.** $7\overline{)98}$

5. $5\overline{)80}$ **6.** $8\overline{)96}$ **7.** $6\overline{)90}$ **8.** $3\overline{)81}$

9. $7\overline{)91}$ **10.** $4\overline{)68}$ **11.** $2\overline{)76}$ **12.** $5\overline{)90}$

13. $3\overline{)54}$ **14.** $6\overline{)84}$ **15.** $3\overline{)78}$ **16.** $7\overline{)84}$

17. $4\overline{)84}$ **18.** $5\overline{)85}$ **19.** $6\overline{)72}$ **20.** $5\overline{)65}$

Supplemental Practice for Lesson 74

1. $2\overline{)110}$ **2.** $9\overline{)126}$ **3.** $3\overline{)222}$ **4.** $8\overline{)432}$

5. $4\overline{)256}$ **6.** $7\overline{)455}$ **7.** $5\overline{)320}$ **8.** $2\overline{)192}$

9. $6\overline{)342}$ **10.** $3\overline{)204}$ **11.** $7\overline{)266}$ **12.** $4\overline{)100}$

13. $8\overline{)456}$ **14.** $5\overline{)365}$ **15.** $9\overline{)468}$ **16.** $6\overline{)162}$

17. $4\overline{)252}$ **18.** $7\overline{)665}$ **19.** $8\overline{)600}$ **20.** $5\overline{)245}$

Supplemental Practice for Lesson 75

1. $4\overline{)93}$ **2.** $2\overline{)115}$ **3.** $5\overline{)182}$ **4.** $3\overline{)173}$

5. $6\overline{)289}$ **6.** $4\overline{)181}$ **7.** $7\overline{)164}$ **8.** $5\overline{)319}$

9. $8\overline{)218}$ **10.** $6\overline{)235}$ **11.** $9\overline{)220}$ **12.** $7\overline{)442}$

13. $2\overline{)189}$ **14.** $8\overline{)595}$ **15.** $3\overline{)109}$ **16.** $9\overline{)892}$

17. $4\overline{)218}$ **18.** $2\overline{)55}$ **19.** $5\overline{)232}$ **20.** $3\overline{)220}$

Supplemental Practice for Lesson 77

1. $\begin{array}{r} 32 \\ \times\ 20 \\ \hline \end{array}$

2. $\begin{array}{r} 43¢ \\ \times\ 30 \\ \hline \end{array}$

3. $\begin{array}{r} 56 \\ \times\ 40 \\ \hline \end{array}$

4. $\begin{array}{r} \$0.68 \\ \times\ \quad 20 \\ \hline \end{array}$

5. $\begin{array}{r} 47 \\ \times\ 60 \\ \hline \end{array}$

6. $\begin{array}{r} \$1.68 \\ \times\ \quad 20 \\ \hline \end{array}$

7. 20×75

8. $30 \times 49¢$

9. 40×87

10. $\$0.97 \times 50$

11. $70 \times \$1.49$

12. 60×38

13. 80×76

14. $48¢ \times 90$

15. 20×89

16. $\$2.25 \times 50$

17. $\$0.39 \times 60$

18. 30×78

19. $40 \times 67¢$

20. 84×70

Supplemental Practice for Lesson 85

1. $3\overline{)700}$

2. $6\overline{)738}$

3. $4\overline{)892}$

4. $7\overline{)868}$

5. $5\overline{)1606}$

6. $8\overline{)915}$

7. $6\overline{)1275}$

8. $9\overline{)1926}$

9. $7\overline{)2415}$

10. $3\overline{)1603}$

11. $8\overline{)1161}$

12. $4\overline{)1111}$

13. $9\overline{)3000}$

14. $5\overline{)625}$

15. $3\overline{)1333}$

16. $6\overline{)1518}$

17. $4\overline{)2250}$

18. $7\overline{)1162}$

19. $8\overline{)1000}$

20. $5\overline{)3743}$

Supplemental Practice for Lesson 89

1. $4\overline{)960}$

2. $5\overline{)1600}$

3. $3\overline{)1206}$

4. $9\overline{)936}$

5. $4\overline{)2082}$

6. $6\overline{)1820}$

7. $7\overline{)2801}$

8. $2\overline{)1819}$

9. $5\overline{)3404}$

10. $3\overline{)2712}$

11. $6\overline{)3000}$

12. $4\overline{)2681}$

13. $7\overline{)5650}$

14. $8\overline{)3275}$

15. $3\overline{)450}$

16. $5\overline{)2001}$

17. $2\overline{)381}$ 18. $8\overline{)6080}$ 19. $9\overline{)3686}$ 20. $6\overline{)4202}$

Supplemental Practice for Lesson 99

1. 12×36
2. 46×15
3. 31×27
4. $\begin{array}{r} 74 \\ \times 16 \\ \hline \end{array}$
5. $\begin{array}{r} 36 \\ \times 63 \\ \hline \end{array}$
6. $\begin{array}{r} 35 \\ \times 35 \\ \hline \end{array}$
7. 14×63
8. 78×22
9. 25×37
10. $\begin{array}{r} 74 \\ \times 58 \\ \hline \end{array}$
11. $\begin{array}{r} 63 \\ \times 49 \\ \hline \end{array}$
12. $\begin{array}{r} 18 \\ \times 65 \\ \hline \end{array}$
13. 96×32
14. 51×76
15. 38×24
16. $\begin{array}{r} 38 \\ \times 47 \\ \hline \end{array}$
17. $\begin{array}{r} 49 \\ \times 86 \\ \hline \end{array}$
18. $\begin{array}{r} 29 \\ \times 31 \\ \hline \end{array}$
19. 33×79
20. 57×42

Supplemental Practice for Lesson 113

Change each improper fraction to a mixed number or a whole number.

1. $\frac{3}{2}$
2. $\frac{9}{3}$
3. $\frac{4}{3}$
4. $\frac{7}{4}$
5. $\frac{12}{5}$
6. $\frac{4}{2}$
7. $\frac{5}{4}$
8. $\frac{7}{5}$
9. $\frac{3}{3}$
10. $\frac{9}{5}$
11. $\frac{5}{2}$
12. $\frac{8}{4}$
13. $\frac{15}{15}$
14. $\frac{5}{3}$
15. $\frac{9}{4}$
16. $\frac{6}{2}$
17. $\frac{6}{3}$
18. $\frac{10}{3}$
19. $\frac{7}{2}$
20. $\frac{7}{3}$

Supplemental Practice for Lesson 117

1. 3.6 + 2.17
2. 5.28 + 12.4
3. 15.4 + 23.56
4. 6.7 + 15.8
5. 16.36 + 14.7
6. 45.3 + 2.91
7. 0.4 + 45.91
8. 3.71 + 6.3
9. 103.7 + 7.41
10. 9.09 + 90.9
11. 1.3 + 4.26 + 2.7
12. 12.4 + 1.15 + 3.3
13. 2.1 + 1.91 + 12.12
14. 6.58 + 3.7 + 0.4
15. 29.6 + 2.96 + 29.62
16. 3.4 + 4.56 + 1.41
17. 36.4 + 6.4 + 0.64
18. 1.2 + 0.21 + 12.1 + 10.21
19. 3.5 + 0.35 + 5.03 + 35.53 + 35.0
20. 2.4 + 4.12 + 20.4 + 42.21 + 1.2

Supplemental Practice for Lesson 118

1. 3.45 − 1.2
2. 23.1 − 2.2
3. 14.25 − 1.6
4. 15.3 − 4.4
5. 7.59 − 1.8
6. 25.34 − 1.21
7. 16.25 − 1.9
8. 8.19 − 0.4
9. 13.26 − 12.2
10. $\begin{array}{r} 3.4 \\ -\ 1.26 \\ \hline \end{array}$
11. $\begin{array}{r} 4.0 \\ -\ 2.14 \\ \hline \end{array}$
12. $\begin{array}{r} 12.4 \\ -\ 1.24 \\ \hline \end{array}$

13. 7.4 − 1.22

14. 3.68 − 1.7

15. 12.1 − 1.21

16. 30.1 − 3.01

17. 34.05 − 6.4

18. 58.0 − 2.14

19. 3.09 − 1.8

20. 20.1 − 3.19

Supplemental Practice for Lesson 125

1. $20\overline{)460}$
2. $30\overline{)630}$
3. $40\overline{)520}$
4. $50\overline{)1600}$
5. $60\overline{)720}$
6. $70\overline{)1470}$
7. $80\overline{)1700}$
8. $90\overline{)1200}$
9. $20\overline{)680}$
10. $40\overline{)1325}$
11. $60\overline{)1450}$
12. $70\overline{)2177}$
13. $80\overline{)2001}$
14. $90\overline{)1359}$
15. $20\overline{)920}$
16. $40\overline{)2088}$
17. $60\overline{)2640}$
18. $70\overline{)1624}$
19. $30\overline{)1680}$
20. $50\overline{)2710}$

Supplemental Practice for Lesson 126

1. 320 × 12
2. 132 × 21
3. 143 × 23
4. $\begin{array}{r} 150 \\ \times\ 32 \\ \hline \end{array}$
5. $\begin{array}{r} 304 \\ \times\ 13 \\ \hline \end{array}$
6. $\begin{array}{r} 315 \\ \times\ 24 \\ \hline \end{array}$
7. 42 × 163
8. 230 × 15
9. 25 × 402
10. $\begin{array}{r} 357 \\ \times\ 34 \\ \hline \end{array}$
11. $\begin{array}{r} 780 \\ \times\ 56 \\ \hline \end{array}$
12. $\begin{array}{r} 406 \\ \times\ 17 \\ \hline \end{array}$
13. 28 × 196
14. 460 × 39
15. 43 × 179

16. $\begin{array}{r} 108 \\ \times\ 39 \\ \hline \end{array}$

17. $\begin{array}{r} 349 \\ \times\ 74 \\ \hline \end{array}$

18. $\begin{array}{r} 470 \\ \times\ 68 \\ \hline \end{array}$

19. 29×357

20. 186×37

Supplemental Practice for Lesson 128

Reduce each fraction.

1. $\frac{5}{10}$
2. $\frac{2}{4}$
3. $\frac{6}{8}$
4. $\frac{2}{6}$
5. $\frac{3}{9}$
6. $\frac{4}{10}$
7. $\frac{3}{6}$
8. $\frac{2}{12}$
9. $\frac{9}{12}$
10. $\frac{4}{6}$
11. $\frac{6}{9}$
12. $\frac{8}{10}$
13. $\frac{2}{8}$
14. $\frac{3}{12}$
15. $\frac{2}{10}$
16. $\frac{4}{8}$
17. $\frac{8}{12}$
18. $\frac{6}{10}$
19. $\frac{4}{12}$
20. $\frac{6}{12}$

Supplemental Practice for Lesson 129

1. $12\overline{)72}$
2. $31\overline{)124}$
3. $11\overline{)100}$
4. $41\overline{)125}$
5. $13\overline{)91}$
6. $21\overline{)107}$
7. $52\overline{)212}$
8. $25\overline{)130}$
9. $32\overline{)130}$
10. $22\overline{)135}$
11. $51\overline{)310}$
12. $14\overline{)80}$
13. $42\overline{)180}$
14. $23\overline{)161}$
15. $34\overline{)175}$
16. $15\overline{)105}$
17. $43\overline{)150}$
18. $24\overline{)200}$
19. $33\overline{)300}$
20. $19\overline{)100}$

Supplemental Practice for Lesson 130

1. $3\frac{1}{2} + 1$
2. $3\frac{1}{3} + 1\frac{1}{3}$
3. $1\frac{1}{5} + \frac{3}{5}$
4. $4 + \frac{1}{2}$
5. $6\frac{3}{5} + 1\frac{1}{5}$
6. $5\frac{5}{8} + 6$

7. $3\frac{3}{7} + 2\frac{2}{7}$

8. $6 + 7\frac{1}{2}$

9. $\frac{5}{9} + 3\frac{2}{9}$

10. $3\frac{3}{10} + 6\frac{6}{10}$

11. $5\frac{2}{3} - 1\frac{1}{3}$

12. $3\frac{3}{4} - 2$

13. $6\frac{1}{2} - \frac{1}{2}$

14. $8\frac{3}{4} - 1\frac{3}{4}$

15. $2\frac{5}{8} - 2\frac{2}{8}$

16. $4\frac{4}{5} - \frac{1}{5}$

17. $4\frac{4}{9} - 3$

18. $1\frac{4}{5} - \frac{4}{5}$

19. $3\frac{1}{2} - 1\frac{1}{2}$

20. $4\frac{5}{7} - 1\frac{3}{7}$

Supplemental Practice for Lesson 131

Simplify each fraction answer.

1. $\frac{1}{2} + \frac{1}{2}$

2. $\frac{1}{3} - \frac{1}{3}$

3. $\frac{1}{4} + \frac{1}{4}$

4. $\frac{3}{8} - \frac{1}{8}$

5. $\frac{1}{6} + \frac{2}{6}$

6. $\frac{5}{6} - \frac{1}{6}$

7. $\frac{2}{3} + \frac{2}{3}$

8. $\frac{7}{8} - \frac{1}{8}$

9. $\frac{4}{5} + \frac{3}{5}$

10. $3\frac{1}{2} - 1\frac{1}{2}$

11. $2\frac{2}{3} + 1\frac{1}{3}$

12. $3\frac{3}{4} - 1\frac{1}{4}$

13. $4\frac{2}{3} + 5\frac{2}{3}$

14. $3\frac{4}{9} - 1\frac{1}{9}$

15. $1\frac{1}{6} + 1\frac{1}{6}$

16. $6\frac{7}{10} - 4\frac{1}{10}$

17. $4\frac{5}{12} - 1\frac{1}{12}$

18. $5\frac{3}{4} + 4\frac{1}{4}$

19. $7\frac{4}{5} + 4\frac{4}{5}$

20. $7\frac{7}{8} - 3\frac{3}{8}$

Supplemental Practice for Lesson 132

In problems 1-8, find the fraction name for 1 used to make the equivalent fraction.

1. $\frac{1}{2} \times M = \frac{3}{6}$

2. $\frac{1}{2} \times Z = \frac{5}{10}$

3. $\frac{1}{2} \times R = \frac{6}{12}$

4. $\frac{1}{3} \times Q = \frac{3}{9}$

5. $\frac{1}{6} \times G = \frac{2}{12}$

6. $\frac{3}{4} \times B = \frac{9}{12}$

7. $\frac{2}{5} \times P = \frac{4}{10}$

8. $\frac{2}{3} \times K = \frac{8}{12}$

In problems 9-20, complete the equivalent fraction.

9. $\frac{1}{4} = \frac{?}{8}$

10. $\frac{1}{3} = \frac{?}{6}$

11. $\frac{1}{2} = \frac{?}{4}$

12. $\frac{1}{2} = \frac{?}{8}$

13. $\frac{3}{4} = \frac{?}{8}$

14. $\frac{2}{3} = \frac{?}{9}$

15. $\frac{2}{5} = \frac{?}{10}$

16. $\frac{1}{2} = \frac{?}{12}$

17. $\frac{5}{6} = \frac{?}{12}$

18. $\frac{1}{2} = \frac{?}{4}$

19. $\frac{3}{4} = \frac{?}{12}$

20. $\frac{2}{3} = \frac{?}{12}$

Supplemental Practice for Lesson 138

1. $11\overline{)253}$
2. $21\overline{)672}$
3. $31\overline{)682}$
4. $12\overline{)504}$
5. $32\overline{)992}$
6. $21\overline{)483}$
7. $11\overline{)165}$
8. $22\overline{)924}$
9. $12\overline{)181}$
10. $31\overline{)963}$
11. $23\overline{)760}$
12. $41\overline{)945}$
13. $15\overline{)375}$
14. $25\overline{)555}$
15. $11\overline{)375}$
16. $21\overline{)924}$
17. $22\overline{)489}$
18. $12\overline{)600}$
19. $33\overline{)1000}$
20. $25\overline{)800}$

Supplemental Practice for Lesson 139

1. $\frac{1}{4} + \frac{1}{2}$
2. $\frac{1}{4} + \frac{1}{8}$
3. $\frac{1}{2} + \frac{1}{8}$
4. $\frac{1}{4} - \frac{1}{8}$
5. $\frac{1}{2} - \frac{1}{8}$
6. $\frac{1}{3} - \frac{1}{6}$
7. $\frac{2}{3} + \frac{1}{6}$
8. $\frac{3}{4} + \frac{1}{8}$
9. $\frac{1}{3} + \frac{2}{9}$
10. $\frac{5}{6} - \frac{2}{3}$
11. $\frac{7}{8} - \frac{3}{4}$
12. $\frac{9}{10} - \frac{4}{5}$
13. $\frac{5}{6} + \frac{1}{12}$
14. $\frac{3}{10} + \frac{2}{5}$
15. $\frac{1}{3} + \frac{1}{12}$
16. $\frac{4}{5} - \frac{1}{10}$
17. $\frac{8}{9} - \frac{2}{3}$
18. $\frac{7}{8} - \frac{1}{4}$
19. $\frac{3}{5} + \frac{3}{10}$
20. $\frac{2}{3} - \frac{7}{12}$

Glossary

addend A number that is added in an addition problem.

angle A corner formed where two lines meet.

area The number of squares of a certain size that it takes to cover a given surface.

average The number that would be in each group if the items were rearranged so that each group had the same number of items.

bar graph A graph that uses bars to show the information.

circle graph *See* pie graph.

circumference The distance around a circle.

common denominator Two fractions have common denominators if the denominators are equal.

decimal number A number that uses digits and a decimal point.

decimal point A dot used to separate the whole part of a decimal number from the fractional part of the decimal number.

denominator The bottom number of a fraction. This number tells the number of parts in the whole.

diameter The distance across a circle through the center.

difference The answer to a subtraction problem.

digit Any of the following symbols: 0, 1, 2, 3, 4, 5, 6, 7, 8, 9.

division bar The bar that separates the numerator of a fraction from the denominator of the fraction. Also called the *fraction bar.*

equal to Has the same value as.

equivalent fractions Two or more fractions that have the same value.

estimate To determine mentally an approximate answer to a problem.

expanded form A way of representing a number which uses the sum of the products of each digit and the place value of each digit.

factor A number that is multiplied in a multiplication problem.

fraction A number used to designate a part of a whole.

fraction bar *See* division bar.

geometric solid A geometric shape that takes up space.

graph A picture that shows number information about a certain topic.

greater than Has a larger value than.

hexagon A six-sided polygon.

improper fraction A fraction that is greater than or equal to 1.

intersecting lines Lines that cross one another.

less than Has a smaller value than.

line graph A graph that displays information by lines that connect data points.

line segment A part of a line.

mixed number A whole number followed by a fraction.

multiple A number found by multiplying a given number by a whole number greater than zero.

number line A line with evenly spaced marks on which a number is associated with each mark.

numerator The top number of a fraction. This number tells the number of parts being described.

octagon An eight-sided polygon.

parallel lines Lines that stay the same distance apart.

pentagon A five-sided polygon.

perimeter The distance around a closed, flat shape.

perpendicular lines Intersecting lines that form right angles.

pictograph A graph that uses pictures of items to show information.

pie graph A graph that represents information using a circle divided into sections that look like slices of pie.

polygon A closed, flat shape whose sides are straight lines.

probability The likelihood that something will happen.

product The answer to a multiplication problem.

quadrilateral A four-sided polygon.

quotient The answer to a division problem.

radius The distance from the center of a circle to the edge of the circle.

reduced fraction An equivalent fraction written with smaller numbers.

right angle A square corner.

Roman numerals Letters used by the ancient Romans to write numbers.

scale A type of number line often used for measuring.

segment Line segment.

sequence An orderly arrangement of numbers determined by following a certain rule.

square unit A square with sides of designated length.

sum The answer to an addition problem.

table A way of presenting information in rows and columns.

tally marks Short, straight marks used for keeping track of the number of times that something occurs.

tally The total of a group of tally marks.

triangle A three-sided polygon.

whole numbers Any of the numbers in this sequence: 0, 1, 2, 3, 4, 5, 6, 7, ...

Index